Lecture Notes in Computer Science 9093

Commenced Publication in 1973
Founding and Former Series Editors:
Gerhard Goos, Juris Hartmanis, and Jan van Leeuwen

More information about this series at http://www.springer.com/series/7409

Mohamed A. Sharaf · Muhammad Aamir Cheema
Jianzhong Qi (Eds.)

Databases Theory and Applications

26th Australasian Database Conference, ADC 2015
Melbourne, VIC, Australia, June 4–7, 2015
Proceedings

 Springer

Editors
Mohamed A. Sharaf
University of Queensland
Brisbane
Queensland
Australia

Jianzhong Qi
The University of Melbourne
Melbourne
Australia

Muhammad Aamir Cheema
Monash University
Clayton
Australia

ISSN 0302-9743   ISSN 1611-3349   (electronic)
Lecture Notes in Computer Science
ISBN 978-3-319-19547-6   ISBN 978-3-319-19548-3   (eBook)
DOI 10.1007/978-3-319-19548-3

Library of Congress Control Number: 2015939808

LNCS Sublibrary: SL3 – Information Systems and Applications, incl. Internet/Web, and HCI

Springer Cham Heidelberg New York Dordrecht London

Printed on acid-free paper

Springer International Publishing AG Switzerland is part of Springer Science+Business Media
(www.springer.com)

Preface

It is our pleasure to present to you the proceedings of the 26th Australasian Database Conference (ADC2015), which took place in Melbourne, Australia. The Australasian Database Conference is an annual international forum for sharing the latest research advancements and novel applications of database systems, data-driven applications, and data analytics between researchers and practitioners from around the globe, particularly Australia and New Zealand. The mission of ADC is to share novel research solutions to problems of today's information society that fulfill the needs of heterogeneous applications and environments and to identify new issues and directions for future research. ADC seeks papers from academia and industry presenting research on all practical and theoretical aspects of advanced database theory and applications, as well as case studies and implementation experiences. All topics related to database are of interest and within the scope of the conference. ADC gives researchers and practitioners a unique opportunity to share their perspectives with others interested in the various aspects of database systems.

ADC 2015 was co-located with ACM SIGMOD 2015 (May 31-June 4, 2015, Melbourne, Australia) and held immediately after it. As in previous years, ADC 2015 accepted all papers that the Program Committee considered as being of ADC quality without setting any predefined quota. The conference received 43 submissions and accepted 29 papers, including 24 full research papers and five demo papers. The Program Committee that selected the papers consisted of 49 members from around the globe, including Australia, China, Germany, Hong Kong, Japan, New Zealand, Switzerland, Taiwan, the UK, and the USA, who were thorough and dedicated to the reviewing process. Each paper was peer reviewed in full by at least three independent reviewers, and in some cases four referees produced independent reviews. A conscious decision was made to select the papers for which all reviews were positive and favorable.

We would like to thank all our colleagues who served on the Program Committee or acted as external reviewers. We would also like to thank all the authors who submitted their papers, and the attendees. We hope that with these proceedings, you can have an overview of this vibrant research community and its activities. We encourage you to make submissions to the next ADC conference and contribute to this community.

June 2015

Mohamed A. Sharaf
Muhammad A. Cheema
Jianzhong Qi

General Chair's Welcome Message

Welcome to the proceedings of the 26th Australasian Database Conference (ADC2015)! ADC is a leading Australia- and New Zealand-based international conference on research and applications of database systems, data-driven applications, and data analytics. In the past 10 years, ADC has been held in Brisbane (2014), Adelaide (2013), Melbourne (2012), Perth (2011), Brisbane (2010), Wellington (2009), Wollongong (2008), Ballarat (2007), Hobart (2006), and Newcastle (2005). This year, the ADC conference came back to Melbourne.

In the past, the ADC conference series was held as part of the Australasian Computer Science Week (ACSW). Starting from 2014, the ADC conferences have departed from ACSW as the database research community in Australasia has grown significantly larger. Now the new ADC conference has an expanded research program and focuses on community-building through a PhD School. ADC 2015 was the second of this new ADC conference series.

The conference this year had three eminent speakers to give keynote speeches: Tamer Özsu from the University of Waterloo, Canada, Gerhard Weikum from the Max Planck Institute for Informatics, Germany, and Bingsheng He from the Nanyang Technological University, Singapore. In addition to 24 full research papers and five demo papers carefully selected by the Program Committee, we were also very fortunate to have two invited talks presented by world-leading researchers: Cyrus Shahabi from the University of Southern California, USA, and Lu Qin from the University of Technology, Sydney, Australia. We had a three-day PhD School program as part of this year's ADC.

We wish to take this opportunity to thank all speakers, authors, and organizers. I would also specially thank our Organizing Committee members: Program Committee Co-chairs Mohamed A. Sharaf and Muhammad A. Cheema, for their dedication in ensuring a high-quality program, Proceedings Chair Jianzhong Qi, for his effort in delivering the conference proceedings timely, and Local Co-chairs Jianxin Li and Jeffrey Chan, for their consideration in covering every detail of the conference logistics. Without them, this year's ADC would not have been a success.

Melbourne is a multicultural city and ADC 2015 was held on the campus of the RMIT University residing at the heart of the City of Melbourne. We trust all ADC2015 participants had a wonderful experience with the conference, the campus, and the city.

Rui Zhang

Organization

General Chair

Rui Zhang   University of Melbourne, Australia

Program Committee Co-chairs

Mohamed A. Sharaf   University of Queensland Australia
Muhammad A. Cheema   Monash University, Australia

Local Co-chairs

Jianxin Li   RMIT University, Australia
Jeffrey Chan   RMIT University, Australia

Proceedings Chair

Jianzhong Qi   University of Melbourne, Australia

Steering Committee

Rao Kotagiri   University of Melbourne, Australia
Timos Sellis   RMIT University, Australia
Gill Dobbie   University of Auckland, New Zealand
Alan Fekete   University of Sydney, Australia
Xuemin Lin   University of New South Wales, Australia
Yanchun Zhang   Victoria University, Australia
Xiaofang Zhou   University of Queensland

Program Committee

Zhifeng Bao   University of Tasmania, Australia
Ljiljana Brankovic   University of Newcastle, Australia
Lijun Chang   University of New South Wales, Australia
Byron Choi   Hong Kong Baptist University, Hong Kong, SAR
           China

Mohamed Eltabakh   WPI, USA
Markus Endres   University of Augsburg, Germany
Junbin Gao   Charles Sturt University, Australia
Dimitrios Georgakopoulos   RMIT University, Australia
Janusz Getta   University of Wollongong, Australia
Yu Gu   Northeastern University, China
Michael Houle   National Institute of Informatics, Japan
Guangyan Huang   Deakin University, Australia
Yoshiharu Ishikawa   Nagoya University, Japan
Jiuyong Li   University of South Australia, , Australia
Ruixuan Li   Huazhong University of Science and Technology,
                                    China
Xue Li   University of Queensland, Australia
Zhixu Li   Soochow University, China
Muhammad Asif Naeem   University of Auckland, New Zealand
Lu Qin   University of Technology, Sydney, Australia
Goce Ristanoski   University of Melbourne, Australia
Uwe Roehm   University of Sydney, Australia
Shazia Sadiq   University of Queensland, Australia
Sherif Sakr   NICTA, Australia
Mohamed Sarwat   Arizona State University, USA
Saket Sathe   EPFL, Switzerland
Shuo Shang   China University of Petroleum, China
Quan Z. Sheng   University of Adelaide, Australia
Laurianne Sitbon   Queensland University of Technology, Australia
Bela Stantic   Griffith University, Australia
Millist Vincent   University of South Australia, Australia
Junhu Wang   Griffith University, Australia
Lei Wang   University of Wollongong, Australia
Yan Wang   Macquarie University, Australia
Chuan Xiao   Nagoya University, Japan
Guandong Xu   University of Technology, Australia
Jian Yang   Macquarie University, Australia
Mi-Yen Yeh   Academia Sinica, Taiwan
Daisaku Yokoyama   University of Tokyo, Japan
Weiren Yu   Imperial College, UK
Nayyar Zaidi   Monash University, Australia
Wenjie Zhang   University of New South Wales, Australia
Xianchao Zhang   Dalian University of Technology, China
Xiuzhen Zhang   RMIT University, Australia
Ying Zhang   University of Technology, Sydney, Australia
Xiang Zhao   National University of Defence Technology, China
Kai Zheng   University of Queensland, Australia
Xiangmin Zhou   Victoria University, Australia
Guido Zuccon   Queensland University of Technology, Australia
Andreas Zufle   Ludwig-Maximilians-Universität, Germany

External Reviewers

Zhan Bu	Nanjing University of Finance and Economics, China
Po-Yu Chen	Imperial College London, UK
Hamidu Abdel-Fatao	University of South Australia
Tao Hoang	University of South Australia
Lili Jiang	Max Planck Institute, Germany
Selasi Kwashie	University of South Australia
Yongrui Qin	The University of Adelaide, Australia
Lie Qu	Macquarie University, Australia
Jie Shao	University of Electronic Science and Technology of China
Xianzhi Wang	The University of Adelaide, Australia
Dong Wen	University of Technology, Sydney, Australia
Zhengyuan Xue	Huazhong University of Science and Technology, China
Diyi Yang	Carnegie Mellon University, USA
Haibin Zhang	Macquarie University, Australia
Cong Zhao	Imperial College London, UK
Xiaoming Zheng	Macquarie University, Australia

Tutorials

An Overview of Graph Data Management and Analysis

M. Tamer Özsu

University of Waterloo

Abstract. Graphs have always been important data types for database researchers. With the recent growth of social networks, Wikipedia, Linked Data, RDF, and other networks, the interest in managing very large graphs have again gained momentum. In this talk I will first present a taxonomy of graph processing systems and then summarize research on querying and analytics over property graphs and management and querying of RDF graphs.

Short Biography. M. Tamer Özsu is Professor of Computer Science at the David R. Cheriton School of Computer Science, and Associate Dean (Research) of the Faculty of Mathematics at the University of Waterloo. His research is in data management focusing on large-scale data distribution and management of non-traditional data. He is a Fellow of the Association for Computing Machinery (ACM), and of the Institute of Electrical and Electronics Engineers (IEEE), an elected member of the Science Academy of Turkey, and member of Sigma Xi and American Association for the Advancement of Science (AAAS). He currently holds a Cheriton Faculty Fellowship at the University of Waterloo.

Knowledge Graphs: From a Fistful of Triples to Deep Data and Deep Text

Gerhard Weikum

Max Planck Institute for Informatics

Abstract. Knowledge graphs (KG's), aka. knowledge bases, are huge repositories of entities, their types, properties, and relationships between entities. KG's have become a key asset for search, analytics, recommendations, and data integration on the Web and in enterprises. Rooted in academic research and community projects such as DBpedia, Freebase, and Yago, KG's are now intensively used at big industrial stakeholders such as Google, Microsoft, Yahoo, Alibaba, Bloomberg, Walmart, and many others.

This talk reviews the knowledge graph technology, discussing strengths and limitations and pointing out opportunities for further research. The talk spans a spectrum of issues that arise in the life-cycle and use-cases of a KG: construction from data and text sources, maintaining over time, extension with common sense knowledge, querying and mining, boosting language understanding and text analytics, and usability issues in interactive exploration.

Short Biography. Gerhard Weikum is a Scientific Director at the Max Planck Institute for Informatics in Saarbruecken, Germany, and also an Adjunct Professor at Saarland University. He graduated from the University of Darmstadt, Germany.

Weikum's research spans transactional and distributed systems, self-tuning database systems, DB&IR integration, and the automatic construction of knowledge bases from Web and text sources. He co-authored a comprehensive textbook on transactional systems, received the VLDB 10-Year Award for his work on automatic DB tuning, and is one of the creators of the YAGO knowledge base.

Gerhard Weikum is an ACM Fellow, a member of the Academia Europaea, and a member of several academies in Germany. He has served on various editorial boards, including Communications of the ACM, ACM TODS and ACM TWEB, and as PC chair of conferences like ACM SIGMOD, Data Engineering, and CIDR. From 2003 through 2009 he was president of the VLDB Endowment. He received a Google Focused Research Award in 2010, the ACM SIGMOD Contributions Award in 2011, and an ERC Synergy Grant in 2013.

Emerging HPC Technologies for Real-Time (Big) Data Analytics: A Tutorial

Bingsheng He

Nanyang Technological University

Abstract. Big data has become a buzz word. Among various big-data challenges, real-time data analytics has been identified as one of the most exciting and promising areas for both academia and industry. We are facing the challenges at all levels ranging from sophisticated algorithms and procedures to mine the gold from massive data to high-performance computing (HPC) techniques and systems to get the useful data in time. In this tutorial, we review the system design and implementation of HPC technologies (including GPUs, FPGAs and RDMA etc) as weapons to address the performance requirement of real-time data analytics. Particularly, we focus on the interplay between HPC and real-time data analytics, where real-time data analytics also poses significant challenges to the design and implementation of HPC technologies. I will also present our recent research efforts in developing real-time data analytics systems by Xtra Computing Group at NTU Singapore (http://pdcc.ntu.edu.sg/xtra/) as well as related research from other groups. Finally, I will outline some open problems in this field.

Short Biography. Dr. Bingsheng He is currently an Assistant Professor at Division of Networks and Distributed Systems, School of Computer Engineering, Nanyang Technological University. Before that, he held a research position in the System Research group of Microsoft Research Asia (2008 - 2010), where his major research was building high performance cloud computing systems for Microsoft. He got the Bachelor degree in Shanghai Jiao Tong University (1999 - 2003), and the Ph.D. degree in Hong Kong University of Science & Technology (2003 - 2008). His current research interests include cloud computing, database systems and high performance computing. His papers are published in prestigious international journals (such as ACM TODS, IEEE TKDE/TPDS/TC) and proceedings (such as ACM SIGMOD, VLDB/PVLDB, ACM/IEEE SuperComputing, PACT, HPDC, ACM SoCC, and CIDR). He has been awarded with the IBM Ph.D. fellowship (2007 - 2008) and with NVIDIA Academic Partnership (2010 - 2011).

Invited Talks

Graph Processing in the Era of Big Data

Lu Qin

University of Technology, Sydney

Abstract. With the emergence and rapid proliferation of applications that deal with big graphs, such as web graphs (Google, Yahoo), social networks (Facebook, Twitter), e-commerce networks (Amazon, Ebay), and road networks, graph processing has become increasingly prevalent and important in recent years. However, in the era of big data, the explosion and profusion of available graph data in a wide range of application domains rise up new challenges and opportunities in graph processing. In this talk, I will first investigate these new challenges. To tickle these challenges, I will then introduce our recent research on big graph processing in terms of new graph query semantics, new query processing algorithms, new graph indexing techniques, and new computing paradigms. Finally, I will discuss our potential future research directions for graph processing.

Short Biography. Dr. Lu Qin received his PhD degree in 2010 from the department of Systems Engineering and Engineering Management (SEEM) in the Chinese University of Hong Kong (CUHK). Dr. Qin is currently a core member in the Centre of Quantum Computation and Intelligent Systems (QCIS) at the University of Technology, Sydney (UTS). Dr. Qin's research interests include algorithm design and analysis on big data, big graph processing in the cloud, and big graph searching and mining. He has published 50+ top-tier conference/journal papers including 7 SIGMOD papers, 9 VLDB papers, and 7 ICDE papers in the top-3 database conferences, and 7 VLDB journal papers, 1 Algorithmica paper, and 2 TKDE papers in the top-ranked database and algorithm journals. His book entitled "Keyword Search in Databases" is the first monograph on keyword search in databases. Dr. Qin served as a program committee member of a lot of top database and data mining conferences. He has received several research funds from Australian government and UTS.

Spatial Indexing and Spatial Crowdsourcing of User-Generated-Video

Cyrus Shahabi

University of Southern California

Abstract. I will start by showing a demo of our MediaQ system prototype. MediaQ is a novel online media management system to collect, organize, share, and search user-generated mobile videos (UGV) from the public. Subsequently, I focus on two of the research challenges underlying MediaQ. First, I discuss our approach to index and search UGVs more effectively by utilizing the smartphone sensors (e.g., GPS locations, compass directions) to geo-tag each video frame by the spatial extent of its coverage area. Next, I introduce our generic framework for spatial crowdsourcing and discuss various techniques for optimal assignment of spatiotemporal tasks (e.g., UGV data collection) to human workers. Finally, I conclude by summarizing our ongoing efforts in spatial crowdsourcing.

Short Biography. Cyrus Shahabi is a Professor of Computer Science and Electrical Engineering and the Director of the Information Laboratory (InfoLAB) at the Computer Science Department and also the Director of the NSF's Integrated Media Systems Center (IMSC) at the University of Southern California (USC). He is also the director of the Informatics Program at USC's Viterbi School of Engineering. He was the CTO and co-founder of a USC spin-off, Geosemble Technologies, which was acquired in July 2012. Since then, he founded another company, ClearPath, focusing on predictive path-planning for car navigation systems. He received his B.S. in Computer Engineering from Sharif University of Technology in 1989 and then his M.S. and Ph.D. Degrees in Computer Science from the University of Southern California in May 1993 and August 1996, respectively. He authored two books and more than two hundred research papers in the areas of databases, GIS and multimedia with an h-index of 41. He also holds more than 12 US Patents.

Dr. Shahabi has received funding from agencies including NSF, NIJ, NASA, NIH, DARPA, AFRL, and DHS as well as industries such as Chevron, Google, HP, Intel, Microsoft, NCR, NGC and Oracle. He was an Associate Editor of IEEE Transactions on Parallel and Distributed Systems (TPDS) from 2004 to 2009 and IEEE Transactions on Knowledge and Data Engineering (TKDE) from 2010 to 2013. He is currently on the editorial board of the VLDB Journal, ACM Transactions on Spatial Algorithms and Systems (TSAS), and ACM Computers in Entertainment. He is the founding chair of IEEE NetDB workshop and also the general co-chair of ACM GIS 2007 - 2009. He chaired the nomination committee of ACM SIGSPATIAL 2011 - 2014. He was a PC co-Chair of DASFAA 2015, IEEE MDM 2013 and IEEE BigData 2013, and regularly serves on the program committee of major conferences such as VLDB, ACM SIGMOD, IEEE ICDE, ACM SIGKDD, and ACM Multimedia.

Dr. Shahabi is a fellow of IEEE, and a recipient of the ACM Distinguished Scientist award in 2009, the 2003 U.S. Presidential Early Career Awards for Scientists and Engineers (PECASE), the NSF CAREER award in 2002, and the 2001 Okawa Foundation Research Grant for Information and Telecommunications.

Contents

Demo Papers

Research Papers

Efficient Discovery of Differential Dependencies Through Association Rules Mining

Selasi Kwashie[1]([✉]), Jixue Liu[1], Jiuyong Li[1], and Feiyue Ye[2]

[1] ITMS, University of South Australia, Adelaide, Australia
selasi.kwashie@mymail.unisa.edu.au
[2] CSE, Jiangsu University of Technology, Changzhou, China

Abstract. Differential dependencies (DDs) extend functional dependencies (FDs) to capture the semantics of distance among data values. To mine DDs in a given relation is, thus, more challenging as the more general definition of DDs creates: a combinatorial large search space; and hugely sized minimal cover sets of DDs. This paper proposes a simple, yet effective and efficient approach to mine DDs in a given relation. We study and present a link between DDs and association rules (ARs): paving way for the adoption of existing ARs mining algorithms in the discovery of DDs. Furthermore, we propose a measure of interestingness for DDs to aid the discovery of essential DDs and avoid mining an extremely large set. Finally, we show the efficiency and scalability of our solution through experiments on three real-world benchmark data sets. The results indicate that our discovery approach is efficient and scalable.

1 Introduction

DDs [15] are recent extension of FDs for data management applications based on the semantic of distance. They replace the strict equality constraint in FDs with a condition stating that the distance between values is within some intervals. That is, a differential dependency (DD) $B[c, d] \rightarrow A[e, f]$ holds on an instance r of a relation if for any two tuples $t_1, t_2 \in r$, if the distance between $t_1[B]$ and $t_2[B]$ is within c and d, then the distance between $t_1[A]$ and $t_2[A]$ is within e and f. For example, consider the relation r in Table 1. A DD $YoS[0,2]Edu[0] \rightarrow Sal[0,3]$ on r states that for any two employees, if the difference between their years of service is within two years ($YoS[0,2]$) and there is no difference in their education levels ($Edu[0]$), then their salary difference must be within \$3K ($Sal[0,3]$).

DDs are, thus, more general than FDs and other metric-based dependencies like metric functional dependencies (MFDs) [8] and matching dependencies (MDs) [5]. Indeed, FDs, MFDs and MDs can be considered as special cases of DDs. An FD is equivalent to a DD with closeness of zero (0) on all attributes. For example, the FD $YoS, Gen \rightarrow Sal$ is the same as the DD $YoS[0]Gen[0] \rightarrow Sal[0]$. Also, since MFDs specify constraint on only the consequent (RHS) attributes while MDs do on only the antecedent (LHS) attributes, they represent special classes of DDs where the constraint on the LHS and the RHS are zero (0) respectively. For example, an MD $YoS[0,2]Gen[0] \rightarrow Sal$ is the same as the

© Springer International Publishing Switzerland 2015
M.A. Sharaf et al. (Eds.): ADC 2015, LNCS 9093, pp. 3–15, 2015.
DOI: 10.1007/978-3-319-19548-3_1

DD $YoS[0,2]Gen[0] \rightarrow Sal[0]$. Therefore, the knowledge representation in DDs subsumes that of FDs, MFDs and MDs.

DDs are different from ARs. Whereas DDs hold at the schema level of a relation, ARs do at the instance (value) level. In other words, an AR is only true for the *specific instance values* specified whiles a DD is true for *any instance values* within the specified distance. Also, as opposed to just the relationship among the raw data values expressed in ARs, DDs encode the distance relationship among the values of attributes. And, more importantly, DDs encode information more compactly.

Table 1. Instance r of a relation R

TID	YoS	Gen	Edu	Sal (\$ K)
1	25	2	4	15
2	25	1	3	18
3	28	1	3	15
4	32	2	2	12
5	30	1	1	15

Edu: 1=Dip.;2=BSc.; 3=M.Phil; 4=PhD.
Gen: 1=Male; 2=Female

Apart from their superior knowledge representation, DDs have many data management applications including violation detection, record linkage, semantic query optimization, among others. In semantic query optimization for instance, integrity constraints are useful tools for reformulating queries into ones that would give the same results more efficiently [3]. DDs can be used to rewrite queries where the semantic of distance is involved. Assume the DD $\sigma : YoS[0,3] \rightarrow Sal[0,5]$ holds on r in Table 1. Then, a query that requires the selection of employees with difference in YoS not more than 3 years and Sal difference within \$5000, can based on σ be efficiently rewritten with only the condition $YoS[0,3]$.

For DD-based data management applications to be effective, it is crucial to have efficient techniques that can automatically discover DDs from existing data. The discovery of DDs in data is, therefore, an important task and the subject of study of this paper. Mining data dependencies is not new. In fact, over the last decade, many algorithms were developed for mining FDs and its various extensions. For example, the works in: [7,12,13] propose methods for FD discovery; [4,6,11] present conditional FD mining techniques; and [14] also propose an algorithm for finding MDs. Unfortunately, these algorithms cannot discover patterns in data specific to DDs.

At the moment, only the works in [15] and [9] study the discovery of DDs. And, this is unsurprising since DD is recently proposed. The discovery technique in [15], on the one hand, relies on reduction algorithms to find DDs and utilises the subsumption order of differential functions to prune the search space. On the other hand, [9] proposes a user-specified constraint on the LHS of DDs and employs a subspace clustering technique to find these LHSs for an efficient discovery. In this paper, we propose a different approach to DD discovery to complement the existing works. We investigate a link between ARs and DDs to allow existing AR mining techniques to be adopted and used in DD discovery.

Our contributions are summarized as follows. First, we establish a relationship between DDs and ARs. This bridges the gap between data dependency (specifically, DDs) discovery in the database community and ARs mining in the data mining community. Second, we propose an algorithm for mining a minimal cover of DDs based on the established link. Furthermore, we propose a measure of interestingness for DDs, reducing a minimal cover of DDs to only essential

DDs. We experimentally evaluate our algorithm on three real-world data sets to demonstrate the feasibility, scalability, and efficiency of our proposition.

2 Preliminaries – DDs and ARs

Differential Dependency. The definitions and concepts concerning DDs here follow those of [15]. Let r be an instance of the relation $R(A_1, \cdots, A_n)$; $dom(A_i)$ be the domain of an attribute $A_i \in R$; $X, Y \subset R$ be subsets of attributes in R.

A *distance metric*, d_A, is a function defined over an attribute A that returns the distance between any two values a_1, a_2 of A in r. d_A is assumed to exhibit: (a) non-negativity: $d_A(a_1, a_2) \geq 0$; (b) identity of indiscernible: $d_A(a_1, a_2) = 0$ if and only if $a_1 = a_2$; and (c) symmetry: $d_A(a_1, a_2) = d_A(a_2, a_1)$, where $a_1, a_2 \in dom(A)$. Given any instance r of R, there exists a finite set $\mathcal{W}_A = \{v_1, \cdots, v_d\}$ of distance values for each attribute $A \in R$. Given that $X = \{A_1, \cdots, A_m\}$, where $m \leq n$, the set of possible distance space of X is $\mathcal{W}_X = \{\mathcal{W}_{A_1} \times \cdots \times \mathcal{W}_{A_m}\}$. Examples of some of the distance metrics include: edit distance, cosine similarity (for textual values); and absolute value of difference (for numeric values).

A *differential function* (DF) of an attribute A w.r.t. the *distance interval* $w = [x, y]$, where $x \leq y \wedge x, y \in \mathcal{W}_A$ is denoted by $A[w]$. A DF, $A[w]$, returns a boolean value indicating whether $d_A(a_1, a_2)$ is within/on w or not for any two values a_1, a_2 of A. If $x = y$, for brevity, we simply denote $A[w]$ as $A[x]$. $A[x]$ is said to have a *point-interval*.

A DF on a set of attributes $X = \{A_1, \cdots, A_m\}$, denoted $X[W_X]$, is: $X[W_X] = A_1[w_1] \wedge \cdots \wedge A_m[w_m]$ where $A_i[w_i]$ is a DF of $A_i \in X$. A tuple pair t_1, t_2 is said to *agree on* (*satisfy*) a DF $X[W_X]$ if for all $A_i[w_i] \in X[W_X]$, $d_{A_i}(t_1[A_i], t_2[A_i])$ is within/on w_i. Let $\mathcal{T}(X[W])$ be *the set of all tuple pairs that agree on* $X[W]$.

Given any two differential functions (DFs) $X[W_X]$ and $Y[W_Y]$: $X[W_X]$ is said to *subsume* $Y[W_Y]$, denoted by $X[W_X] \succeq Y[W_Y]$, if and only if for all $A_i[w_i] \in X[W_X]$, there exists $A_i[w'_i] \in Y[W_Y]$ such that $w'_i \subseteq w_i$.

The distance intervals specified by DFs serve as the constraints in DDs.

Definition 1. *A DD is a statement* $\sigma : X[W_L] \rightarrow Y[W_R]$ *between two DFs* $X[W_L], Y[W_R]$. σ *holds over* r *of* R *if and only if for any two tuples* $t_1, t_2 \in r$, *if* $X[W_L]$ *returns true,* $Y[W_R]$ *returns true.* $X[W_L], Y[W_R]$ *are termed the LHS (determinant/antecedent) and RHS (dependent/consequent) of* σ *respectively.*

Given the sets Σ_1, Σ_2 of DDs, Σ_1 is said to be a *logical implication* of (or to *imply*) Σ_2 if for all instances, if Σ_1 holds, then Σ_2 holds. For example, if $\Sigma_1 = \{A[0,1] \rightarrow B[0,3], B[0,4] \rightarrow C[0,2]\}$, and $\Sigma_2 = \{A[0,1] \rightarrow C[0,2]\}$, then Σ_1 implies Σ_2.

Definition 2. *Let* Σ *be a set of DDs in* r. *A DD* $X[W_L] \rightarrow Y[W_R] \in \Sigma$ *is* **minimal** *if and only if the following conditions are satisfied: (a) there does not exist any DD* $X'[W_X] \rightarrow Y[W_R] \in \Sigma \wedge X'[W_X] \succeq X[W_L]$. *(b) there does not exist any DD* $X[W_L] \rightarrow Y'[W_Y] \in \Sigma \wedge Y[W_R] \succeq Y'[W_Y]$.

The set Σ_1 is a **cover** of the set Σ if every DD in Σ is in or is implied by DDs in Σ_1. A **minimal cover** Σ_c is a cover set of Σ such that there does not exist a cover $\Sigma' \subset \Sigma_c$.

For example, given the relation in Table 1, let $d_A(a_1, a_2) = |a_1 - a_2|$ for all $A \in R$. According to Definition 1, the DD $YoS[0] \to Edu[1]$ holds in r since $\mathcal{T}(YoS[0]) = \{1, 2\} \subseteq \mathcal{T}(Edu[1]) = \{\{1, 2\}, \{1, 3\}, \{2, 4\}, \{3, 4\}, \{4, 5\}\}$. Also, assuming $\Sigma = \{YoS[0, 5]Gen[0] \to Edu[0], YoS[0, 5] \to Edu[0, 2], YoS[0, 7] \to Edu[0], Gen[0]Sal[0] \to YoS[3, 5], Gen[0]Sal[0] \to YoS[3, 7]\}$, then a minimal cover of Σ is $\Sigma_c = \{YoS[0, 7] \to Edu[0], Gen[0]Sal[0] \to YoS[3, 5]\}$.

Association Rule. We recall here the basic concepts and definition of ARs [1]. Given a relation schema $R = \{A_1, \cdots, A_n\}$, let $dom(A_i)$ represent the domain of the attribute $A_i \in R$ for all $i \in [1, \cdots, n]$.

An *item*, $A\langle w \rangle$[1], refers to an attribute $A \in R$ with a value $w \in dom(A)$. The set of items with distinct attributes is known as an *itemset* $X\langle W \rangle = A_1\langle w_{A_1} \rangle \cdots A_m \langle w_{A_m} \rangle$. A *transaction* t is a sequence of values

Table 2. Distance relation $\mathcal{D}$ of r

TID	YoS	Gen	Edu	Sal ($K)
1	0	1	1	3
2	3	1	1	0
3	7	0	2	3
4	5	1	3	0
5	3	0	0	3
6	7	1	1	6
7	5	0	2	3
8	4	1	1	3
9	2	0	2	0
10	2	1	1	3

$v_1, \cdots, v_n$, where $v_i \in dom(A_i); i \in [1, \cdots, n]$. The set of all transactions under R represents an instance r of R.

A transaction $t \in r$ supports an itemset $X\langle W \rangle$ iff: for all $A_i \langle w_i \rangle \in X\langle W \rangle, w_i = v_i \ \forall \ i \in [1, \cdots, n]$. The *support*, $s(X\langle W \rangle)$ of an itemset $X\langle W \rangle$, is the proportion of transactions $t \in r$ that supports $X\langle W \rangle$. Thus, $s(X\langle W \rangle) = freq(X\langle W \rangle)/|r|$, where $freq(X\langle W \rangle)$ is the number of transactions in r that support $X\langle W \rangle$.

Given two itemsets $X\langle W_X \rangle, Y\langle W_Y \rangle$ such that $X\langle W_X \rangle \cap Y\langle W_Y \rangle = \emptyset$, an AR is an implication F of the form $X\langle W_X \rangle \xrightarrow{s,c} Y\langle W_Y \rangle$, where s, c are the support and confidence of F respectively. The AR F is said to be satisfied by the set of transactions in r if s and c are not less than their respective specified minimum.

The *support* of F is given as $s(F) = s(X\langle W_X \rangle \cup Y\langle W_Y \rangle)$; the *confidence* of F is the conditional probability of a transaction $t \in r$ to support $Y\langle W_Y \rangle$, given that it supports $X\langle W_X \rangle$. Thus, the confidence of F, $c(F) = s(X\langle W_X \rangle \cup Y\langle W_Y \rangle)/s(X\langle W_X \rangle)$. $X\langle W_X \rangle, Y\langle W_Y \rangle$ are termed LHS and RHS respectively.

For instance, Table 2 shows a relation on schema $R = \{YoS, Gen, Edu, Sal\}$. An example of an item is $YoS\langle 0 \rangle$; $Gen\langle 1 \rangle Edu\langle 1 \rangle$ is an itemset with frequency 5 and support of $\frac{5}{10}$. Given the minimum thresholds of both support and confidence to be 0.1, $YoS\langle 7 \rangle \xrightarrow{0.1, 0.5} Sal\langle 3 \rangle$ is an AR since $s(YoS\langle 7 \rangle Sal\langle 3 \rangle) = 0.1 \geq 0.1$ and $c(YoS\langle 7 \rangle \to Sal\langle 3 \rangle) = \frac{s(YoS\langle 7 \rangle Sal\langle 3 \rangle)}{s(YoS\langle 7 \rangle)} = \frac{0.1}{0.2} = 0.5 \geq 0.1$.

3   Problem Formulation

Below, we present a relationship between DDs and ARs: allowing existing ARs mining techniques in efficient discovery of DDs. Next, to avoid mining a large

---

[1] We note that we use $A\langle w \rangle$ for an item and $A[w]$ for a DF of A.

number of DDs, we restrict our problem to finding minimal DDs that represent *interesting* patterns in data; and state the formal problem definition.

Relationship Between DDs and ARs. Given an instance r of a relation schema R, there exists a finite set $\mathcal{W}_A$ of distance values for each attribute $A \in R$. To discover $\mathcal{W}_A$ for each $A \in R$, and more importantly to relate ARs and DDs, we generate the *distance relation* of the given r of R. Given a relation r on R, a distance relation $\mathcal{D}$ on R is a set of tuples defined as $\mathcal{D} = \{u \mid \forall A \in R, u[A] = d_A(t_1[A], t_2[A]) \wedge t_1, t_2 \in r\}$.

Given $\mathcal{D}$, the set $\mathcal{W}_A$ of distance values of an attribute A in r is given as: $\mathcal{W}_A = \{v \mid v \in dom'(A)\}$, where v is a value of A in $\mathcal{D}$, and $dom'(A)$ is the domain of A in $\mathcal{D}$. Furthermore, we can relate the ARs in $\mathcal{D}$ to the DDs in r. From the definitions in Section 2, we note the following: (a) each tuple in $\mathcal{D}$ represents a transaction; (b) each item $A\langle w\rangle$ in $\mathcal{D}$ is a point-interval DF $A[w]$ in r; and (c) an itemset $X\langle W_X\rangle$ represents a set of point-interval DFs $X[W_X]$.

From the foregoing, given the distance relation, $\mathcal{D}$, of an instance r of a relation R, the relationship between the ARs in $\mathcal{D}$ and the DDs in r is formally presented in Lemma 1 as follows.

Lemma 1. *Given the distance relation $\mathcal{D}$ of a relation r on schema R, if the DD $X[W_X] \to A[v_1, v_h]$ is satisfied by r, then, the set of ARs: $X\langle W_X\rangle \xrightarrow{\sigma, c_1} A\langle v_1\rangle$, $\cdots$, $X\langle W_X\rangle \xrightarrow{\sigma, c_h} A\langle v_h\rangle$ will hold in $\mathcal{D}$ and $\Sigma_{i=1}^{h} c_i = 100\%$, and vice versa.*

We show how ARs in $\mathcal{D}$ can be mapped to DDs in r by Lemma 1 in Table 3. Consider the relation, r, in Table 1 and its distance relation, $\mathcal{D}$, in Table 2. Any AR F with confidence $c(F) = 1.0$ (100%) is transformed into DDs σ such

Table 3. An illustration of Lemma 1

ARs in $\mathcal{D}$ (Table 2)	DDs in r (Table 1)
$YoS\langle 3\rangle Sal\langle 0\rangle \xrightarrow{0,1,1.0} Edu\langle 1\rangle$	$YoS[3]Sal[0] \to Edu[1]$
$Edu\langle 1\rangle \xrightarrow{0.5,1.0} Gen\langle 1\rangle$	$Edu[1] \to Gen[1]$
$Gen\langle 1\rangle Sal\langle 0\rangle \xrightarrow{0.2,0.5} YoS\langle 5\rangle$	
$Gen\langle 1\rangle Sal\langle 0\rangle \xrightarrow{0.2,0.5} YoS\langle 3\rangle$	$Gen[1]Sal[0] \to YoS[3,5]$
$Edu\langle 2\rangle Sal\langle 3\rangle \xrightarrow{0.2,0.5} YoS\langle 7\rangle$	
$Edu\langle 2\rangle Sal\langle 3\rangle \xrightarrow{0.2,0.5} YoS\langle 5\rangle$	$Edu[2]Sal[3] \to YoS[5,7]$

that, an item $A\langle w\rangle \in F$ corresponds to a point-interval DF $A[w]$ in σ (see 1st row of Table 3). Let Ω_A be a set of ARs with same LHS $X\langle W_X\rangle$, and same RHS attribute A. Given that $c(F) < 1.0$ (100%) for every AR $F \in \Omega_A$, and the sum of the confidence of all ARs in Ω_A is 100%, then by Lemma 1, Ω_A can be transformed into a DD $\sigma : X[W_X] \to A[w_a]$. Where (a) $X[W_X]$ is the corresponding DF of the common LHS $X\langle W_X\rangle$, of Ω_A and (b) $w_a = [v_1, v_h]$; v_1, v_h being the smallest and largest values of the RHSs in Ω_A (see 2nd & 3rd rows of Table 3).

From Lemma 1, given the ARs in a distance relation $\mathcal{D}$ of a given relation r, we can deduce the DDs that hold in r. This forms the foundation for the DD discovery approach presented in Section 4.

Problem Statement. The general goal of data dependency discovery is to find a *minimal cover* of all dependencies that hold in a given relation. That is, since the number of dependencies in a given relation can be extremely large, only an irreducible set of dependencies is of interest.

However, the more general definition of DDs makes finding even a minimal cover of DDs for a given relation more challenging. This is because, although the task of finding DDs has the same combinatorial explosion of attribute set as other data dependencies, the search space of DDs is significantly larger as different DFs may hold over one attribute. This leads to another complication: the gigantic size of a minimal cover set of DDs – making further evaluation, analysis and use of discovered DDs difficult.

On account of the huge size of a minimal cover of DDs in a given relation, we mine only DDs that: (a) are minimal; (b) represent persistent patterns in data; and (c) are reliable. We term such DDs as *interesting*, defined as follows.

Definition 3. *Let* ms, mr, *be the minimum support and recall thresholds. A DD* $\sigma : X[W_X] \to A[w_a]$ *is interesting iff.:* σ *is minimal;* $sup(\sigma) \geq ms$; $rec(\sigma) \geq mr$.

Where $sup(\sigma) = \frac{|\mathcal{T}(X[W_X]) \cap \mathcal{T}(A[w_a])|}{|\mathcal{T}(r)|} = \frac{|\mathcal{T}(X[W_X])|}{|\mathcal{T}(r)|}$ is the probability of occurrence of σ in r; and $rec(\sigma) = \frac{|\mathcal{T}(X[W_X]) \cap \mathcal{T}(A[w_a])|}{|\mathcal{T}(A[w_a])|} = \frac{|\mathcal{T}(X[W_X])|}{|\mathcal{T}(A[w_a])|}$ is the conditional probability of a tuple pair $t_1, t_2 \in r$ agreeing on $X[W_X]$ given that it agrees on $A[w_a]$. $|\mathcal{T}(r)|$ is the total number of tuple pairs in r.

From the above definition, a minimal DD (Definition 2) is only considered interesting if it has high support and recall values. Note that, *all valid DDs have 100% confidence*. Therefore, we use the support of a minimal DD to measure its persistence in data, and recall to judge its reliability.

Definition 4 (Problem Definition). *Given an instance* r *of a relation* R, *a minimum support,* ms, *and a minimum recall,* mr: *we find a minimal cover* Σ_c^i *of all interesting DDs in* r.

4   The DD Discovery Approach and Algorithms

The Approach. We propose a simple but effective and efficient process for mining a minimal cover Σ_c^i of all interesting DDs based on Lemma 1. The pseudo-code of this process is captured in Algorithm 1. Given a relation r on

---
Algorithm 1 AR-DDMiner

Input: The relation r on schema R, minimum support, ms, minimum recall, mr
Output: A minimal cover Σ_c^i of interesting DDs in r
1: generate the distance relation $\mathcal{D}$ of r
2: mine ARs in $\mathcal{D}$, setting support and confidence to $1/|\mathcal{D}|$
3: transform the set of ARs into DDs
4: prune the set of DDs
5: **return** minimal cover of interesting DDs

---

schema R, we generate the distance relation $\mathcal{D}$ of r (line 1). Next, we find the set Ψ of all ARs in $\mathcal{D}$ by setting both the minimum support and confidence thresholds to $1/N$ ($N = |\mathcal{D}|$) to ensure that we do not miss any DD (line 2). The third step involves the transformation of ARs to DDs (line 3). Finally, all implied and trivial DDs are pruned out to output a minimal cover Σ_c^i of all interesting DDs (lines 4, 5).

Generating the distance relation $\mathcal{D}$ of the given relation r on schema R is a simple yet essential step in the discovery solution. For each attribute $A \in R$, any distance function d_A can be adopted according to the domain of A to measure the closeness among the instance values of each attribute A in r. $\mathcal{D}$ can then be generated by a straight-forward process: a pairwise difference of all tuples in r. We therefore do not present any algorithm for this process. As an example, Table 2 is the distance relation of Table 1, assuming the absolute value of difference as the distance metric for all attributes.

Mining ARs in $\mathcal{D}$. Numerous ARs mining algorithms have been proposed in the literature over the past decades. We present here a discussion on the choice of appropriate ARs mining algorithm for our discovery process.

The efficiency and effectiveness of algorithms are two instinctive factors that influence the choice of one algorithm over another. However, in recent times, most data mining methods are comparable with respect to these factors. Given two rules $f_x : X \rightarrow A$ and $f_y : Y \rightarrow A$, such that $X \subseteq Y$; f_x is said to be a more general rule than f_y, and f_y more specific that f_x. Unlike ARs discovery [1], the concern in data dependency discovery is finding the most *general* rules. To reduce the cost of pruning the set of rules generated by the classical ARs mining algorithms, we choose the algorithm presented in [10] for mining a class of non-redundant ARs from $\mathcal{D}$ namely, *optimal rules*. An optimal ARs set [10] is the set of all ARs except the set of more specific rules with no greater confidence/support than any of their more general rules. Experiments to show the advantages of using non-redundant ARs mining algorithms (e.g. [10]) rather than the traditional ARs mining algorithms are reported in Section 5.

Transforming ARs to DDs. Here, the ARs mined from $\mathcal{D}$ are transformed into DDs of the given relation r. The procedure is presented in Function 1. The transformation consists of two processes namely: the point-interval (PI) and the multiple-interval (MI) transformation processes.

Function 1 Transforms ARs in $\mathcal{D}$ to DDs in r

1: **function** ARsToDDs(Ψ)
2:     **for each** $AR\ F \in \Psi$ **do**
3:         **if** $(con(F) == 100\% \wedge sup(F) \geq ms \wedge rec(F) \geq mr)$ **then**
4:             transform AR F to DD σ by Lemma 1; add σ to Σ_p
5:         **else if** $(con(F) < 100\%)$ **then** add F to Ψ'
6:     **for all** ARs $X\langle W_X \rangle \rightarrow A\langle w \rangle \in \Psi'$ **do**
7:         $\Upsilon_{X\langle W_X \rangle} \leftarrow comLHS(X\langle W_X \rangle)$; $\Upsilon \dashv = \Upsilon_{X\langle W_X \rangle}$
8:     **for each** $\Upsilon_{X\langle W_X \rangle} \in \Upsilon$ **do**
9:         **for all** AR $X\langle W_X \rangle \rightarrow A\langle w \rangle \in \Upsilon_{X\langle W_X \rangle}$ **do**
10:            $\Omega_A \leftarrow comRHSAtt(A)$
11:            **if** $(\Sigma_{i=1}^{h} con(F_i) == 100\% \ \forall\ AR\ F_i \in \Omega_A)$ **then**
12:                transform the set Ω_A of ARs to a DD σ by Lemma 1
13:                **if** $(sup(\sigma) \geq ms \wedge rec(\sigma) \geq mr)$ **then** add σ to Σ_m
14:    $\Sigma = \Sigma_p \cup \Sigma_m$
15:    **return** Σ

The PI transformation (lines 2–5) is a straight-forward process: a 100% confidence AR F is transformed into a DD σ if its support and recall are no less than the specified minimum ms, mr respectively. All DDs in this transformation have point-interval DFs. The set Σ_p contains all such DDs (line 4).

The MI transformation entails the conversion of the set Ψ' of ARs with confidence less than 100% to a set Σ_m of valid DDs (lines 6–13). Firstly, the set $\Upsilon_{X\langle W_X \rangle}$ of ARs in Ψ' with same LHS $X\langle W_X \rangle$ is formed for all LHSs of ARs in Ψ'. That is, the set Ψ' is reorganised into the set $\Upsilon = \{\Upsilon_{X\langle W_X \rangle} | \ X\langle W_X \rangle \in LHS(ARs)\}$ (lines 6,7). For each set $\Upsilon_{X\langle W_X \rangle} \in \Upsilon$, there exist subsets $\Omega_A \subseteq \Upsilon_{X\langle W_X \rangle}$ such that all ARs in Ω_A have the same consequent attribute A. Next, line 10 of Function 1 finds Ω_A for each attribute $A \in R$ in each $\Upsilon_{X\langle W_X\rangle}$. For all $\Omega_A \in \Upsilon$, if Ω_A the total confidence of all ARs in Ω_A is 100%: the set Ω_A is transformed into a DD σ, according to Lemma 1. σ is added to the set Σ_m if and only if the support and recall of σ is more than the specified minima (lines 11–13). Finally, the set of all transformed DDs Σ_m and Σ_p is returned in line 14 to Algorithm 1 as Σ. Pruning of Σ is the next step, performed by Function 2.

Pruning and Generating a Minimal Cover. The final step in the discovery process is the generation of a minimal cover Σ_c^i of interesting DDs (Definition 3) from the set Σ of valid DDs obtained from the previous stage. Note that, for all $\sigma \in \Sigma$, $sup(\sigma) \geq ms$ and $rec(\sigma) \geq mr$. Hence, we only need to find a minimal cover for the minimal DDs (Definition 2) in Σ. The process is described in Function 2 and consists of three key steps: left-reduction (lines 2–5); right-subsumption (lines 6–9); elimination of *transitively-implied* and *distance-trivial* DDs (line 10) to give Σ_c^i.

Left-reduction pruning. This pruning ensures that the first condition of Definition 2 is satisfied by all DDs in Σ_c^i. Given a DF $A[w]$, let $\Sigma_{det(A[w])} \subseteq \Sigma$ such that for all $\sigma \in \Sigma_{det(A[w])}, LHS(\sigma) \to A[w]$. Thus, $\Sigma_{det(A[w])}$ is the set of all DDs in Σ that deter-

Function 2 Prunes non-minimal DDs
1: **function** PRUNEDDs(Σ)
2: **for all** $X[W_X] \to A[w] \in \Sigma$ **do**
3: $\Sigma_{det(A[w])} \leftarrow$ find DDs with common RHS ($A[w]$)
4: $\Sigma'_{det(A[w])} \leftarrow$ find left-reduced DDs in the set $\Sigma_{det(A[w])}$
5: $\Sigma'+ = \Sigma'_{det(A[w])}$
6: **for all** $X[W_X] \to A[w] \in \Sigma'$ **do**
7: $\Sigma_{dep(X[W_X])} \leftarrow$ find DDs with common LHS ($X[W_X]$)
8: $\Sigma'_{dep(X[W_X])} \leftarrow$ find DDs in $\Sigma_{dep(X[W_X])}$ with smallest RHS
9: $\Sigma''+ = \Sigma'_{dep(X[W_X])}$
10: $\Sigma_c^i \leftarrow pruneF(\Sigma'')$
11: **return** Σ_c^i

mine $A[w]$. We are interested in the set $\Sigma'_{det(A[w])} \subseteq \Sigma_{det(A[w])}$ such that for all $X[W_X] \to A[w] \in \Sigma'_{det(A[w])}$ there does not exist a $Y[W_Y] \to A[w] \in \Sigma_{det(A[w])}; Y \subset X$, and $Y[W_Y] \succeq X[W_X]$. For example, given that $\Sigma_{det(Gen[0])} = \{Edu[0] \to Gen[0], Edu[0]Sal[0,2] \to Gen[0], Edu[0,1] \to Gen[0]\}$, it easy to derive the set $\Sigma'_{det(Gen[0])} = \{Edu[0,1] \to Gen[0]\}$. Thus, since $Edu[0] \subseteq Edu[0]Sal[0,2]$ we eliminate $Edu[0]Sal[0,2] \to Gen[0]$ from $\Sigma'_{det(Gen[0])}$ and $Edu[0,1] \succeq Edu[0]$, means $Edu[0] \to Gen[0]$ must be excluded from $\Sigma'_{det(Gen[0])}$.

Right-subsumption pruning. The right-subsumption pruning fulfils the second requirement of Definition 2. Given a set of DF $X[W_X]$, let the set of DDs in Σ with $X[W_X]$ as LHS function be $\Sigma_{dep(X[W_X])}$. Here, we are interested in the set $\Sigma'_{dep(X[W_X])} \subseteq \Sigma_{dep(X[W_X])}$ where for all $X[W_X] \to A[w_i] \in \Sigma'_{dep(X[W_X])}$, there does not exist any DD $X[W_X] \to A[w_j] \in \Sigma_{dep(X[W_X])}$ such that $A[w_i] \succeq A[w_j]$. As an example, let $\Sigma_{dep(Sal[0,2])} = \{Sal[0,2] \to Edu[0], Sal[0,2] \to Edu[0,1], Sal[0,2] \to Edu[0,3]\}$. We are interested in the set $\Sigma'_{dep(Sal[0,2])} = \{Sal[0,2] \to Edu[0]\}$ since $Edu[0,3] \succeq Edu[0,2] \succeq Edu[0,1] \succeq Edu[0]$.

Further pruning. The function *pruneF* in line 10 removes all transitively implied and distance-trivial DDs from Σ_c^i. Given that $YoS[0] \to Edu[1], Edu[1] \to Gen[1]$ hold in the relation in Table 1, by the property of transitivity, $YoS[0] \to$

$Gen[1]$ must hold in Table 1. Pruning out all transitively implied DDs ensures that the set Σ_c^i of interesting DDs is a minimal cover.

A DD $X[W_L] \rightarrow A[w]$ where $w = v_x, v_y \in \mathcal{W_A}$ is distance-trivial if v_x, v_y are the smallest and largest intervals values in $\mathcal{W_A}$ respectively. We consider these kinds of DDs non-interesting and prune them out because they always hold in any instance. For example, in Table 1, any DD with $YoS[0,7]$, $Gen[0,1]$, $Edu[0,3]$, or $Sal[0,6]$ as dependent DFs is distance-trivial and always holds. This is because the said DFs cover the entire distance space of each attribute.

5   Empirical Evaluations

Experimental Set-up and Data sets. The proposed algorithm in this paper is implemented in Java. The experiments were conducted on a 2.0 GHz AMD Athlon 64 processor computer with 2.0 GB of memory running Windows XP OS. Various experiments have been conducted on three real-world data sets namely: Adult, USair and DBLP. These data sets are processed: all tuples with missing values and repetitive columns are removed. The Adult data set is the 1992 US Census data with 15 attributes and 32,000 tuples. The USair is a data set on the on-time performance of US airlines. It contains 20 attributes and 500,000 tuples. The DBLP data set is a collection of conference/journal proceedings. We collect about 40,000 tuples and transform them into a 6 attributes relation after the elimination of columns with null values.

We defined various distance metrics to measure the distance among tuples in each data set. For example, word difference count for attributes with textual values, absolute values of difference for some numeric attributes and a special taxonomical distance for attributes with hierarchical values were used.

Results and Analysis. The experimental results and analysis are presented here.

Choice of AR mining algorithm. In this experiment, we show the time-benefit of adopting a non-redundant AR mining algorithm as opposed to a classical AR mining technique. We compare the time performance of our proposition when the two classes of AR miners are used. The results of our choice of ARs mining algorithm in [10] (Optimal AR) and a well-known efficient classical ARs mining algorithm implementation in [2] (Classical AR) are presented in Fig. 1.

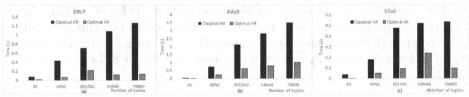

Fig. 1. Transformation & Pruning Time of Classical vs Optimal AR sets

The ordinate and abscissa of each graph (a, b, c) for the three data sets are the time performance in seconds (s) and the number of tuples in each $\mathcal{D}$ of r

respectively. A fixed schema size of 6 attributes for all data instances was used. It is clear that, for all instance sizes of all data sets, the time performances of our algorithm is better when the optimal AR method (gray bars) is used than when the classical AR method (black bars) is used. This is simply because the classical AR method returns more redundant rules. Hence, more time is required for transformation and pruning. This explains the patterns shown by Fig. 1 and validates the argument on choice of ARs mining algorithms in Section 4.

Time performance and scalability. Here, we show the practical feasibility of our algorithm (AR-DDMiner) w.r.t.: (a) existing methods; (b) varying number of tuples; and (c) varying schema size. The results are captured in Fig. 2.

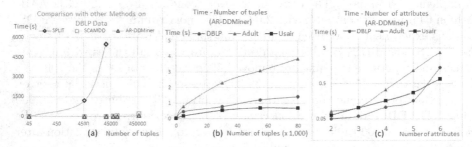

Fig. 2. Time performance

Part (a)[2] of Fig. 2 shows the performance of AR-DDMiner, SCAMDD in [9] and SPLIT[3] in [15] on the DBLP data set. It is noteworthy that although the three algorithms are different in approach and search different sections of the search space of DDs, they are complementary. Indeed, the SPLIT method finds minimal DDs with a fixed lower-limit (of zero) on DFs; SCAMDD mines minimal DDs with user-specified constraint on the upper-limit of LHS DFs; and AR-DDMiner discovers minimal DDs with point-interval LHS DFs. Therefore, a direct comparison of the time performance of the three algorithms is not useful. However, we included the existing works in our experiments to honour them and show the performance of AR-DDMiner w.r.t them. For a fair comparison, no minimum support and recall is set for our algorithm. Thus, we find a minimal cover for minimal DDs *not* interesting DDs. From the plot, it can be seen that AR-DDMiner and SCAMDD (constraint is set to strictest (0) for SCAMDD) have a comparable performance and are more efficient than SPLIT. This is because, it is often practically infeasibility to mine DDs in the entire search space even when the lower-limit of DFs is fixed to a minimum – like in SPLIT.

---

[2] Due to page limit, we show only the best performance of the other methods – DBLP.
[3] We name it Split since it 'splits' the search space.

In part (b) of Fig. 2, the graphs show how the AR-DDMiner algorithm performs on all three data sets. The number of tuples on this x-axis is scaled down by 1,000. From the performance graphs of the three data sets, we note that AR-DDMiner has a comparable the time performance on the USair and DBLP data sets, whereas the time performance on the Adult data set is incomparable to any of the other two data sets. This is explained by the nature of the data sets. The

Table 4. k in varying $\mathcal{D}$

$\mathcal{D}$	USair	DBLP	Adult
0.4	18	18	20
5.0	31	44	51
30.1	31	44	62
54.9	31	44	83
79.8	31	44	83

space of candidate DDs $\Phi(r)$ in a given instance r of a relation R with n attributes is given by $\Phi(r) \leq nk \times (1 + k)^{n-1}$, where k is the largest number of DFs on an attribute $A \in R$. In this experiment, n is fixed at six. Table 4 reveals that, the USair and DBLP data sets have similar k values and less than those of the Adult data set. Hence, time performance on the Adult data set is higher – bigger $\Phi(r)$.

The next category of experiments performed is on the time performance of AR-DDMiner on varying attribute sizes of each of the three data sets. For a fixed $\mathcal{D}$ size of 79,800 tuples, the projection of $n = (2 - 6)$ attributes of each data set was used to generate a relation. In Fig. 2 (c), the time taken to find a minimal cover of DDs (mine ARs, transform ARs to DDs, and prune the set of valid DDs) is shown on the y-axis against an x-axis of varying attribute sizes for each data set. Once more, the search space complexity relation $\Phi(R)$ can account for the patterns of the graphs in Fig. 2 (c) as follows. In this experiment, both k and n are variable. For $n = 2$, all three data sets have com-

Table 5. k in varying R

R	USair	DBLP	Adult
2	31	20	15
3	31	20	83
4	31	20	83
5	31	20	83
6	31	44	83

parable k values as seen in Table 5, explaining the closeness of the three graphs at $n = 2$. A critical look at Table 5 for $2 \leq n$ (or $|R|$) ≤ 5 clarifies the performances on the 3 data sets for the range. For instance, in this range, the DBLP data set has the smallest k values; clearly affecting the time performance (the best for range).

Effect of the parameters. In the evaluation of the time performance of our algorithms in the previous experiments, no ms and mr thresholds were set. The results, therefore, represent the worse execution time of AR-DDMiner. Setting ms, mr will obviously improve efficiency. We investi-

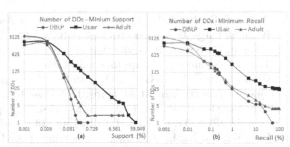

Fig. 3. Effect of minimum support and recall

gate here, the effect of ms and mr thresholds on the number of DDs discovered – we used 79,800 distance tuples and 6 attributes of each data set.

Parts (a) and (b) of Fig. 3 show the effects of setting only ms and only mr thresholds on the number of DDs mined respectively. In both experiments, the number of DDs mined drastically reduces for increasing ms and mr values. Indeed, from an ms value of 0.5% and an mr of 1.0%, significant amounts of

DDs were pruned. This shows the effectiveness of both measures in pruning DDs that may exist in data by chance or caused by noise (errors).

The plots of the support and recall values of all minimal DDs (Definition 2) for all three data sets are presented in Fig. 4. This clearly reveals that although there may be many minimal DDs, most have small support and recall values. Thus, not all of them necessarily capture prevalent patterns in data. This observation substantiates the need for measures to prune the set of minimal DDs: hence, our definition of *interesting* DDs (Definition 3).

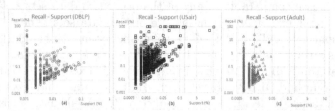

Fig. 4. Recall - Support Dist. of Minimal DDs of the Datasets

6   Conclusion

This paper introduces a new approach to DD discovery based on a link between DDs and ARs. Empirical evaluations show that the proposed algorithm, AR-DDMiner, is efficient and scalable; nonetheless, naturally there is still work to be done. In our next studies, we shall address two challenges to applying AR-DDMiner to very large datasets: (a) the cost of generating the distance relation $\mathcal{D}$ and (b) the need to set minimum support and confidence during AR mining phase low. Indeed, we shall explore techniques in sampling to allow parts of very large datasets to be accurately drawn to represent the whole and investigate the discovery of approximate DDs. Another line of work is in the application of discovered DDs to data management applications like semantic query optimization.

References

1. Agrawal, R., Imieliski, T., Swami, A.: Mining Association Rules between Sets of Items in Large Databases. SIGMOD Rec. **22**(2), 207–216 (1993)
2. Borgelt, C., Kruse, R.: Induction of association rules: apriori implementation. In: 15th Conference on Computational Statistics, pp. 395–400 (2002)
3. Chakravarthy, U.S., Grant, J., Minker, J.: Logic-based Approach to Semantic Query Optimization. ACM Trans. Database Syst. **15**(2), 162–207 (1990)
4. Fan, W., Geerts, F., Li, J., Xiong, M.: Discovering Conditional Functional Dependencies. IEEE Trans. Knowl. Data Eng. **23**(5), 683–698 (2011)
5. Fan, W., Jia, X., Li, J., Ma, S.: Reasoning about Record Matching Rules. PVLDB **2**(1), 407–418 (2009)
6. Golab, L., Karloff, H., Korn, F., Srivastava, D., Yu, B.: On Generating Near-optimal Tableaux for Conditional Functional Dependencies. Proc. VLDB Endow. **1**(1), 376–390 (2008)
7. Huhtala, Y., Krkkinen, J., Porkka, P., Toivonen, H.: Tane: An Efficient Algorithm for Discovering Functional and Approximate Dependencies. The Computer Journal **42**(2), 100–111 (1999)

8. Koudas, N., Saha, A., Srivastava, D., Venkatasubramanian, S.: Metric functional dependencies. In: 25th International Conference on Data Engineering, pp. 1275–1278. IEEE Computer Society (2009)
9. Kwashie, S., Liu, J., Li, J., Ye, F.: Mining differential dependencies: a subspace clustering approach. In: Wang, H., Sharaf, M.A. (eds.) ADC 2014. LNCS, vol. 8506, pp. 50–61. Springer, Heidelberg (2014)
10. Li, J.: On optimal Rule Discovery. IEEE Trans. on Knowledge and Data Engineering 18(4), 460–471 (2006)
11. Li, J., Liu, J., Toivonen, H., Yong, J.: Effective Pruning for the Discovery of Conditional Functional Dependencies. Computer Journal 56(3), 378–392 (2013)
12. Liu, J., Ye, F., Li, J., Wang, J.: On Discovery of Functional Dependencies from Data. Data & Knowledge Engineering 86, 146–159 (2013)
13. Novelli, N., Cicchetti, R.: FUN: an efficient algorithm for mining functional and embedded dependencies. In: Van den Bussche, J., Vianu, V. (eds.) ICDT 2001. LNCS, vol. 1973, pp. 189–203. Springer, Heidelberg (2000)
14. Song, S., Chen, L.: Discovering matching dependencies. In: 18th ACM Conference on Information and Knowledge Management, pp. 1421–1424 (2009)
15. Song, S., Chen, L.: Differential Dependencies: Reasoning and Discovery. ACM Trans. Database Syst. 36(3), 16:1–16:41 (2011)

Ontology Augmentation via Attribute Extraction from Multiple Types of Sources

Xiu Susie Fang$^{(\boxtimes)}$, Xianzhi Wang, and Quan Z. Sheng

School of Computer Science, The University of Adelaide, Adelaide, SA 5005, Australia
{xiu.fang,xianzhi.wang,michael.sheng}@adelaide.edu.au

Abstract. A comprehensive ontology can ease the discovery, mainte-nance and popularization of knowledge in many domains. As a means to enhance existing ontologies, *attribute extraction* has attracted tremen-dous research attentions. However, most existing attribute extraction techniques focus on exploring a single type of sources, such as structured (e.g., relational databases), semi-structured (e.g., Extensible Markup Language (XML)) or unstructured sources (e.g., Web texts, images), which leads to the poor coverage of knowledge bases (KBs). This paper presents a framework for *ontology augmentation* by extracting attributes from four types of sources, namely existing knowledge bases (KBs), query stream, Web texts, and Document Object Model (DOM) trees. In partic-ular, we use query stream and two major KBs, DBpedia and Freebase, to seed the attribute extraction from Web texts and DOM trees. We specially focus on exploring the extraction technique from DOM trees, which is rarely studied in previous works. Algorithms and a series of fil-ters are developed. Experiments show the capability of our approach in augmenting existing KB ontology.

Keywords: Knowledge base · Information extraction · Dom tree · Web data

1 Introduction

With the sheer amount of data produced and communicated over the Inter-net and the Web in the last few years, the Web has gradually evolved into a huge information repository with hidden knowledge. To explore this knowl-edge, researchers have developed various extraction techniques (i.e., *extractors*) to augment ontologies or enhance existing knowledge bases (KBs). While the existing ontologies have already included a wide range of entities, the number of attributes contained in these KBs is still small (see Table 1 for some statistics we have done). For example, Freebase has 25 million entities, but only 4,000 attributes. The type *University* in Freebase (note that in Freebase, classes are referred to as *types* and attributes are referred to as *properties*)[1] has only 9 prop-erties, while in reality we can easily identify more attributes. For this reason, it becomes urgent to find more attributes of classes for ontology augmentation.

---

[1] Hereafter, we will use the terms class and type, attribute and property interchange-ably.

© Springer International Publishing Switzerland 2015
M.A. Sharaf et al. (Eds.): ADC 2015, LNCS 9093, pp. 16–27, 2015.
DOI: 10.1007/978-3-319-19548-3_2

Table 1. Statistics of Representative KBs

KB	# Entities(million)	# Attributes
YAGO	10	100
DBpedia	4	6,000
Freebase	25	4,000
NELL	0.3	500

Although tremendous previous efforts have been conducted, they mostly extract attributes from a single type of Web sources, such as Web texts (e.g., [5]), Web tables (e.g., [9]) DOM trees (e.g., [3,12]), or relational databases. These approaches often lead to a poor coverage of the results.

There are generally three challenges. First, the single pattern (e.g., "*what is the A of E*" or "*the A of (the/a/an) E*") used by previous attribute extraction systems [6,11] no longer applies due to the poor coverage and incapability of filtering noisy inputs (such as "*what is the plural of apple*"). Second, since the same entity may have different attributes in different KBs, the attributes should be consolidated for better usage. To the best of our knowledge, there is no previous work on merging the attributes from different KBs. Third, as the DOM trees of Web pages may differ from one and another, it is tricky to develop an generic solution to exploring new attributes from such sources.

In our work, we propose more comprehensive attribute extraction from four types of sources, namely existing KBs (Freebase and DBpedia in our case), query stream, Web texts, and DOM trees, to resolve the above challenges. In a nutshell, this paper makes the following contributions:

- We propose a novel framework that extracts and merges the attributes from four types of sources, KBs (Freebase and DBpedia) and query streams, Web texts, and DOM trees, for comprehensive ontology augmentation.
- We develop an improved query stream extraction technique that adopts new patterns and filtering rules to improve the coverage and quality of the extractions.
- We develop an algorithm for extracting attributes from DOM trees. We use the attributes extracted from the query stream and existing KBs as seeds to learn the tag path patterns from Web pages, and use these patterns to extract more attributes from the DOM trees.

The remainder of this paper is organized as follows. Section 2 gives a brief overview of the related work. Section 3 introduces our framework and particularly compares with a recent research work named Biperpedia [6]. The methods for merging the attribute extractions from Freebase and DBpedia are also described in this section. Section 4 presents our method for query stream extraction. Section 5 describes the algorithm for extracting attributes from DOM trees using the above extractions as seeds. Finally, Section 6 reports the experimental results, and Section 7 provides some concluding remarks.

2   Related Work

In this section, we overview the related work on attribute extraction, particularly the approaches on multiple types of sources and DOM trees.

2.1   Extracting Attributes from Multiple Types of Sources

While attribute extraction has been widely studied in recent years, few works have been conducted to extract attributes from multiple types of data sources. Pasca et al. [15] are the first to exploit attributes from query streams. By using a head-to-head qualitative comparison, they conclude that extracting attributes from query stream achieves 45% higher accuracy than that from Web texts. Based on this insight, Pasca et al. [14] extract attributes from both query logs and query sessions, and Kopliku et al. [9] combine extractions from structured data sources including Web tables, search hit counts, Wikipedia, and DBpedia.

Comparing with above efforts, our approach is the first to extract attributes from four different types of sources, namely *query stream*, *Web texts*, *DOM trees*, and *existing KBs*. Our work is inspired by a very recent work conducted by Gupta et al. [6], which proposes a novel ontology named Biperpedia. We will discuss the differences between Biperpedia and our work in Section 3. Another related work is proposed by Lee et al. [11], which extracts attributes from query logs, Web documents, and external KBs independently to compute the typicality for a class (resp. attribute) given an attribute (resp. class). In contrast, our system extracts attributes from DOM trees and Web texts seeded by the attributes extracted from query stream and two major KBs (i.e., Freebase and DBpedia).

2.2   Extracting Attributes from DOM Trees

Extracting attributes from DOM trees is not completely new. Early supervised approaches [1,13] use manually defined wrappers to extract attributes from each Website, which are time-consuming and non-scalable. Wrapper learning techniques (e.g., [16] proposed by Turmo et al.) can help reduce human intervention, but additionally requires labeled data for the training, and are inapplicable to new websites that have not been handled before. Generative models designed in [17] alleviate this problem by segmenting and labeling the training samples, but they can only extract the attributes that are predefined in the training data. Interactive learning techniques developed by Irmak et al. [8] and Kristjansson et al. [10] can also help reduce human efforts on preparing the training data, but they are still not automated.

Unsupervised methods include *template-based* methods and *pattern-based* methods. The template-based methods, represented by RoadRunner (designed by Crescenzi et al. [4]) and EXALG (developed by Arasu et al. [2]), detect website-specific templates to extract attribute values. The pattern-based methods proposed by Liu et al. [12] and Bing et al. [3] extract data records from a single list page, based on some patterns that repeatedly occur in multiple

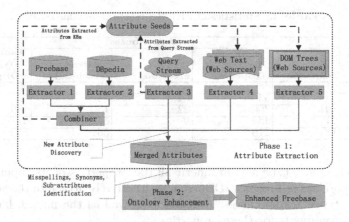

Fig. 1. The Framework for Ontology Augmentation

data records. Both methods however require re-implementation for new websites. Comparing with previous works, our approach enables more accurate and extensive attribute extraction from DOM trees automatically.

3 The Framework for Ontology Augmentation

Our framework (see Figure 1) contains two main phases: *attribute extraction* and *ontology enhancement*. At the *attribute extraction* phase, we extract attributes from DBpedia and Freebase, which are then combined. The details will be introduced in the remainder of this section. We also extract attributes from query stream (see Section 4). The resulting attributes will be used as seed to extract more attributes from the open Web. We adopt different methods to deal with Web texts and DOM trees. For Web texts, we first perform standard natural language processing (NLP), and then apply distant supervision to induce lexical patterns and use these patterns to extract more attributes. For DOM trees, we learn the tag path patterns and define several filters to extract attributes (see Section 5). At the *ontology enhancement* phase, we identify the misspellings, synonyms, and sub-attributes within the extracted attributes. Because our work focuses on the first phase, we simply reuse the methods in Biperpedia [6] for the ontology enhancement. Specifically, we use search engines to identify misspellings, a Support Vector Machine (SVM) to identify synonyms, and two heuristics[2] to identify sub-attributes.

Attribute extraction from Freebase and DBpedia. We use two dominant KBs, Freebase and DBpedia, for attribute extraction. Since Freebase contains the largest number of entities and isA pairs, we use Freebase as the basis for ontology augmentation.

---

[2] i) The former includes a modifier over the latter, such as *English teacher* and *teacher*.
  ii) The two attributes have relation like "*A1 is a A2*", e.g., *supervisor* is a *teacher*.

Table 2. Statistics of Five Representative Classes

Class	# Attributes				
	DBpedia	Extrac. (DBpedia)	Freebase	Extrac. (Freebase)	Combine (Freebase &DBpedia)
Book	21	48	5	19	60
Film	53	53	54	54	92
Country	191	360	22	150	489
University	21	484	9	57	518
Hotel	18	216	7	56	255

Intuitively, a sub-type should inherit all the properties of its super-type, so for every type/class in Freebase/DBpedia, we iteratively attach to the type/class all its super-types'/super-classes' properties, as well as the names, labels, and descriptions/comments for these properties.

Combining attribute extractions from DBpedia with Freebase. We combine the two KBs by attaching the attributes of every DBpedia class to its similar types in Freebase. By similar types, we mean the types with synonymous names of the class, or the types that have high overlaps (e.g., more than 50%) with the class in their contained entities. To avoid redundancy, we compare the attributes of Freebase and DBpedia in terms of name, label, and comment to determine if they are actually the same. For the attributes that have no comments/descriptions, we solely rely on their names, but leave the development of methods for annotating these attributes as our future work. Table 2 shows that our approach obtains more attributes for all the five representative classes in Freebase.

Comparison with Biperpedia. Our work distinguishes from the most similar approach, Biperpedia [6], in three aspects. First, we fuse existing KBs (Freebase and DBpedia), instead of using a single KB—as what Biperpedia does, for attribute extraction. Second, we define filters and more practical patterns for query stream extraction. Third, while Biperpedia regards Web tables meaningless, the value of Web tables for attribute extraction has been proved by many works (e.g., [9]). For this reason, besides Web texts, we additionally extract attributes from DOM trees of which Web tables are regarded as a sub-type.

4   Seed Extraction from Query Stream

Query stream is useful for attribute extraction because it naturally reflects users' collective convictions on possible attributes of entities. To extract attributes from query stream, for each type, T, in Freebase, we conduct an iterative procedure consisting of five steps, which will be discussed in this section.

The first step focuses on identifying relevant query stream. To do so, we apply an *entity recognizer* to identify the queries that contain entities of T and these queries as regarded as the relevant query stream of T. Then in the second step, we identify attribute candidates. We particularly exploit a set of predefined patterns, such as *"what/how/when/who is the A of (the/a/an) E"*, *"the A of (the/a/an) E"*, and *"E's A"*, to extract attribute candidates from the relevant query stream. For example, given relevant queries, namely "who is the director of Taken 3", "the release date of Taken 3", and "Taken 3's box office", we can identify that *director, release date* and *box office* as the attributes of type *Film* in Freebase according to above patterns. We denote the set of all the identified attributes as *Attri*, and the set of queries that match the above patterns as *Selected_QUERY*.

The third step focuses on filtering out non-attributes because the *Attri* set may be noisy. For example, the query text *"The University of Adelaide"* matches one of our patterns, but *"University"* is not actually an attribute of *"Adelaide"*. Similarly, for the query *"the plural of country"*, *"plural"* is not an attribute of *"country"*. We develop the following two rules to handle such cases.

Rule 1: We use a *blacklist* to exclude the generic attributes that appear in multiple different types. First, we select a set of dissimilar types from the query stream, and rank the attributes by the number of dissimilar types they belong to. We then add the top attributes to the blacklist to avoid their appearance in *Attri*. In this way, the words such as *"lack"*, *"rest"*, *"best"*, *"meaning"*, *"summary"*, *"definition"* and *"plural"* can be excluded from *Attri*.

Rule 2: We match each query in *Selected_QUERY* with the entities of *T* in Freebase to resolve the false negatives caused by long named entities. Specially, some queries with capitalized initials can be directly recognized as named entities and removed directly from *Selected_QUERY*. For example, *"Toyota"* is not an attribute of *"Glendale"* in *"the Toyota of Glendale"*. For this case, we directly remove this query from *Selected_QUERY*.

The fourth step is to identify entity-attribute pairs. For the relevant query stream of *T*, if a query contains both entity, denoted as *E*, of *T* in Freebase and attribute (denoted as *A*, $A \in Attri$), we keep the corresponding (A, E) pair in a set denoted as *AEpair*. For now, we only consider simple queries containing single entity and attribute. We will deal with complex queries with multiple entities and multiple attributes in our future work. The final step focuses on identifying credible attributes. To further improve the quality of the extractions, we first define the following three functions that will be used:

- *EntityNumber(T)*: The number of entities contained by *T* in Freebase.
- *EntityDiversity(T, A)*: The number of distinct entities co-appears with an attribute *A* in *AEpair* . We employ the standard co-reference resolution algorithm [7] to identify all the (A, E) ($\in AEpair$) pairs where *A* and *E* co-refer to the same entity, and delete all such pairs to reduce the redundancy.
- *EntityFrequency(E)*: The number of queries that contain *E* in the relevant query stream of *T*.

We develop two more rules to further clean the extractions as the following:
Rule 3: Given a type *T* containing entities $\{E_1, E_2, \ldots, E_k\}$ and an attribute *A*, *A* will not be attached to *T*, if $\exists E_j \in \{E_1, E_2, \ldots, E_k\}$, $EntityFrequency(E_j)$ $\geq \max_{(A, E*) \in AEpair} (EntityFrequency (E^*))$ and $(A, E_j) \notin AEpair$.

Rule 4: For each *EntityDiversity(T, A)* $\neq$ 0, we remove the (T, A) pair, if $\frac{EntityDiversity(T,A)}{EntityNumber(T)} \leq \alpha$ (a pre-defined threshold).

After the filtering, we finally obtain the credible attributes of type T in the form of $(A, EntityDiversity(T, A))$.

5   Extraction from DOM Trees

Web pages are typically semi-structured and described by nested HTML tags. The tree-like structures can be commonly found in Web pages that contain Web lists, Web tables, as well as deep-Web sources, and are referred to as DOM trees[3]. Traditional extractors simply remove the tags and extract data from the plain texts. Thus, they fail to exploit the knowledge contained in the DOM trees. In this paper, we introduce a two-step approach to extract attributes from DOM trees for ontology augmentation. We first extract additional attributes from the DOM trees seeded by the attributes extracted from query stream and existing KBs (denoted by *SEED_SET(T)*). We then define a set of filters to refine and differentiate the attributes. These two steps will be described in details in Section 5.1 and Section 5.2, respectively.

5.1   Original Extraction from DOM Trees

Different from attribute extraction from Web texts, where lexical and parse patterns can be learned and used all over the Web, extracting attributes from DOM trees is more tricky because different Websites have different styles and formats, and the tag path patterns extracted from one Web page can hardly be applied to another page. To resolve this challenge, our approach alternatively extract attributes and learns tag path patterns through an iterative process. The detailed procedure is described in Algorithm 1.

Briefly, given a type T, the algorithm first identifies the Websites related to T (e.g., http://www.imdb.com/ for type *Film*). For each Web page, the algorithm analyzes the DOM structure and classifies the text nodes into *entity node* (the texts represent the name of an entity E of T) and *non-entity node*. The tag paths between each *entity node* and their corresponding *non-entity node* are then extracted, removed of noisy tags, and kept in a *tag path set*. For each Website, the algorithm iteratively finds out Web pages that contain at least one (A, E) pair, where E is an *entity node*, A is the content of a *non-entity node* and $A \in SEED\_SET(T)$. For each Web page, the algorithm traverses the *tag path set* for this Web page to obtain the tag paths between the seed A and E, and transfers these tag paths from the *tag path set* to an *induced tag path pattern set* for this Web page. We next compare all the tag paths in the *tag path set* with the patterns in the *induced tag path pattern set*. Those *non-entity nodes* with tag paths that are similar with the induced patterns are finally recognized as new attributes, and are added to *SEED_SET(T)*, with the corresponding tag paths removed from the *tag path set*.

---

[3] http://www.w3.org/DOM

Algorithm 1. Algorithm for DOM Tree Extraction

Input: *Type T_k in Freebase; a set of Websites regarding to T_k, $S=\{S_1, S_2, \ldots, S_n\}$, for each Website $S_j \in S$, it contains a set of Web pages, $P_j=\{P_{j_1}, P_{j_2}, \ldots, P_{j_m}\}, j_m$ is the number of Web pages belong to S_j; the entity set Set_E of T_k in Freebase; the seed attribute set A_{T_k} extracted from query stream and existing KBs for T_k*

Output: *Original attributes for Type T_k in Freebase (i.e., enriched A_{T_k}).*

1 **Initialization**: identify all the *entity node* and *non-entity node* in every Web pages, and obtain *tag path set*(denote as Tagpath) for each Web page, e.g., for $P_{j_l} \in P_j$, we keep a set of tag paths Tagpath(P_{j_l}).

2 **for** *each $S_j \in S, j = 1, 2, \ldots, n$* **do**

3   **for** *each $P_{j_i} \in P_j, i = 1, 2, \ldots, j_m$, and P_{j_i} contains at least an entity $E \in Set_E$ and an attribute $A \in A_{T_k}$* **do**

         `/* if `$|A_{T_k}|$` is increased, the algorithm continues the loop for this Website;`

               `else the algorithm begins to traverse another Website            */`

4   extract the tag path(s) between E and A, and transfer them to the *induced tag path pattern set*;

5   compare all the other tag paths $\in$ Tagpath(P_{j_i}) with the induced tag path(s) in *induced tag path pattern set*;

6   **if** *(a tag path is similar to the induced tag path(s))* **then**

7   add the text of that *non-entity node* to A_{T_k};

8   remove the tag path from $Tagpath(P_{j_i})$;

The algorithm turns to another Website when the number of attributes in *SEED_SET(T)* reaches a certain threshold. Since the number of Web pages and text nodes in a Web page are limited, the algorithm can always terminate with an output.

5.2 Extraction Filtering

Similar to extractions from query stream, the attributes obtained by the first step in Section 5.1 may contain noises due to the open nature of Web content. We therefore employ the following three types of features to refine the extracted attributes:

- The inherent features of attribute: A node that denotes an attribute in a DOM tree always follows some inherent rules, e.g., the text node always contains a colon as the end of the string, and the length of the text is always limited to a certain number.
- The intra-site features of attribute: If a Website contains an attribute, the attribute tends to appear frequently in a considerable number of pages of this Website.
- The inter-site features of attribute: Attributes tend to appear in multiple Websites instead of very few Websites.

We can simply remove the attributes that mismatch these features, but this may result in some loss of recall. For example, the number of movies that win an Oscar award are quite limited. Thus, the attribute *"winner in Oscar"* would not appear frequently in the Web pages of a movie Website, which could surely dissatisfy the second feature. We sequentially use three filters to deliver three

Table 3. Inherent Features of DOM Trees

Feature	Description
Word count	The number of words of an attribute should be no more than 10 (we discover that almost all attributes in Freebase is described by less than 10 words)
End with colon	The text ends with a colon should be the name of an attribute
The first letter of every word is in the upper-case	Web page always capitalizes the first letter for each word of the name of an attribute

attribute sets in turn, i.e., potential attributes, attribute candidates and credible attributes. Each set represents a different balance between the precision and recall of the extraction results and can be used by knowledge-driven applications based on their own requirements. We exploit the *inherent feature filter* to obtain the set of potential attributes by using the specific rules followed in the DOM trees, see Table 3 for some examples.

For the *intra-site feature filter*, we remove all attributes with intra-site frequency lower than a predefined threshold β to obtain the attribute candidate set. We calculate the frequency of an attribute A_i in a Website $S_j (i = 1, 2, \ldots, j = 1, 2, \ldots)$ by $f_j(A_i) = \frac{N(A_i)}{N(S_j)}$, where $N(S_j)$ is the number of Web pages of each Website, and $N(A_i)$ is the number of Web pages that contain attribute A_i.

Finally, the intra-site feature filter may incorrectly take some Web site-specific terms as attributes. For example, "*edit*" appears frequently in IMDb (a famous movie Website), but seldom contained by other Web sites. We remove such terms by examining the inter-site frequency feature of each attribute. Based on above discussion, we can obtain the credible attribute set by keeping only the attributes that appear evenly and frequently in many different websites. Specifically, we calculate the inter-site frequency of an attribute, and use a predefined threshold γ to exclude the attributes with low inter-site frequency.

6   Experiments

We implemented the proposed approach in Java and conducted some preliminary experimental studies using an ASUS P550C computer with a 2.5 GHz i7 processor and 8 GB RAM. In this section, we report our experimental results on attribute extraction from query stream and DOM trees.

6.1   Experiments on Query Stream Extraction

We conducted experiments on five representative types in Freebase, namely *Book*, *Film*, *Country*, *University*, and *Hotel*, to validate the capability of our approach for extracting attributes from query streams. For the entity recognition, each class is specified as a set of representative entities of Freebase (Table 4). Since our goal is to extract attributes rather than attribute values and the entities of the same class should share the same attributes, pre-specifying the target class by a set of entities will not be a limiting factor. Based on a query stream of 29,283,918

Table 4. Entities in Representative Classes

Class	# Representative Entities	Examples of Entities
Book	1200	Asia Grace, Cool Tools
Film	1000	A Christmas Story, A Chump at Oxford
Country	727	Germany, Australia, Iran
University	1000	Brandeis University, Maynooth University
Hotel	1000	Hotel Sacher, Hotel Georgia

Table 5. Query Stream Extraction Results

Class	Relevant Query Records	Credible Attributes	Precision (%)			
			Top-10	Top-20	Top-50	Top-100
Book	259,556	96	80	65	62	N/A
Film	403,672	59	100	75	66	N/A
Country	393,244	182	100	96	95	93
University	24,633	20	100	100	N/A	N/A
Hotel	15,544	N/A	N/A	N/A	N/A	N/A

query records (which is the combination of two real-world datasets, Google[4] and AOL[5]), we finally obtained the extraction results as shown in Table 5.

We took the voting of three volunteers to determine the precision of the results. Volunteers manually gave their opinions on whether each attribute is reasonable for a class. The precision was calculated as the fraction of attributes that were labeled as reasonable. To measure the *precision*, we ranked the attributes of a class by *EntityDiversity(T, A)* attached to each attribute. Specifically, we evaluated the top-k (k=10, 20, 50, 100) attributes for each class. The evaluation results (see Table 5) indicate that more relevant query records lead to more reasonable attributes. The precision of the top-k attributes peaks at k=10, but decreases as k increases. This is consistent with our assumption, that the attributes appear with various entities would be more credible. Although the results have shown good precision (60%~100%), the query stream used for the experiments is still relatively small. For this reason, we can hardly obtain good attributes for some classes. For example, the class of *Hotel* has only 15,544 relevant query records, and no reasonable attribute could be found for it. On the other hand, as relevant query dataset of *Country* was much larger, where we successfully obtained 182 credible attributes. It is reasonable to anticipate that more attributes can be extracted if larger datasets are available.

6.2   Experiments on DOM Tree Extraction

We also conducted experiments for the same five classes to study the extraction performance of our approach for DOM trees. As inputs, we used the merged extractions from existing KBs (Table 2) and query stream (Table 5). For the entity recognition, we also used the same entity dataset for each class (Table 4). We exploited crawler4j[6] to craw the Web and jsoup 1.8.1 to reformat the

---

[4] https://code.google.com/p/hypertable/downloads/detail?name=query-log.tsv.gz
[5] http://www.cim.mcgill.ca/~dudek/206/Logs/AOL-user-ct-collection/
[6] http://code.google.com/p/crawler4j/

Table 6. DOM Tree Extraction Results

Class	# Attributes				Precision (%)
	Query Stream	Existing KBs	Seed Attributes	DOM Trees	
Book	96	60	118	168	81.5
Film	59	92	121	329	88.6
Country	182	489	621	725	92.7
University	20	518	536	539	93.3
Hotel	N/A	255	255	312	79.8

collected Web pages. We filtered out all the nodes with long text (more than ten words in our case) to avoid tackling too many non-attribute nodes. Similarly, we used voting of three volunteers to determine the quality of resulting attributes (Table 6).

From Table 6, we can see that more attributes were extracted from DOM trees than from either query stream or existing KBs. More seeds tend to lead to more attributes extracted from DOM trees. The results also demonstrate a high precision achieved by our approach. This is reasonable because the information contained in DOM trees are often more structured and cleaner than that of Web texts. Clearly, DOM trees are a high-quality source for attribute extraction, which unfortunately are not considered in many recent research efforts in knowledge base construction such as Biperpedia [6].

7   Conclusion and Future Work

In this paper, we have proposed a framework for ontology augmentation by extracting attributes from four types of sources (existing KBs, query stream, Web texts, and DOM trees). We combine the attribute extractions from existing KBs (Freebase and DBpedia) and improve the existing query stream extraction methods by introducing new extraction patterns and filtering rules. We then apply these attribute extractions as seeds to induce extractions from the open Web (Web texts and DOM trees). While Web texts extraction has been widely studied, we focus on attribute extraction from DOM trees. To the best of our knowledge, this is the first approach to extracting attributes from DOM trees using open information extraction techniques. Experimental results show that our system achieves more comprehensive yet still accurate ontology augmentation. Our future work will focus on the enhancement of attribute description and the further improvement for the DOM tree extraction.

References

1. Adelberg, B.: NoDoSE - A Tool for Semi-automatically Extracting Structured and Semistructured Data from Text Documents. ACM SIGMOD Record **27**(2), 283–294 (1998)

2. Arasu, A., Garcia-Molina, H.: Extracting structured data from web pages. In: Proceedings of ACM SIGMOD Conference (SIGMOD 2003), New York, USA (2003)
3. Bing, L., Lam, W., Gu, Y.: Towards a unified solution: data record region detection and segmentation. In: Proceedings of the 20th ACM Intl. Conf. on Information and Knowledge Management (CIKM 2011), New York, NY, USA (2011)
4. Crescenzi, V., Mecca, G., Merialdo, P.: RoadRunner: automatic data extraction from data-intensive web sites. In: Proceedings of the 2002 ACM SIGMOD Conference (SIGMOD 2002), New York, NY, USA (2002)
5. Grishman, R.: Information extraction: capabilities and challenges. In: Notes for the 2012 International Winter School in Language and Speech Technologies. Rovira i Virgili University, Tarragona (2012)
6. Gupta, R., Halevy, A., Wang, X., Whang, S., Wu, F.: Biperpedia: An Ontology for Search Applications. The VLDB Endowment (PVLDB) 7(7), 505–516 (2014)
7. Haghighi, A., Klein, D.: Simple coreference resolution with rich syntactic and semantic features. In: Proceedings of the 2009 Conference on Empirical Methods in Natural Language Processing (EMNLP 2009), Singapore (2009)
8. Irmak, U., Suel, T.: Interactive wrapper generation with minimal user effort. In: Proceedings of the 15th International Conference on World Wide Web (WWW 2006), New York, NY, USA (2006)
9. Kopliku, A., Boughanem, M., Pinel-Sauvagnat, K.: Towards a framework for attribute retrieval. In: Proceedings of the 20th ACM International Conference on Information and Knowledge Management (CIKM 2011), New York, NY, USA (2011)
10. Kristjansson, T., Culotta, A., Viola, P., McCallum, A.: Interactive information extraction with constrained conditional random fields. In: Proceedings of the 19th National Conf. on Artifical Intelligence (AAAI 2004), San Jose, California (2004)
11. Lee, T., Wang, Z., Wang, H., won Hwang, S.: Attribute extraction and scoring: a probabilistic approach. In: Proceedings of 29th International Conference on Data Engineering (ICDE 2013), Brisbane, Australia (2013)
12. Liu, B., Grossman, R., Zhai, Y.: Mining data records in web pages. In: Proceedings of the 9th ACM SIGKDD International Conference on Knowledge Discovery and Data Mining (KDD 2003), New York, NY, USA (2003)
13. Liu, L., Pu, C., Han, W.: XWRAP: an XML-enabled wrapper construction system for web information sources. In: Proceedings of the 16th International Conference on Data Engineering (ICDE 2000), San Diego, California, USA (2000)
14. Paşca, M., Alfonseca, E., Robledo-Arnuncio, E., Martin-Brualla, R., Hall, K.: The role of query sessions in extracting instance attributes from web search queries. In: Gurrin, C., He, Y., Kazai, G., Kruschwitz, U., Little, S., Roelleke, T., Rüger, S., van Rijsbergen, K. (eds.) ECIR 2010. LNCS, vol. 5993, pp. 62–74. Springer, Heidelberg (2010)
15. Pasca, M., Durme, B.V.: What you seek is what you get: extraction of class attributes from query logs. In: Proceedings of the 20th International Joint Conference on Artificial Intelligence (IJCAI 2007), Hyderabad, India (2007)
16. Turmo, J., Ageno, A., Català, N.: Adaptive Information Extraction. ACM Computing Surveys (CSUR) 38(2), 4-es (2006)
17. Zhu, J., Nie, Z., Wen, J.R., Zhang, B., Ma, W.Y.: Simultaneous record detection and attribute labeling in web data extraction. In: Proceedings of the 12th ACM SIGKDD Conference (KDD 2006), New York, NY, USA (2006)

Predicting Passengers in Public Transportation Using Smart Card Data

Mengyu Dou[1], Tieke He[1], Hongzhi Yin[2], Xiaofang Zhou[2],
Zhenyu Chen[1(✉)], and Bin Luo[1]

[1] State Key Laboratory for Novel Software Technology,
Nanjing University, Nanjing 210093, China
zychen@software.nju.edu.cn
[2] School of Information Technology and Electrical Engineering,
The University of Queensland, Brisbane, St. Lucia, QLD 4072, Australia

Abstract. Transit prediction has long been a hot research problem, which is central to the public transport agencies and operators, as evidence to support scheduling and urban planning. There are several previous work aiming at transit prediction, but they are all from the macro perspective. In this paper, we study the prediction of individuals in the context of public transport. Existing research on the prediction of individual behaviour are mostly found in information retrieval and recommender systems, leaving it untouched in the area of public transport. We propose a NLP based back-propagation neural network for the prediction job in this paper. Specifically, we adopt the concept of "bag of words" to build user profile, and use the result of clustering as input of back-propagation neural network to generate predictions. To illustrate the effectiveness of our method, we conduct an extensive set of experiments on a dataset from public transport fare collecting system. Our detailed experimental evaluation demonstrates that our method gets good performance on predicting public transport individuals.

Keywords: Transportation · Prediction · Smart card · Bag-of-words · Back-propagation neural network

1 Introduction

While much computing research has been concerned with theory and algorithms, as well as their improvements, the focus has recently been shifting toward human beings. Urban computing is one of these new research areas, the notion was first brought up in 2004 by Eric Paulos and introduced in his paper on Familiar Strangers [9]. It is an emerging field of study that focuses on taking advantage of technology in public circumstances, thus to facilitate cities, parks, forests and suburbs. It also studies the activities of humans in these circumstances, with the help of various devices that collect sensor data. It is well acknowledged that research in urban computing can be useful for extending existing theories in relevant fields and for further blending of these fields into a coherent understanding

© Springer International Publishing Switzerland 2015
M.A. Sharaf et al. (Eds.): ADC 2015, LNCS 9093, pp. 28–40, 2015.
DOI: 10.1007/978-3-319-19548-3_3

of public social life. Also, with the concept of "Intelligent City" becoming reality, digital "sensors" are distributed in every corner of the urban areas, such as WiFi hotspots and fare collecting systems. These technologies automatically record the human mobilities and the interactions without additional observers or interviewers and avoid build-in errors caused by human memory [2]. In our work, we use data from an automated fare collection public transportation system, it contains precise timing and location information for both boarding and alighting, which serves as a basis to generate information on spatial-temporal properties and passengers' spatial trajectories, also, transfer information can be derived from that.

Urban computing regards heterogeneous data sources in the first place due to its nature, from geospatial to temporal, as well as social network, text, images and so on. Major research issues in the area of urban computing include sensing city dynamics, computing with heterogeneous data sources and blending the physical and virtual worlds. To sense city dynamics, we are mainly talking about capturing traffic flow, human mobility, environment, populations and so on so forth, and the corresponding available data are from GPS traces of vehicles and people, ticketing data from public transportation systems, environmental sensor network as well as transaction records of credit cards, etc. Computing with heterogeneous data sources means a synergy of various data and discovery of collective knowledge, in the area of urban computing the most commonly adopted dimension of data is geospatial data, nevertheless, bringing in other dimensions of data such as temporal, text and images often invokes many interesting findings. While blending the physical and virtual worlds mainly concerns building systems to serve both people and cities as well as modelling city traffics and environments [14], which is mostly application driven.

Our intuition comes from two assumptions, 1) each passenger has his or her specific inherent regularity and mobility pattern encoded in the historical records, 2) more or less you are sharing some mobility patterns with your "friends" when using public transportation services. The word "friends" here is an obscure notion. It could be some real friends that you share the same trips, or school mates that follow similar timetables, or colleagues working at the same company. It can also be guys you are not even aware of, but you are actually following similar mobility patterns. Afterwards, we can use mobility patterns of a passenger's "friends" to complete his or her own ones. Then, we can move to the next stage, i.e., prediction. In the bus service, there are limited boarding and alighting stops. Each trip consists of one boarding stop and one alighting stop, and a passenger may have many trips within a given time range. We adopt the **bag-of-words model** to measure user similarity, i.e., we take each trip as a word, and trips of a passenger as the text, in that way we can classify or cluster these texts or say passengers, those fall into the same class can then be viewed as "friends". While for the prediction phase, we mainly adopt the back-propagation neural network, which we will discuss in detail in following sections.

To the best of our knowledge, our work is the first to consider predicting individuals in public transportation services. We explore this problem in the context of a real bus service, with the smart card data collected by a fare collection system in one month. Our proposed framework provides new sights on prediction in urban computing. Our main contributions are the following:

- We propose a NLP based back-propagation Neural Network which can effectively predict individuals using public transportation service in selected time range. Which provides another option for security department to deal with emergencies. Also, it is significant to transportation agencies in order to improve their service level by designing timetables, adjusting velocity and choose suitable dwell time at each stop.
- Our two step framework is of great value in many other related urban computing problems, especially for the prediction tasks. It also opens many opportunities of pushing this research on predicting individuals to the limit.
- We perform an extensive set of experiments to validate the effectiveness of our proposed framework. The experimental results is pretty good to support our assumptions, and we analyzed the results in depth.

The rest of this paper is organized as follows. First, we review the related work In Section 2. Then, we introduce our notations and define the problem formally in Section 3. Next, we present our solutions in Section 4. In Section 5, we evaluate our framework on real data, detailed experimental settings is also presented in this Section, together with a simple analysis of our dataset. Finally, we conclude our paper in Section 6.

2   Related Work

We study the related work in this Section, positioning our work in the research community.

Most previous work on prediction in the area of urban computing is about the road traffic and destination. In their work, Wang and Mao et al. [11] uncover the potential daily predictability of urban traffic patterns, using data of daily vehicle mobility based on GPS positioning device installed in taxis, their result indicates that each day has its specific inherent regularity and traffic pattern encoded in the historical records. There is another work which is also based on taxi data that deals with inferring human mobility [3], they mainly use a Hidden Markov Model based algorithm to identify trips, which gets relatively high precision and recall.

Based on human mobility and **POIs**, Yuan and Zheng et al. [13] proposed a framework to discover regions of different functions in a city. Others studied the composition of cities [1] and urban spatial-temporal structure [7]. He and Li et al. [4] proposed an algorithm to mine regular route. For the prediction, Xue and Zhang et al. [12] developed a system for destination prediction. Min and Wynter studied the real-time road traffic prediction in their work [8]. Also, Ishak and Kotha et al. [6] presented an approach for optimizing short-term traffic-prediction.

Our work on the individual-level prediction makes some use of the result of [11], i.e., each person has his or her specific inherent regularity and mobility pattern encoded in the historical records.

3 Problem Definition

We first introduce the preliminary concepts, and then we introduce the problem this paper seeks to solve. For convenience, we summarize the major notations used throughout this paper in Table 1.

Table 1. A List of Notations

Notation	Meaning
TR_a	A trip
T_i	The ith time frame of the day
D_i	The ith day of the week
S_i	The ith stop of a certain bus route
P_i	Denote a passenger i
F_i	Denote "friends" of passenger i
J_i	Journey of passenger P_i within given time period
n	The number of passengers
m	The number of stops
k	The number of time frames of operation hours
l	Total number of trips
$C_{i,p}$	Count of the pth trip in joueney J_i
$D_{i,j}$	The distance between journey i and journey j
S_i	The standard deviation of vector

3.1 Preliminary Concepts

We first provide the definitions for trip and journey in the context of public transportation service, in which we mean the bus service.

Trip. In our case trips are confined to time frame and a combination of boarding stop and alighting stop. Specifically, we use each hour as a time frame, denoted as T_i, also, this time frame is related to the day of the week, here we use D_i to stand for that, that is, D_1 represents *Monday* and D_2 means *Tuesday* and so on so forth.

Journey. We define a journey as a group of trips of a passenger. In which, we use P_i to represent passenger i, and correspondingly, $J_i = \{TR_a, TR_b, TR_c, \dots\}$ contains all the trips of passenger P_i within a given time period, note that TR_a means a trip.

3.2 The Problem

Problem: Given journeys of all the passengers $\{P_1, P_2, \ldots, P_n\}$ within historical time period, i.e., $\{J_1, J_2, \ldots, J_n\}$, predict whether these passengers will take the bus in the coming hours/days in future. That is, for each passenger, we will get a 1 if he/she will take the bus route 169 between 8 and 9 in the coming morning, and 0 if he/she won't.

4 NLP Based BP NN Prediction

We will describe in detail our proposed framework in this section. Fig. 1 illustrates the workflow of the framework.

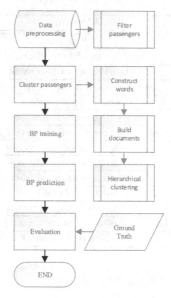

Fig. 1. Framework of Passenger Prediction

4.1 Construct Words and Documents

Given a certain bus route, which have m stops, each trip consists of one boarding stop and one alighting stop, so we get C_m^2 combinations in total. In our assumption, we think a passenger's mobility pattern is decided or affected by many hidden factors, such as time of the day, day of the week, weekday and weekend, as well as social factors and environments, and so on so forth. Because these social factors and environment elements are hard to collect, we only consider the time factor in this paper, specifically, we only consider the operation time and day of the week. We divide the operation time by hour, say, we have k hours every day of bus operation. Also, for every week, we have 7 days of operation. By definition, for a certain bus route we will have a total number of different trips as follows:

$$l = C_m^2 \cdot k \cdot 7$$

By now, we've got the words to construct the documents for each passenger, i.e., $\{TR_a, TR_b, \ldots\}$. Then, for passenger P_i, we have journey J_i, as defined above, composed of all these trips, or words. In that way, for each passenger, we get a document like $\{TR_a, TR_a, TR_b, \ldots\}$, which can be further converted as a vector of count of every trip, in the way of $\{C_1, C_2, \ldots\}$. That is the phase of constructing words and documents of our framework.

4.2   Cluster Passengers

In this step, we mainly adopt the hierarchical clustering to get groups of passengers, thus to put "friends" into the same group. As we don't know which passenger is similar to whom beforehand, or how many passengers would fall into the same group, hierarchical clustering is ideal for our task. Key steps of our applied hierarchical clustering is:

1. Build vector for each document, elements of the vector is the count of each word appeared in the document, such that if TR_a appeared twice in a journey, we have a number 2 for the corresponding element.
2. Journey of each passenger is regarded as a cluster in the first place, say, we have n clusters in the beginning. Compute the distance between every two journeys (vectors), here we adopt the weighted euclidean distance as Equation 1.
3. Find the closest pair of journeys (smallest weighted euclidean distance), and put them into one cluster, and in that way we accomplish one iteration.
4. Compute the distance between the new cluster and all the other old ones.
5. Iterate step 3) & 4) until we get to the final cluster with all journeys included.

$$D_{i,j} = \sqrt{\sum_{k=1}^{l} (\frac{C_{i,k} - C_{j,k}}{S_k})^2} \tag{1}$$

4.3   Back-Propagation Neural Network

Back-propagation is a method that monitors learning, it utilizes the method of mean square error and gradient descent to realize the modification to the connection weight of network. The modification to the connection weight of network is aimed at achieving the minimum error sum of squares. The initial weight of network is generally generated at random in certain interval, the training starts with an initial point and reaches gradually to a minimum of error along the slope of error function.

In this paper, we adopt the three-layer back-propagation network, suppose its input node is x_i, the node of hide layer is y_j, and the node of output layer is z_h. Weight value of network between the input node and node of hide layer is w_{ji}, and weight value of network between the nodes of hide layer and output layer is v_{hj}. And the expected value of the output node is t_h, $f(\cdot)$ is the active function. The details of the computational formula of the model can be referred in [5].

5　Experiments

In this section, we first introduce the dataset we are experimenting on, with a simple statistic analysis. Then we describe in detail how we set up the experiments and evaluate the experimental results.

5.1　Dataset

Smart card automated fare collection systems are being used more and more by public transit agencies [10]. On one hand, their main purpose is to collect revenue, on the other hand, they produce large quantities of very detailed data on boarding and alighting transactions. These data can be very useful to both transit planners and researchers. In our experiment, we collaborate with *Translink*[1] to perform such individual prediction in the context of public transportation service. *Translink* is a division of the department of transport and main roads, who is responsible for leading and shaping *Queensland*'s overall passenger transport system. It facilitates passenger transport service for *Queenslanders* and aims to provide a single integrated transport network accessible to everyone. We use data of a specific bus route, from **1 November 2012** to **29 November 2012**, of the "Inbound" direction. Totally, we get 28, 559 traffic transaction records for that period of 9968 passengers. For each tuple, we have format like:
< *RECORDID, OPERATIONS_DATE, ROUTE,*
DIRECTION, SMARTCARD_ID, BOARDING_TIME,
ALIGHTING_TIME, BOARDING_STOP,
ALIGHTING_STOP, JOURNEY_ID, TRIP_ID >
The meaning of value in each field is explained in Table 2.

Table 2. Meaning of Each Field

Field	Meaning
RECORDID	Key value specifies a transaction
OPERATIONS_DATE	Date the transaction took place
ROUTE	Route number of the bus
DIRECTION	Inbound/Outbound
SMARTCARD_ID	Encrypted unique id of passenger
BOARDING_TIME	Date/Time touch on a card
ALIGHTING_TIME	Date/Time touch off a card
BOARDING_STOP	Boarding Stop (ID & Description)
ALIGHTING_STOP	Alighting Stop (ID & Description)
JOURNEY_ID	Identify a journey by passenger
TRIP_ID	Identify the trip within a journey

Some statistical characteristics of the dataset we are using are presented in Table 3. Table 4 is about the number of passengers boarding and alighting along

---

[1] http://translink.com.au/

the sequence of all the stops, of direction *"Inbound"*, in which, $Stop_1$ stands for the first stop, and so on so forth. According to common sense, there are no passengers alighting at the first stop, and no passengers boarding at the last stop, we use *"N/A"* to represent that in Table 4.

Table 3. Statistical Characteristics

Parameter	Value
#Records	28559
#Passengers	9968
#Avg.records	2.865
#Max.records	26
#Min.records	1
#Busstops	10

Table 4. Passengers Boarding & Alighting

StopSequence	Boarding	Alighting
$Stop_1$	1227	N/A
$Stop_2$	3356	256
$Stop_3$	2061	668
$Stop_4$	553	193
$Stop_5$	442	185
$Stop_6$	1387	448
$Stop_7$	605	955
$Stop_8$	1649	485
$Stop_9$	120	94
$Stop_{10}$	N/A	7452

Fig. 2 illustrates the number of trips passengers take during the selected time period of our experiment. As we are performing experiments on dataset which records human activities, some social characteristics can also be seen from it. From Fig. 2, we can see that the distribution of number of trips passengers take follows the so-called "power law" phenomenon.

5.2   Finding "Friends"

In this step, we cluster all the passengers in order to ensure that every passenger have some "friends", or say passengers that share some mobility patterns. As described above, trips are divided by operation of hours and day of week, as well as the combination of boarding stop and alighting stop, specifically, we have operation hours between 6.am and 10.pm every day, that's 16 hours in total, and 7 days of operation, 10 bus stops, in that case we have a total number of words as follows:

$$l = C_m^2 \cdot k \cdot 7 = C_{10}^2 \cdot 16 \cdot 7 = 5040$$

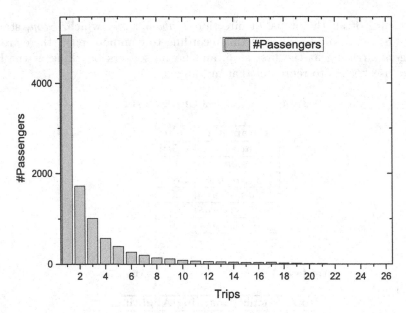

Fig. 2. Number of trips

Then, we can build documents or texts for each passenger from all these words. In the long run, passengers shall have multiple records of trips, for our case of one month's data, we filter passengers with less than 5 trips in order to gain more precise mobility patterns of passengers. After this preprocessing, we have 1581 passengers left. In our experiment, we use the first four weeks' data as the training data, and the records of the last day as ground truth test data.

After we performed the hierarchical clustering described in 4.2, we got 414 clusters at last, all clusters have more than 2 passengers which ensures each passenger have at least one "friend". Fig. 3 depicts the result of hierarchical clustering, there are 156 clusters with only 2 passengers and cluster with the largest number of passengers is 25.

5.3   BP NN Prediction

Now that we've got the clusters of passengers. In this paper, we want to make use of passengers' inherent regularity and mobility pattern, as well as the influence of his/her "friends" on him/her. For each passenger, we assume 9 parameters that have connection with his/her mobility pattern based on our data at hand, which are activities of the previous three weeks of the same hour, activities of the previous three days of the same hour, and activities of the previous three hours. The structure of our applied back-propagation neural network is shown in Figure 4. For the input layer, we have 9 nodes, as described above. Through the back-propagation neural network, we explore the influencing factor of each parameter.

The 9 input nodes for the back-propagation neural network are the activities in public transportation of the corresponding hours as well as the activities of

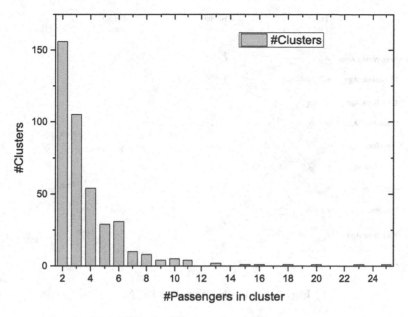

Fig. 3. Clusters of passengers

his/her "friends. We use α to represent the weight of his/her own activities and β to represent those of his/her "friends, in which α add β makes 1. It corresponds to our intuition that "more or less you are sharing some mobility patterns with your friends when using public transportation services". Also, if a passenger had a record in the corresponding hour, we use 1 to represent that, and 0 if none. For the part of his/her "friends, we use the accumulative value. In that way, the input node can be represented as follows:

$$Node_{input} = \alpha \cdot [0,1] + \beta \cdot \frac{Friends have activities}{Number of friends} \tag{2}$$

In the training process, BPN constructs an input-output mapping, adjusting the weights and biases at each iteration based on the minimization of some error between predicting values and desired outputs.

5.4   Evaluation

We mainly use *Precision*, *Recall* and *F1* to evaluate the effectiveness of our proposed method.

Precision is defined as the ratio of relevant items that are predicted to all the predicted ones. In our experiments, relevant items are passengers taking bus in the predicting hour, so the *Precision* can be calculated by following equation:

$$Precision = \frac{Relevant \cap PredictedOnes}{Number of Predicted}$$

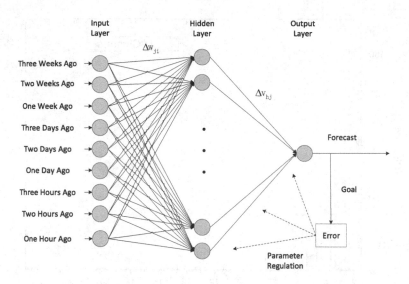

Fig. 4. Structure of applied BP NN

Recall is the defined as the ratio of relevant items that are predicted to all the relevant ones. In our experiments, for one specific hour, we have ground truth data of whom taking the bus, so *Recall* can be defined as follows:

$$Recall = \frac{Relevant \cap PredictedOnes}{Number of Relevant}$$

We expect both *Precision* and *Recall* to be good. However, they usually conflict with each other, improving one is usually at the expense of the other. Thus, F_1 measure is introduced to combine *Precision* and *Recall*. F_1 measure is calculated as follows:

$$F_1 = \frac{2 \times Precision \times Recall}{Precison + Recall}$$

The results of the three measurements are presented in Fig. 5, Fig. 6 and Fig. 7. In Equation 2, we have the constraint that $\alpha + \beta = 1$, in which α means how much the input weights on his/her own activities, and vice versa. From Fig. 5 we can see that our method gets good performance on *Precision*, in which the X-axis stands for the weight of one's own activities when preparing input for the training process. The *Precision* are all above 0.56 under different ratios, it increases when adding up the ratio till $\alpha = 0.5$, it stays stable around 0.63 after that. Fig. 6 depicts the result of *Recall*, which illustrates that we get *Recall* around 0.3. From Fig. 6 we can see that when taking both his/her own activities and those of his/her "friends" ($\alpha = 0.7$) we can get the peak value of *Recall*. It proves that incorporating the records of his/her "friends" can improve the effectiveness of our prediction. Fig. 7 is the result of *F1* measure, which mainly follows the shape of *Recall*, it vibrates around 0.42 under all ratios.

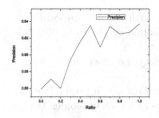
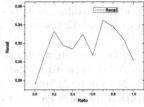
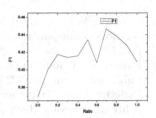

Fig. 5. Precision   **Fig. 6.** Recall   **Fig. 7.** F1 measure

6 Conclusions

In this paper we address the problem of prediction on individuals in the context of public transportation services. As far as we know, we are the first to mention such problem in the area of urban computing. We propose a NLP based back-propagation neural network to deal with the problem, specifically, we integrate the bag-of-words model and back-propagation pattern recognition. Results show our proposed method can effectively fulfil this job, which also validate our assumptions that passengers have their inner mobility patterns and more or less they are sharing these patterns with their "friends".

Acknowledgments. This is supported in part by the National Basic Research Program of China (973 Program 2014CB340702), the National Natural Science Foundation of China (Grant No. 11171148).

References

1. Cao, Z., Wang, S., Forestier, G., Puissant, A., Eick, C.F.: Analyzing the composition of cities using spatial clustering. In: UrbComp 2013: Proceedings of the 2nd ACM SIGKDD International Workshop on Urban Computing, New York, USA, pp. 14:1–14:8 (2013)
2. Eagle, N., Pentland, A.S., Lazer, D.: Inferring friendship network structure by using mobile phone data. PNAS **106**(36), 15274–15278 (2009)
3. Ganti, R., Srivatsa, M., Ranganathan, A., Han, J.: Inferring human mobility patterns from taxicab location traces. In: Proceedings of the 2013 ACM International Joint Conference on Pervasive and Ubiquitous Computing, pp. 459–468. ACM (2013)
4. He, W., Li, D., Zhang, T., An, L., Guo, M., Chen, G.: Mining regular routes from gps data for ridesharing recommendations. In: UrbComp 2012: Proceedings of the ACM SIGKDD International Workshop on Urban Computing, New York, USA, pp. 79–86 (2012)
5. Hecht-Nielsen, R.: Theory of the backpropagation neural network. In: International Joint Conference on Neural Networks, IJCNN, pp. 593–605. IEEE (1989)
6. Ishak, S., Kotha, P., Alecsandru, C.: Optimization of dynamic neural network performance for short-term traffic prediction. Transportation Research Record: Journal of the Transportation Research Board **1836**(1), 45–56 (2003)
7. Jiang, S., Jr., J.F., Gonzalez, M.C.: Discovering urban spatial-temporal structure from human activity patterns. In: UrbComp 2012, pp. 95–102, August 2012

8. Min, W., Wynter, L.: Real-time road traffic prediction with spatio-temporal correlations. Transportation Research Part C: Emerging Technologies **19**(4), 606–616 (2011)
9. Paulos, E., Goodman, E.: The familiar stranger: anxiety, comfort, and play in public places. In: pp. 223–230. ACM, New York (2004)
10. Pelletier, M.-P., Trépanier, M., Morency, C.: Smart card data use in public transit: A literature review. Transportation Research Part C: Emerging Technologies **19**(4), 557–568 (2011)
11. Wang, J., Mao, Y., Li, J., Li, C., Xiong, Z., Wang, W.-X.: Predictability of road traffic and congestion in urban areas. arXiv preprint arXiv:1407.1871 (2014)
12. Xue, A.Y., Zhang, R., Zheng, Y., Xie, X., Yu, J., Tang, Y.: Desteller: a system for destination prediction based on trajectories with privacy protection. In: Proceedings of the VLDB Endowment, vol. 6 (2013)
13. Yuan, J., Zheng, Y., Xie, X.: Discovering regions of different functions in a city using human mobility and pois. In: KDD 2012, pp. 186–194, August 2012
14. Zheng, Y., Liu, Y., Yuan, J., Xie, X.: Urban computing with taxicabs. In: UbiComp 2011, pp. 89–98, September 2011

Unifying Spatial, Temporal and Semantic Features for an Effective GPS Trajectory-Based Location Recommendation

Hamidu Abdel-Fatao[✉], Jiuyong Li, and Jixue Liu

School of Information Technology & Mathematical Sciences, University of South
Australia, Adelaide, Australia
hamidu.abdel-fatao@mymail.unisa.edu.au

Abstract. Location recommendation aims at providing personalized
suggestions of a set of new and potentially interesting locations to a
target user. The underlying principle of this problem is to predict the
Degree of Relevance of candidate locations to the user and make recommendations accordingly. Enormous attention has been devoted to this
problem by research and industrial community lately due to its applicability in numerous applications. In this work we develop an effective GPS
trajectory-based location recommendation framework for *Location Based
Social Networks*. We propose an algorithm, *STS Location Recommender*,
to leverage unique properties of GPS trajectories namely spatial, temporal and semantic features for recommendation. Our algorithm specifically
exploits temporal and semantic influence on users' mobility fused with
spatial properties of locations to model relevance of locations to users.
Prior to our work, no existing studies based on GPS trajectories simultaneously used all of these features for location recommendation. We
experiment on real-world GPS datasets to show that our approach provides more precise recommendations compared with baseline approaches.

Keywords: GPS trajectory data mining · Collaborative filtering · Location recommendation systems · Location-based social networks

1 Introduction

Location Based Social Networks (LBSNs) are new wave of interactive platforms
which allow users to share their geospatial locations alongside other location-
related contents such as comments, experiences etc. The advent of LBSNs is a
direct consequence of recent advances and ubiquity of Web and mobile technologies [1–3]. Since their evolution, LBSNs have received considerable attention from
research and industry because they play crucial role in the development of many
important applications [2,4]. Popular among these are location recommendation
systems, urban planning, counter-terrorism etc.

The object of this study is to develop a GPS trajectory-based Location Recommendation System (LRS) for LBSNs. This problem fundamentally aims at

© Springer International Publishing Switzerland 2015
M.A. Sharaf et al. (Eds.): ADC 2015, LNCS 9093, pp. 41–53, 2015.
DOI: 10.1007/978-3-319-19548-3_4

exploiting historical GPS trajectory data to provide personalized suggestions of previously unvisited locations to users in LBSNs. The core issue is to predict, with a reasonable level of accuracy, the *Degree of Relevance (DoR)* of candidate locations to users and to make recommendations accordingly. For example if a user highly favours a particular type of restaurant e.g. Asian restaurants, LRS can infer this and make recommendations accordingly.

The major challenges that confront LRSs entail how to effectively model the DoR of locations to users that is - (i) able to leverage spatial features, temporal features and semantic features of locations for recommendation, (ii) personalized and diverse i.e. capable of recommending truly relevant locations which represent all possible interests to users. This is challenging considering the number of factors that must by taken into account.

In a bid to address these challenges a number of techniques have been proposed. For example Ye et al [5] exploit three factors namely user preference, social influence and geographical influence for *PoI* recommendations based on user check-in datasets. Yuan et al [3] is similar to [5] except that, [3] considers temporal factor as an additional constraint for time-specific PoI recommendation. Zheng et al al. [6] proposed a matrix decomposition technique using GPS trajectories combined with activity information, for global location and activity recommendations. As an extension to [6], Zheng et al [1] proposed *UCLAF* that incorporates a user dimension as an additional entity for personalized location and activity recommendations using Tensor decomposition technique.

These existing works are riddled with either one or both of two major flaws. These flaws include the fact that (*i*) *they are geographically constrained* − cannot make recommendations when there exist no geographical overlap between a target user's location history and candidate locations. (*ii*) *they are time-unaware* − only recommend locations globally and cannot tell where a user will like to be at a specific time. This is because majority of the works [2,5,7] only rely on spatial features for recommendation. Few others [1,6,8] consider semantic features in addition to spatial features. Even fewer works [3,4] consider temporal features in addition to spatial features. To the best of our knowledge, no existing work based on GPS trajectories simultaneously used all the three features for location recommendation. Our work bridges these knowledge gaps by considering all three features simultaneously to provide more precise time-aware and semantically meaningful recommendations.

We summarize the main contributions of this paper as follows.

- We develop a novel approach for modeling DoR of locations to users that simultaneously exploits semantic, temporal and spatial features of locations.
- We develop an effective algorithm *STS Location Recommender*, the provides more precise, time-aware and semantically meaningful locations recommendations based on historical GPS trajectory data.

We evaluate our proposed location recommendation framework by conducting experiments using real-world GPS dataset. Experimental results show that our proposed LRS is more precise compared with baseline approaches.

The rest of this work is organized as follows. We explain relevant concepts and our problem statement in Section 2. In Section 3, we detail our proposed technique, present our experiments in Section 4 and conclude in Section 5.

2 Statement of Problem

In this section, we clarify relevant concepts and present formal statement of the problem addressed in this paper.

2.1 Preliminaries

Historical human mobility behaviour can be reconstructed using traces of geographic points known as trajectory points. A *Trajectory Point* p is a spatial point associated with a timestamp, denoted by the triple $p = (x, y, t)$ where x and y are the latitude and longitude respectively of point p at timestamp t.

UserID	Date	Time	Latitude	Longitude
1509	2008-02-06	16:29:31	123.5322	42.30713
1509	2008-02-06	16:29:36	123.53218	42.30713
1509	2008-02-06	16:29:41	123.53217	42.30713
1509	2008-02-06	16:29:46	123.53217	42.30714
1509	2008-02-06	16:29:51	123.53217	42.30713

Table 1. Sample Trajectory Dataset

Definition 1. *Trajectory* denoted by $P = \langle p_1, p_2, ..., p_z \rangle$ is a sequence of trajectory points organised in ascending order of timestamps, where $\{p_i \in P : p_i = (x_i, y_i, t_i)\}$ is a trajectory point and $t_i < t_{i+1}, \forall i \in [1, z]$.

Typically, raw trajectory datasets (see sample in Table 1) are available as very large volumes of geographically close points. Analysing such datasets directly can introduce significant computational overheads. To curb this problem, we follow the intuition that when people visit places of interest, they stay within nearby areas for significant periods of time. For example, in a cinema people usually stay within the Cinema Hall for considerably periods of time watching movies. We therefore extract such significant areas called *Stay Points* from trajectories.

Definition 2. *Stay Point* denoted by $s = [(x, y), t_a, t_s]$ is a geographical area characterized by a maximum distance threshold δ_d where a user stayed for at least a minimum threshold period of time δ_t, and x, y, t_a and $(t_s \geq \delta_t)$ are respectively latitude, longitude, arrival time and stay time of s.

To extract stay points, we implement an existing algorithm by Zheng et al [2]. We chose this approach for its intuitive and consistency with our definition. Figure 1a illustrates the approach diagrammatically (please refer to [2] for depths).

Transformation of trajectory points to stay points drastically

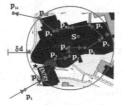

(a) Stay Point   (b) Reference Points

Fig. 1. Trajectory Transformation

reduces the trajectory data. However, the volume of stay points can still be reduced by clustering close and adjacent stay points representing the same location but having slightly different coordinates. To achieve this we employ a density-based clustering algorithm *OPTICS* [9] to cluster geographically close stay points into non-overlapping clusters. We represent each discovered cluster with a single point called *Reference Point* (see Figure 1b) defined below.

Definition 3. *A Reference Point denoted by* $r = [(x_r, y_r), t_a, t_s]$, *is a representative of a cluster of stay points* $S_c = \{s'_1, s'_2, ..., s'_q\}$, *where* (x_r, y_r) *is the average coordinate of the stay points* $s'_i \in S_c$, t_a *and* t_s *are respectively the earliest arrival time and mean stay time of the stay points in* S_c.

Since our core objective in this study is to recommend semantically meaningful locations to users, it is necessary to enrich reference points discovered with their underlying semantic tags such as gym, restaurant, park etc. Fortunately, many places in LBSNs have been labeled with semantic tags [8]. We therefore take steps to annotate each reference point with its corresponding semantic tag and call it a *Semantic Location*.

Definition 4. *Semantic Location L, denoted by* $L = [(x_r, y_r) : l_f, t_a, t_s]$ *symbolizes a reference point annotated with a semantic tag where* l_f *represents semantic tag, and* $(x_r, y_r), t_a, t_s$ *have their usual meanings.*

To date, there exists no universally accepted standard for assigning precise semantic tags to geographic points even though there has been attempts by some studies [10,11]. In this work, we use of a *Point of Interest (PoI)* database for this task. A PoI database is a corpus of PoIs such as restaurant, shop, cinema, etc associated with geographic coordinates. Specifically, we use Foursquare[1] category database because it's PoI data is highly reliable [12] and used by popular search engines such as Microsoft Bing.

Having obtained semantic locations from reference points we model users' mobilities as sequences of semantic locations called semantic trajectories.

Definition 5. *Semantic Trajectory denoted by* $T_s = \langle L_1, L_2, ..., L_m \rangle$ *is a time-ordered sequence of semantic locations where* $L_i \in T_s$ $(1 \leq i \leq m)$ *is a semantic location.*

As an example, a semantic trajectory of a user can be represented by the sequence $\langle [(39.9993, 116.3269) : \text{Hotel}, 13:30, 72 \text{ min}] \rightarrow [(39.9952, 116.3272) : \text{Park}, 14:43, 54 \text{ min}] \rightarrow [(39.8201, 116.2766) : \text{Restaurant}, 17:02, 32 \text{ min}] \rangle$.

2.2 Semantic Trajectory Pattern Mining

A *Semantic Trajectory Pattern* represents routine mobility behaviour of a user discovered from his/her semantic trajectories. For example, if on most days a user goes to work at 9 am, restaurant at 12 pm and gym at 6 pm we call such a mobility behaviour a semantic trajectory pattern.

---

[1] www.foursquare.com

A number of works [8,13] have tackled the problem of mining mobility patterns. These studies, inspired by the fact that user mobility is typically sequential in nature, perform sequential pattern mining on preprocessed mobility histories to discover frequent mobility behaviours. However, most of these works neglect temporal information in their approaches. For example Ying et al [8] represent a semantic trajectory pattern in the form ⟨{Hospital}{Park}{Bank}⟩, which clearly does not contain temporal information. We argue that, temporal information is crucial in understanding mobility behaviours because users' mobilities typically exhibit temporal patterns. For example, a user might visit a restaurant at 12 pm everyday, but he/she may typically visit other places at nighttime.

In our work, we take temporal information into account in mining semantic trajectory patterns. To facilitate this, we encode each semantic location with an equivalent numerical representation in a three-step process. Firstly, we assign the semantic tag associated with each semantic location a unique integer identifier. For example (Restaurant $\Rightarrow$ 103) denotes a transformation of *Restaurant* to an integer value 103. Secondly we split each day into six equal four-hourly non-overlapping time slots e.g. [00:00 − 03:59]$\Rightarrow$ 1, [04:00 − 7:59]$\Rightarrow$ 2 etc. Finally we identify each stay time by integers 1, 2 or 3 denoting short, medium and long stay times respectively. As an example, the semantic location [(39.8201, 116.2766) : Restaurant, 17:02, 32 mins] is transformed to [103, 4, 1] meaning the user visited semantic feature 103 during time slot 4 and stayed for short time period. We then perform sequential pattern mining on the transformed dataset to extract frequent semantic trajectory patterns.

2.3   Problem Formalization

Having explained basic concepts, we now detail our problem definition.

Problem Definition Given a target user $u_i \in U$ such that $U = \{u_1, u_2, ..., u_n\}$ is a set of users in a city, the problem is to recommend *top k previously unvisited and semantically meaningful locations* that u_i might be interested in at time t, based on his/her location preferences and current location in the city.

To address this problem there is a need to : (i) precisely model u_i's preferences for past locations using his/her location histories, (ii) accurately estimates the DoR of each previously unvisited location to u_i based on his/her preferences. We propose an algorithm, *$\underline{S}$patio-$\underline{T}$emporal and $\underline{S}$emantic-Aware (STS) Location Recommender* to solve the problem. The novelty of our algorithm lies in its ability to leverage semantic, temporal and spatial feature for time-aware and semantically meaningful location recommendation even where there exist no geographical overlap between unvisited locations and users' location histories.

3   The STS Location Recommender

In this section, we elaborate on the STS Location Recommender. The recommender takes a three-step approach summarized as follows: (i) Location preference estimation (ii) User similarity estimation (iii) User-based collaborative location recommendation. We explain each step in the following subsections.

3.1   User Preference Estimation

A User's preference for a semantic location is a measure of interestingness of the location to the user. We express this measure mathematically in terms of *Preference Score* defined and formulated as follows.

Definition 6. *Preference Score* of a location with respect to a target user during a specified time slot is a numerical estimate of the likelihood of the user visiting the location during the given time slot.

Preference score comprises two component probabilities viz *Visit Probability* in terms of *(i) Semantic Influence; (ii) Temporal Influence*, derived as follows.

Suppose $T_s(u) = \{L_1, L_2, ..., L_m\}$ denotes a set of semantic locations visited by a user u and let $\tau(u) = \{l_1, l_2, ..., l_m\}$ be corresponding set of semantic tags of $T_s(u)$, where each $l_i \in \tau(u)$ is the semantic tag of $L_i \in T_s(u)$ $(1 \leq i \leq m)$. Let ϕ_{l_i} be a binary variable denoting a visit to a location tagged by l_i. *Visit Probability in terms of semantic influence* $P(l_i)$ of l_i is expressed as $P(l_i) = w_s \times c(\phi_{l_i})/|\tau(u)|$, where $c(\phi_{l_i})$ is total count of ϕ_{l_i} and w_s is a weight expressed in terms of popularity of l_i. That is, $w_s = 1 + log(N_u/N_{l_i})$ where N_u is total number of users, and N_{l_i} is the number of users who visited a location tagged by l_i.

Also, suppose $T_s^t(u) = \{L_1^t, L_2^t, ..., L_k^t\}$ denotes a set of semantic locations visited by u during a specific time slot t and let $\tau^t(u) = \{l_1^t, l_2^t, ..., l_k^t\}$ be corresponding set of semantic tags of $T_s^t(u)$, where each $l_i^t \in \tau^t(u)$ is the semantic tag of $L_i^t \in T_s^t(u)$ $(1 \leq i \leq m)$. Let $\phi_{l_i^t}$ be a binary variable denoting a visit to a location tagged by l_i by u during t. *Visit probability in terms of temporal influence* $P(l_i^t)$ of l_i is expressed as $P(l_i^t) = w_t \times \sum \phi_{l_i^t}/|T_s^t(u)|$, where $w_t = 2^{\overline{\delta_t(l_i^t)}/\sum \overline{\delta_t(l_j^t \in \tau^t(u))}}$ such that $\overline{\delta_t(l_i^t)}$ is the mean stay time at l_i during t and $\sum \overline{\delta_t(l_j^t \in \tau^t(u))}$ is the sum of mean stay times for all locations visited during t.

Finally, *Preference Score* for a semantic location with semantic tag l_i with respect u during time slot t expressed in terms of $P(l_i)$ and $P(l_i^t)$ is given by

$$P_u(l_i, t) = \lambda P(l_i) + (1 - \lambda)P(l_i^t) \tag{1}$$

where λ in Equation 1 is a tuning parameter that controls the influence of semantic and temporal factors on $P_u(l_i, t)$.

3.2   User Similarity Estimation

User similarity measures the extent to which two users share preferences for semantic locations visited during specified time slots. Intuitively, users who share similar lifestyles will exhibit similar preferences for semantic locations during specific time slots. We estimate similarity between any two users using Pearson's correlation coefficient similarity metric explained as follows.

Let $\tau^t = \tau^t(u) \cup \tau^t(v)$ be a set of semantic tags corresponding to semantic locations visited by users u and v during a given time slot t. For u, there exists a preference score $P_u(l_i, t)$ for $l_i \in \tau^t$ if u has visited l_i. Similarly for v, there exist a preference score $P_v(l_i, t)$ if v has visited $l_i \in \tau^t$. Given vectors of preference scores for u and v corresponding to semantic locations visited during t, the similarity between u and v during t is given by

$$sim^t(u,v) = \frac{\sum\limits_{l_i \in \tau^t} [P_u(l_i,t) - \overline{P_u(\tau^t(u))}][P_v(l_i,t) - \overline{P_v(\tau^t(v))}]}{\sqrt{\sum\limits_{l_i \in \tau^t} [P_u(l_i,t) - \overline{P_u(\tau^t(u))}]^2} \sqrt{\sum\limits_{l_i \in \tau^t} [P_v(l_i,t) - \overline{P_v(\tau^t(v))}]^2}} \qquad (2)$$

where $\overline{P_u(\tau^t(u))}$ and $\overline{P_v(\tau^t(v))}$ are the mean preference scores of locations visited by u and v during t respectively.

Given a set of time slots T, the overall similarity between the two users measured over all time slots is given by

$$Sim(u,v) = \frac{\sum\limits_{t \in T} sim^t(u,v)}{|T|} \qquad (3)$$

3.3 Location Recommendation

In this subsection we present our algorithm for location recommendation.

Our idea follows user-based Collaborative Filtering (CF) [14] model. We utilize the intuition that like-minded individuals who share similar lifestyles are most likely to visit similar semantic locations at similar times. For example, two users who exhibit similar levels of preferences for nightlife are likely to visit say, a cinema for movies at night time.

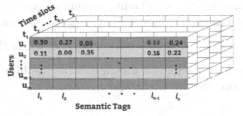

Fig. 2. Aggregated Users' Location Histories

Specifically, we treat each semantic location visited at a specific time as an "item" and users' preference scores for these locations as their implicit ratings on the locations. We then build user-location matrix for each time slot, where entries are preference scores for corresponding user-location pairs. For all time slots the user-location matrices are

---

Algorithm 1: STS Location Recommender

Input: Target user u, set of users U, user-location matrices M for predefined time slots T

Output: *Top-N* recommended locations for u at $t \in T$

foreach $v \in U$ **do**
  | $Sim(u,v) \leftarrow$ compute similarity between u and v using Equation 3
$U_K \leftarrow$ select K users with the largest $Sim(u,v)$ values
foreach $v_K \in U_K$ **do**
  | **foreach** $l_j \in M$ *for which u has no preference for but visited by v_K at t* **do**
  | | $R^t(u,l_j) \leftarrow$ compute DoR for l_j using to Equation 4
  | | $\omega_{l_j} \leftarrow$ compute willingness measure according to Equation 5
  | | $\hat{R}^t_{u,l_j} \leftarrow$ compute u's recommendation score for l_j
return *Top-N* $l_j \in M$ with the highest $\hat{R}^t_{u,l_j}$ for u at t

---

Fig. 3. STS Location Recommender

aggregated into a three dimensional representation (see Figure 2). Based on this model STS Location Recommender algorithm 3 is executed for recommendation. We summarize the steps in the algorithm as follows.

Firstly, given a target user u during a specific time slot $t \in T$ and a set of LBSN users U, the similarity $Sim(u, v)$ between u and each user v ($v \in U$) is computed. Secondly, a set of K users U_K, with the highest similarity score is selected. For each $v_K \in U_K$, a set of semantic locations L_r, previously unvisited by u but visited by v_K are extracted as candidates. For each semantic location in this set, the DoR of its semantic tag l_j to u given by

$$R^t(u, l_j) = \overline{P_u(\tau^t(u))} + \frac{\Sigma_{v_K \in U_K} Sim(u, v_K).P_{v_K}(l_j, t)}{\Sigma_{v_K \in U_K} Sim(u, v_K)} \tag{4}$$

is computed as u's implicit preference for the l_j, where $\overline{P_u(\tau^t(u))}$, $P_{v_K}(l_j, t)$ and $Sim(u, v_K)$ have their usual meanings. Finally, the recommendation score for the semantic location tagged by l_j during time slot t is obtained by taking into account the geospatial distance of u's current location to the semantic location corresponding to l_j. We use a *Proximity Measure* ω_{l_j} defined in terms of geospatial distance to implicitly determine u's willingness to visit the candidate locations. That is $\omega_{l_j} = \dfrac{1}{log(dist(l_c, l_j))}$, where $dist(l_c, l_j)$ is the geospatial distance between the current semantic location of u having semantic tag l_c and semantic location having semantic tags l_j. Using the proximity measure, u's recommendation score for l_j is computed as

$$\hat{R}^t_{u, l_j} = \omega_{l_j} \times R^t(u, l_j) \tag{5}$$

The *top-N* semantic locations with the highest $\hat{R}^t_{u, l_j}$ are returned as recommendations for the user u during time slot t.

4   Experiments

In this section, we evaluate the effectiveness of STS Location Recommender through experiments. We present a description of our dataset, then discuss metrics employed for evaluation and compare our approach with baseline models.

4.1   Description of Dataset and Experimental Settings

In this work, we utilized *GeoLife*[2] real-world GPS trajectory dataset collected from 182 individuals over a period of 5 years (April 2007 to August 2012). We chose individuals with sufficiently large number trajectories (i.e. having trajectories spanning a period of at least one week) in order to increase our chances of finding trajectories which exhibit routine mobility cycles. We found that trajectories of 149 users satisfied this requirement and processed their datasets accordingly. To evaluate our approach we compared with the methods in [15] herein abbreviated as $UBCF$ and [1] abbreviated as $UCLAF$. We describe how we adapt our dataset to perform these comparisons as follows.

---

[2] http://research.microsoft.com/en-us/downloads/b16d359d-d164-469e-9fd4-daa38f2b2e13/

UBCF by Herlocker et al [15] is a benchmark for most conventional user-based CF approaches. The idea behind UBCF is to perform location recommendations based on users' implicit ratings on locations estimated from their location histories, regardless of temporal information. For each user we used visit probability in terms of semantic influence as his/her implicit rating on corresponding locations. We then construct a user-location matrix to perform UBCF.

In *UCLAF* [1], Zheng et al employed a *User-Location-Activity tensor* in addition to *User-User, Location-Feature, User-Location* and *Activity-Activity* matrices for collaborative location and activity recommendation using a tensor decomposition technique. Since our dataset lacks activity information and we did not have access to their dataset, we adapt our dataset to conform to their settings. To do this, we utilized activity information mentioned in their experiments namely *Food & Drink, Shopping, Movies & Shows, Sports & Exercise* and *Tourism & Amusement* in addition to our user-location information to generate User-Location-Activity tensor. Further we considered the first five closest semantic features to each reference point to construct location-feature matrix. We utilized Pearson correlation similarity metric to compute similarity between users to construct user-user matrix. For activity-activity matrix, we employed the web in the same manner as their work, using activity information mentioned earlier to get the entries. Finally, we use frequency of visit of each semantic location to generate entries for user-location matrix .

4.2   Evaluation Methodology

Our STS Location Recommender under investigation estimates a recommendation score for each candidate location and returns the *top-N* highest ranked locations to a target user as recommendations, given his/her current location and time of the day. To study the effectiveness and prediction accuracy of our proposed approach, we evaluate our LRS in terms of (i) Precision and Recall; (ii) Root Mean Square Error *(RMSE)*.

Precision and Recall investigate how many locations marked off in our test dataset during the processing step are recovered in the returned recommended locations. More specifically, we examine

1. Precision@N: how many locations in the top-N recommended locations during timeslot t correspond to the hold-off locations in the testing data.
2. Recall@N: how many locations in the hold-off locations in the testing set are returned as top-N recommended locations during timeslot t.

We tested the performance when $N = 5, 10, 15$ with 5 as default value.

Root Mean Squared Error (RMSE) measures deviation of generated recommendation scores $\hat{R}_{u,l_j}.predicted$, for each user-location pair (u, l_j) from the actual values $\hat{R}_{u,l_j}.actual$. $RMSE$ between predicted and actual scores is given by

$$RMSE = \sqrt{\frac{1}{|Test_{set}|} \sum_{(u,l_j) \in Set} (\hat{R}_{u,l_j}.predicted - \hat{R}_{u,l_j}.actual)^2} \qquad (6)$$

Note that low values of $RMSE$ indicate a good quality of prediction.

Since we do not have ground truth evaluation of our work due to the source of our dataset, we only rely on these metrics for evaluations. For each user, we randomly mark-off 30%, 40% and 50% of all locations visited for testing.

4.3   Experimental Results and Comparison

Firstly, we investigate the impact of semantic and temporal information on the performance of STS Location Recommender. We then compare the effectiveness of our algorithm with a conventional approaches namely $UBCF$ [15]. Finally, we compare our accuracy with $UCLAF$ [6] and $UBCF$ [15] in terms of RMSE.

Influence of Semantic and Temporal Information. We study influence of temporal and semantic features on recommendations by varying tuning parameter λ in Equation 1. A high value of λ indicates semantic information has a higher impact on the accuracy of recommendation. On the other hand, a low value of λ means accuracy of recommendation improves

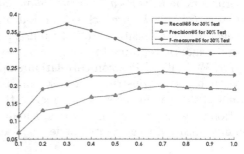

Fig. 4. Variations in Tuning Parameter

with increase in temporal information. Our findings are shown in Figure 4.

As shown in the Figure 4, the best precision was achieved at $\lambda = 0.7$ and the best recall at $\lambda = 0.3$. This shows that semantic information has more effect on determining precision whiles recall is predominantly determined by temporal factor. Since the best value of λ is not the same for precision and recall, we take the harmonic mean of precision and recall. In particular, for each value of λ, we find the F-measure given by $F\text{-}measure = 2 * \frac{precision*recall}{precision+recall}$. We found that λ value of 0.7 gave the best value for F-measure as shown in Figure 4. We therefore set our $\lambda = 0.7$ in our experiments to measure precision and recall.

Comparison of Precision and Recall with Conventional Approach. We investigate the impact of temporal information on the effectiveness of STS Location Recommender in comparison with a conventional approach. Specifically, we compare *Precision* and *Recall* for various percentages of test dataset using our algorithm in comparison with UBCF. UBCF is purely based on conventional user-based collaborative filtering technique which does not take temporal information of users' movement into account. Our algorithm on the other hand uses temporal information in terms of time of visit weighted by stay time at visited locations. Figure 5 shows the results obtained. From the results, both STS Location Recommender and UBCF show similar trends in performances in terms of both precision and recall. That is, both precision and recall generally increase with increase in percentage test of datasets using the two approaches.

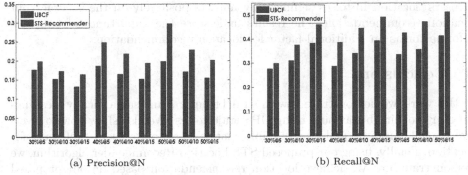

(a) Precision@N   (b) Recall@N

Fig. 5. Comparison of Precision and Recall

Also increase in the number of top recommended locations leads to decrease in precision but increase in recall in both cases. However, in spite of the similarities in the trends, STS Location Recommender significantly outperforms UBCF in terms of precision and recall in all cases. This indicates that incorporating temporal information improves both the quality and accuracy of location recommendation.

We record low values for both precision and recall using STS Location Recommender and UBCF. This is not surprising because GPS trajectory datasets are typically sparse in nature and thus contribute to the low values of precision and recall. However, it is much better than random prediction. In this work, we emphasize on the relative improvements achieved instead of the absolute values we obtained. Note that we do not compare precision and recall with UCLAF because we run UCLAF on our pre-processed dataset using the authors code which does not consider these metrics in their evaluations.

Comparison of RMSE Evaluation with Baseline Methods. We investigate the accuracy of our algorithm in comparison with UCLAF and UBCF using RMSE evaluation. We run each method 5 times for various percentages of test dataset and report the mean values and standard deviations in Table 2.

The results obtained show that STS Location Recommender outperforms both UCLAF and UBCF for all percentages of test data. Note that UCLAF also

Method	RMSE@50%	RMSE@40%	RMSE@30%
UBCF	0.009962 ± 0.001	0.008146 ± 0.003	0.006443 ± 0.001
UCLAF	0.00824 ± 0.001	0.006889 ± 0.002	0.00423620 ± 0.001
STS Location Recommender	0.005511 ± 0.002	0.003567 ± 0.001	0.002252 ± 0.001

Table 2. Root Mean Square Error Evaluation

performs better than UBCF. This can be attributed to the fact that UBCF is purely based on conventional user-based collaborative filtering that relies only on user location features. UCLAF on the other hand employs additional information mentioned earlier to improve accuracy of recommendation. STS Location Recommender outperforms both approaches because incorporating temporal dimension of users' location histories significantly impacts positively on the accuracy our location recommendation. The results underscore the importance of temporal information as an additional factor for location recommendation.

5 Conclusions

In this work, we demonstrate how to model temporal and semantically meaningful user mobility behaviours using GPS trajectories. We also show how to leverage spatial, temporal and semantic information to estimate users' preference for location. Finally, using our proposed STS Location Recommender algorithm, we demonstrate how we achieve location recommendation based on our proposed mobility model. Through experimental evaluation, we show that our approach improves location recommendation compared with the baseline approaches.

References

1. Zheng, V.W., Cao, B., Zheng, Y., Xie, X., Yang, Q.: Collaborative filtering meets mobile recommendation: a user-centered approach. In: Proceedings of the 24rd AAAI Conference on Artificial Intelligence, pp. 236–241, (2010)
2. Zheng, Y., Zhang, L., Xie, X., Ma, W.-Y.: Mining interesting locations and travel sequences from GPS trajectories. In: Proceedings of the 18th International Conference on World Wide Web, pp. 791–800 (2009)
3. Yuan, Q., Cong, G., Ma, Z., Sun, A., Thalmann, N.M.: Time-aware point-of-interest recommendation. In: Proceedings of the 36th International ACM SIGIR Conference on Research and Development in Information Retrieval, pp. 363–372 (2013)
4. Ying, J.J.-C., Lee, W.-C., Tseng, V.S.: Mining geographic-temporal-semantic patterns in trajectories for location prediction. ACM Trans. Intell. Syst. Technol. 5(1), 2:1–2:33 (2014)
5. Ye, M., Yin, P., Lee, W.-C., Lee, D.-L.: Exploiting geographical influence for collaborative point-of-interest recommendation. In: Proceedings of the 34th International ACM SIGIR Conference on Research and Development in Information Retrieval, pp. 325–334 (2011)
6. Zheng, V.W., Zheng, Y., Xie, X., Yang, Q.: Collaborative location and activity recommendations with GPS history data. In: Proceedings of the 19th International Conference on World Wide Web, pp. 1029–1038 (2010)
7. Takeuchi, Y., Sugimoto, M.: CityVoyager: an outdoor recommendation system based on user location history. In: Ma, J., Jin, H., Yang, L.T., Tsai, J.J.-P. (eds.) UIC 2006. LNCS, vol. 4159, pp. 625–636. Springer, Heidelberg (2006)
8. Ying, J.J.-C., Lu, E.H.-C., Lee, W.-C., Weng, T.-C., Tseng, V.S.: Mining user similarity from semantic trajectories. In: Proceedings of the 2Nd ACM SIGSPATIAL International Workshop on Location Based Social Networks, pp. 19–26 (2010)
9. Ankerst, M., Breunig, M.M., Kriegel, H.-P., Sander, J.: OPTICS: Ordering points to identify the clustering structure. In: Proceedings of the 1999 ACM SIGMOD International Conference on Management of Data, vol. 28, pp. 49–60 (1999)
10. Ye, M., Shou, D., Lee, W.-C., Yin, P., Janowicz, K.: On the semantic annotation of places in location-based social networks. In: Proceedings of the 17th ACM SIGKDD International Conference on Knowledge Discovery and Data Mining, pp. 520–528 (2011)

11. Xiao, X., Zheng, Y., Luo, Q., Xie, X.: Finding similar users using category-based location history. In: Proceedings of the 18th SIGSPATIAL International Conference on Advances in Geographic Information Systems, pp. 442–445 (2010)
12. Chon, Y., Lane, N.D., Li, F., Cha, H., Zhao, F.: Automatically characterizing places with opportunistic crowdsensing using smartphones. In: Proceedings of the 2012 ACM Conference on Ubiquitous Computing, UbiComp 2012, pp. 481–490 (2012) doi:10.1145/2370216.2370288
13. Alvares, L.O., Bogorny, V., Kuijpers, B., de Macelo, J., Moelans, B., Palma, A.T.: Towards semantic trajectory knowledge discovery, Data Mining and Knowledge Discovery
14. Spertus, E., Sahami, M., Buyukkokten, O.: Evaluating similarity measures: a large-scale study in the orkut social network. In: KDD 2005: Proceedings of the Eleventh ACM SIGKDD International Conference on Knowledge Discovery in Data Mining, pp. 678–684 (2005)
15. Herlocker, J.L., Konstan, J.A., Borchers, A., Riedl, J.: An algorithmic framework for performing collaborative filtering. In: Proceedings of the 22Nd Annual International ACM SIGIR Conference on Research and Development in Information Retrieval, pp. 230–237 (1999)

A Cache-Based Semi-Stream Join
to deal with Unmatched Stream Data

M. Asif Naeem[1]([✉]), Imran Sarwar Bajwa[2], and Noreen Jamil[1]

[1] School of Computer and Mathematical Sciences,
Auckland University of Technology, Private Bag 92006, Auckland, New Zealand
mnaeem@aut.ac.nz, njam031@aucklanduni.ac.nz
[2] Department of Computer Science & IT, The Islamia University of Bahawalpur,
Bahawalpur, Pakistan
imran.sarwar@iub.edu.pk

Abstract. In Data Stream Management System (DSMS) semi-stream processing has become a popular area of research due to the high demand of applications (e.g. real-time data warehousing) for up-to-date information. One common operation in semi-stream processing is joining of incoming stream with disk-based master data. A recent algorithm called CACHEJOIN was proposed to implement this join operation. However, CACHEJOIN loads entire stream data into join module and consumes all its resources without eliminating those stream tuples which have no relevant tuples in disk-based master data. Due to this, the performance of CACHEJOIN remains suboptimal. In this paper we present a revised version of CACHEJOIN called Improved CACHEJOIN which removes this limitation. This reduces the processing cost for the new algorithm and as a consequence, the new algorithm outperforms existing CACHEJOIN significantly. In order to quantify the performance differences, we compare both algorithms using both synthetic and real datasets with a known skewed distribution. We also present the cost model for our new algorithm.

Keywords: Unmatched stream data · Semi-Stream joins · Performance optimization · Data transformation

1 Introduction

In real-time data warehousing the changes occurring at source level are reflected in data warehouses without any delay. Extraction, Transformation, and Loading (ETL) tools are used to access and manipulate transactional data and then load them into the data warehouse. An important phase in the ETL process is a transformation where the source level changes are mapped into the data warehouse format. Common examples of transformations are units conversion, removal of duplicate tuples, information enrichment, filtering of unnecessary data, sorting of tuples, and translation of source data key. A particular type of stream-based joins

© Springer International Publishing Switzerland 2015
M.A. Sharaf et al. (Eds.): ADC 2015, LNCS 9093, pp. 54–65, 2015.
DOI: 10.1007/978-3-319-19548-3_5

called semi-stream joins are required to implement these transformations examples. In this particular type of stream-based joins, a join is performed between a single stream and a slowly changing table. In the application of real-time data warehousing [4,5], the slowly changing table is typically a disk-based master data table (we denote it here by R) while incoming real-time sales data may comprise the stream.

In the past, the algorithm CACHEJOIN [6] was proposed for joining a stream with a slowly changing table with limited main memory requirements. This algorithm was particularly designed to exploit the skew in stream data as the skew is common in a wide range of applications [1]. Based on that the algorithm introduced a new cache module to holds a particular part of R (which is frequent in stream) into memory permanently.

Although this improves the performance of the algorithm however, there is a key limitation that need to be addressed. It is possible that there can be some stream tuples which do not have any matching tuple in R. So these stream tuples can simply be eliminated without loading them into the join module. Contrarily, CACHEJOIN does not take it into account and loads all these tuples into memory and process them similar to the other tuples which have corresponding matches in R. This creates an overhead for the algorithm in terms of its processing cost and ultimately affects the performance of the algorithm negatively.

In this paper we present a revised version of CACHEJOIN with name "Improved CACHEJOIN" which addresses the stated limitation. In the new algorithm we develop an affective strategy to deal with unmatched stream tuples and the algorithm does not load these tuples into join module. Hence the performance of the new algorithm improves in case if stream contains unmatched tuples. Further details about the new algorithm are provided in Section 4.

2   Related Work

In this section, we present an overview of the previous work that has been done in this area, focusing on those which are closely related to our problem domain.

A seminal algorithm MESHJOIN [9,10] has been designed especially for joining a continuous stream with R, like in the scenario of active data warehouses. The MESHJOIN algorithm is a hash join, where the stream serves as the build input and R serves as the probe input. A characteristic of MESHJOIN is that it performs a staggered execution of the hash table build in order to load stream tuples more steadily. To implement this staggered execution the algorithm uses a queue. After each iteration the algorithm removes stream tuples from the end of the queue, as they have been matched with all R partitions. The algorithm makes no assumptions about data distribution or the organization of R, hence does not perform well for skewed distributions.

R-MESHJOIN (reduced Mesh Join) [7] clarifies the dependencies among the components of MESHJOIN. As a result the performance has been improved slightly. However, R-MESHJOIN implements the same strategy as in the MESHJOIN algorithm for accessing R.

One approach to improve MESHJOIN has been a partition-based join algorithm [3] which can also deal with stream intermittence. It uses a two-level hash table in order to attempt to join stream tuples as soon as they arrive, and uses a partition-based waiting area for other stream tuples. For the algorithm in [3], however, the time that a tuple is waiting for execution is not bounded.

Another approach, Semi-Streaming Index Join (SSIJ) [2], was developed recently to join stream data with R. In general, the algorithm is divided into three phases: the pending phase, the online phase and the join phase. In the pending phase, the stream tuples are collected in an input buffer until either the buffer is larger than a predefined threshold or the stream ends. In the online phase, stream tuples from the input buffer are looked up in cached disk blocks. If the required disk tuple exists in the cache, the join is executed. Otherwise, the algorithm flushes the stream tuple into a stream buffer. When the stream buffer is full, the join phase starts where partitions of R are loaded from disk using an index and joined until the stream buffer is empty. This means that as partitions are loaded and joined, the join becomes more and more inefficient: partitions that are joined later can potentially join only with fewer tuples because the stream buffer is not refilled between partition loads.

CACHEJOIN [6] adds an additional cache module to cope with Zipfian stream distributions. This is similar to Partitioned Join and SSIJ, but a tuple-level cache is used instead of a page-level cache to use the cache memory more efficiently. However, CACHEJOIN loads all stream tuples into join module without taking care of unmatched stream tuples. Contrarily, these unmatched stream tuples can be filtered out and removed before the join operation which is implemented in this paper. This will save a significant processing cost and eventually will improve the performance of the join operator.

3 Problem Definition

This section presents the working of existing CACHEJOIN with problem definition.

The execution architecture for CACHEJOIN is shown in Figure 1. The largest components of CACHEJOIN with respect to memory size are two hash tables, one storing stream tuples denoted by H_S and the other storing tuples from R denoted by H_R. The other main components of CACHEJOIN are a disk buffer, a queue and a stream buffer. Relation R and stream S are the external input sources. Hash table H_R, for R contains the most frequently accessed part of R and is stored permanently in memory. The CACHEJOIN algorithm possesses two complementary hash join phases, somewhat similar to Symmetric Hash Join. One phase uses R as the probe input; the largest part of R will be stored on disk. This phase is called the disk-probing phase. The other join phase, called the stream-probing phase, uses the stream as the probe input, but will deal only with a small part of relation R. For each incoming stream tuple, CACHEJOIN first uses the stream-probing phase to find a match for frequent requests quickly, and if no match is found, the stream tuple is forwarded to the disk-probing phase.

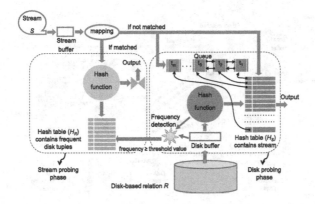

Fig. 1. Data structures and architecture of CACHEJOIN

CACHEJOIN alternates between the stream-probing and disk-probing phases. According to the procedure described above, the hash table H_S is used to store only that part of the update stream which does not match tuples in H_R. A stream-probing phase ends if H_S is completely filled or if the stream buffer is empty. Then the disk-probing phase becomes active. The length of the disk-probing phase is determined by the fact that a few disk pages of R have to be loaded at a time in order to amortize the costly disk access. In the disk-probing phase of CACHEJOIN, the oldest tuple in the queue is used as an index to determine the partition of R that is loaded for a single disk-probing phase into the disk buffer.

After presenting the existing algorithm now we highlight our observations about CAHEJOIN. We have seen from the architecture of CACHEJOIN that the algorithm does not consider the filtering of unmatched stream tuples before their processing while it loads all unmatched stream tuples into the join module. This creates a significant overhead for the algorithm in terms of processing cost. In practice we can easily find such examples where stream data does not has matching with R e.g. in online trading system there can be fake transaction without any real product [8]. One of our aspect to consider this issue is to filter out the fake transactions from stream data. As a solution this gives us two-fold benefits. On one hand it reduces the processing cost for the algorithm. On the other hand the fake transactions can be eliminated and not to be loaded into data warehouse.

4   Improved CACHEJOIN

We propose a revised version of existing CACHEJOIN with name "Improved CACHEJOIN". The new algorithm addresses the issue, described in problem definition section, about the existing approach. This section presents an overview of the new algorithm while its detailed walk trough is presented in Section 4.1.

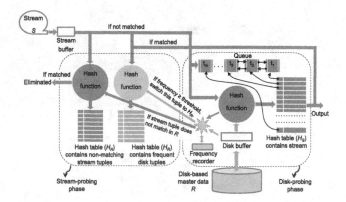

Fig. 2. Data structures and execution architecture of Improved CACHEJOIN

The data structures and execution architecture of the proposed algorithm is shown in Figure 2. From the figure, although Improved CACHEJOIN also has same two phases as in existing CACHEJOIN however, from the architectural point of view the new algorithm has an additional hash table which is denoted by H_N in the figure. It contains those stream tuples which do not have related matching tuples in R.

4.1   Algorithm

The pseudo-code for Improved CACHEJOIN is shown in Algorithm 1. The outer loop of the algorithm is an endless loop, as it is natural for a stream processing algorithms (line 2). The body of the outer loop has two main parts, the stream-probing phase and the disk-probing phase. Due to the endless loop, these two phases are executed alternatingly.

Line 3 to 13 comprises the stream-probing phase. The stream-probing phase has to know the number of empty slots in H_S. This number is kept in variable *hSavailable*. At the start of the algorithm, all slots in H_S are empty (line 1). The stream-probing phase has an inner loop that continues while stream tuples as well as empty slots in H_S are available (line 3). In the loop, the algorithm reads one input stream tuple t at a time (line 4). The algorithm first looks up t in H_N and if t matches in H_N the algorithm simply discards t as t does not has matching tuple in R (lines 5, 6). In case if t does not match in H_N then the algorithm looks up t in H_R and if it matches, the algorithm generates the join output without storing t in H_S (lines 7, 8). In the case t does not match in H_R, the algorithm loads t into H_S along with enqueuing its key attribute value in the queue (line 10). The counter of empty slots in H_S has to be decreased (line 11).

Line 14 to 32 comprises the disk-probing phase. At the start of this phase, the algorithm reads the join attribute value g from the end of the queue (line 14). After reading g the algorithm finds the relative index value in R. If the relative

Algorithm 1. Pseudo-code for Improved CACHEJOIN

Input: A disk based relation R with index on join attribute and a stream of updates S.

Output: $R \bowtie S$

Parameters: w (where $w=w_N+w_R+w_S$) tuples of S and d number of tuples of R.

Method:

```
 1: hSavailable ← h_S
 2: while (true) do
 3:     while (stream available AND hSavailable > 0) do
 4:         READ a stream tuple t from stream buffer
 5:         if t ∈ H_N then
 6:             DISCARD t
 7:         else if t ∈ H_R then
 8:             OUTPUT t ⋈ H_R
 9:         else
10:             LOAD t into H_S along with placing its join attribute value into Q
11:             hSavailable ← hSavailable − 1
12:         end if
13:     end while
14:     READ a join attribute value g from the end of Q
15:     if g does not match with index value in R then
16:         SWITCH the stream tuple(s) with key g from H_S to H_N
17:         DELETE g from Q with corresponding stream tuple(s) from H_S
18:         Go to line 14
19:     else
20:         READ a segment of R into the disk buffer using g as an index look-up
21:     end if
22:     for each tuple r in disk buffer  do
23:         if r ∈ H_S then
24:             OUTPUT r ⋈ H_S
25:             f(r) ← number of matching tuples found in H_S
26:             DELETE all matched tuples from H_S along with the corresponding nodes
                 from Q
27:             hSavailable ← hSavailable + f(r)
28:             if f(r) ≥thresholdvalue then
29:                 SWITCH the tuple r into hash table H_R
30:             end if
31:         end if
32:     end for
33: end while
```

index value does not exist in R, the algorithm marks g as an unmatched tuple and switches the stream tuple(s) against key value g from H_S to H_N (lines 15, 16). Since H_S is a multi-hash-map, there can be more than one values against g. If H_N is already filled then the algorithm overwrites the least frequent tuple in H_N with new one(s). To determine the least frequent tuple in H_N the algorithm uses an extra variable as a counter in the frequency recorder which counts the frequency of occurrence for each tuple of H_N in stream data. The algorithm then deletes g from the queue with all its corresponding values from H_S and jumps back to line 14 (lines 17, 18). Otherwise, if the relative index value exists in R, the algorithm loads a segment of R into disk buffer using g as an index (line 20). In an inner loop, the algorithm looks up one-by-one all tuples r from the disk buffer in hash table H_S. In case of a match, the algorithm generates the join output (lines 22 to 24). As described before due to multi-hash-map, there can be more than one match in H_S against r, the number of matches be $f(r)$ (line 25). The algorithm removes all matching tuples from H_S along with deleting the corresponding nodes from the queue (line 26). This creates empty slots in H_S (line 27).

Lines 28 to 30 in disk-probing phase are concerned with the frequency detection. In line 28 the algorithm tests whether the matching frequency $f(r)$ of tuple r is larger than a pre-set threshold. If yes, it means r is a frequent tuple and therefore the algorithm entered r into H_R. If there is no empty slot in H_R the algorithm overwrites the least frequent tuple in H_R. To determine the least frequent tuple the algorithm uses the same phenomenon as described above i.e. frequency recorder keeps the record of all tuples in H_R with respect to their frequency in stream data.

The threshold is a flexible barrier. Initially, an appropriate value is assigned to it while later on this value can be varied up and down depending on available size of H_R and the rate of matching the disk tuples in the disk buffer.

4.2 Cost Model

In this section we develop the cost model for the new algorithm. The cost model presented here follows the style used for MESHJOIN [9,10]. Equation 1 represents the total memory used by the algorithm (except the stream buffer), and Equation 2 describes the processing cost for each iteration of the algorithm. The notations we used in our cost model are given in Table 1.

Memory Cost. The major portion of the total memory is assigned to the hash table H_S together with the queue while a comparatively much smaller portion is assigned to H_R, H_N, the disk buffer, and the frequency recorder. The memory for each component can be calculated as follows:

Memory for the disk buffer $(bytes) = d \cdot v_R$
Memory for frequency recorder $(bytes) = 8h_R$
Memory for H_R $(bytes) = h_R \cdot v_R$
Memory for H_N $(bytes) = h_N \cdot v_R$
Memory for H_S $(bytes) = \alpha[M - 8h_R - (d + h_R + h_N)v_R]$
Memory for Q $(bytes) = (1 - \alpha)[M - 8h_R - (d + h_R + h_N)v_R]$
By aggregating the above, the total memory M for Improved CACHEJOIN can be calculated as shown in Equation 1.

$$M = 8h_R + (d + h_R + h_N)v_R + \alpha[M - 8h_R - (d + h_R + h_N)v_R] + \\ (1 - \alpha)[M - 8h_R - (d + h_R + h_N)v_R] \tag{1}$$

Table 1. Notations used in cost estimation of Improved CACHEJOIN

Parameter name	Symbol
Number of stream tuples discarded in each iteration through H_N	w_N
Number of stream tuples processed in each iteration through H_R	w_R
Number of stream tuples processed in each iteration through H_S	w_S
Size of disk tuple (*bytes*)	v_R
Disk buffer size (*tuples*)	d
Size of H_N (*tuples*)	h_N
Size of H_R (*tuples*)	h_R
Size of H_S (*tuples*)	h_S
Memory weight for the hash table	α
Memory weight for the queue	$(1-\alpha)$
Cost to read d disk tuples into the disk buffer (*nanosecs*)	$c_{I/O}(d)$
Cost to look-up one tuple into a hash table (any of H_N, H_R or H_S) (*nanosecs*)	c_H
Cost to generate the output for one tuple (*nanosecs*)	c_O
Cost to remove one tuple from hash table H_S and the queue (*nanosecs*)	c_E
Cost to read one stream tuple into the stream buffer (*nanosecs*)	c_S
Cost to append one tuple in hash table H_S and the queue (*nanosecs*)	c_A
Cost to compare the frequency of one disk tuple with the threshold value (*nanosecs*)	c_F
Total cost for one loop iteration (*secs*)	c_{loop}

Currently, the memory for the stream buffer in not included because it is small (0.05 MB is sufficient in our experiments).

Processing Cost. In this section we calculate the processing cost for the algorithm. To make it simple we first calculate the processing cost for individual components and then sum these costs to calculate the total processing cost for one iteration.

Cost to load d tuples from disk to the disk buffer (*nanosecs*)—$c_{I/O}(d)$

Cost to look-up w_N tuples in H_N (*nanosecs*)=$w_N \cdot c_H$

Cost to look-up w_R tuples in H_R (*nanosecs*)=$w_R \cdot c_H$

Cost to look-up disk buffer tuples in H_S (*nanosecs*)=$d \cdot c_H$

Cost to compare the frequency of all the tuples in disk buffer with the threshold value (*nanosecs*)=$d \cdot c_F$

Cost to generate the output for w_R tuples (*nanosecs*)=$w_R \cdot c_O$

Cost to generate the output for w_S tuples (*nanosecs*)=$w_S \cdot c_O$

Cost to read the $w = w_N + w_R + w_S$ tuples from the stream buffer (*nanosecs*)=$w \cdot c_S$

Cost to append w_S tuples into H_S and Q (*nanosecs*)=$w_S \cdot c_A$

Cost to delete w_S tuples from H_S and Q (*nanosecs*)=$w_S \cdot c_E$

Since the loading of stream tuples in hash tables H_N and H_R is rare once the algorithm gets streamlined, we ignore the costs of loading operation for both of the hash tables. By aggregating the above costs the total cost of the algorithm for one iteration can be calculated using Equation 2.

$$c_{loop}(secs) = 10^{-9}[c_{I/O}(d) + d(c_H + c_F) + w \cdot c_S + w_N \cdot c_H + \\ w_R(c_H + c_O) + w_S(c_A + c_E + c_O)] \qquad (2)$$

The term 10^{-9} is a unit conversion from nanoseconds to seconds. Since in c_{loop} seconds the algorithm processes $w = w_N + w_R + w_S$ tuples of the stream S, the service rate μ can be calculated using Equation 3.

$$\mu = \frac{w}{c_{loop}} \tag{3}$$

5   Experiments

5.1   Experimental Setup used in all Experiments

We performed our experiments on a *Pentium-i5* with 8GB main memory and 500GB hard drive as secondary storage. We implemented our experiments in Java using the Eclipse IDE. We analyzed the service rate of both the algorithms using synthetic, TPC-H, and real-life datasets.

Synthetic Data. The stream dataset we used is based on a Zipfian distribution with exponent value is equal to 1. R was stored on disk using a MySQL database and it has index on it. To measure the I/O cost more accurately, we set the fetch size for `ResultSet` equal to the disk buffer size. The detailed specifications of our synthetic dataset are shown in Table 2.

TPC-H. We also analyzed the service rate of both the algorithms using the TPC-H dataset, which is a well-known decision support benchmark. We created the datasets using a scale factor of 100. More precisely, we used the table `Customer` as R and the table `Order` as stream data. In table `Order` there is one foreign key attribute `custkey`, which is a primary key in the `Customer` table, so the two tables can be joined. Our `Customer` table contained 20 million tuples, with each tuple having a size of 223 bytes. The `Order` table contained the same number of tuples, with each tuple having a size of 138 bytes. The plausible scenario for such a join is to add customer details corresponding to an order before loading the order into the warehouse.

Real-life Data. We also compared the service rate of the both algorithms using a real-life dataset[1]. This dataset basically contains cloud information stored in a summarized weather report format. We consider R by combining meteorological data corresponding to months April and August, while consider the stream data by combining data files from December. R contains 20 million tuples, while the streaming data table contains 6 million tuples. The size of each tuple in both R and the stream data table is 128 bytes. Both tables are joined using a common attribute, longitude (`LON`). The domain of the join attribute is the interval [0,36000].

---

[1] This dataset is available at: http://cdiac.ornl.gov/ftp/ndp026b/

Table 2. Data specification

Parameter	value
Total allocated memory M	50MB to 250MB
Size of R	0.5 *million* to 8 *million tuples*
Size of each disk tuple	120 *bytes* (similar to CACHEJOIN)
Size of each stream tuple	20 *bytes* (similar to CACHEJOIN)
Size of each node in the queue	12 *bytes* (similar to CACHEJOIN)
Data set	based on Zipf's law (exponent is equal to 1)

5.2 Performance Evaluation

In this section we evaluate the performance of both the algorithms. We have identified three parameters, for which we want to understand the behavior of the algorithms. The three parameters are: the total memory available, the size of R, and the percentage of unmatched stream data in R . For the sake of brevity, we restrict the discussion for each parameter to a one-dimensional variation, i.e. we vary one parameter at a time.

Memory Size. In our first experiment we test the performance of all algorithms using different memory budgets while keeping the size of R fixed (*2 million tuples*). Figure 3(a) presents the comparisons of both approaches. From the figure it is clear that Improved CACHEJOIN performs better than existing CACHEJOIN.

Size of R. In our second experiment we test the performance by varying the size of R, We choose the discrete sizes of the parameter, the size of R, from a simple geometric progression. The performance results are shown in Figure 3(b). In our experiments, the performance of Improved CACHEJOIN is again better than existing CACHEJOIN.

Percentage of Unmatched Stream. We also test the performance of both the algorithms by considering different percentages of unmatched stream in R. The results of this experiment are presented in Figure 3(c). From the figure it can be observed that Improved CACHEJOIN performs significantly better than existing CACHEJOIN except one setting of 0% unmatched stream. Under this setting hash table H_N does not add any value in the performance while it increases a little processing cost. However, for other settings the improvement in performance becomes more pronounced for increasing the value of percentage of unmatched stream data.

TPC-H and Real-life Datasets. In these experiments we measured the service rate produced by both the algorithms at different memory settings. We allocate the size of memory in percentages with respect to the size of R. The results of using TPC-H and real-life datasets are shown in Figure 3(d) and Figure 3(e) respectively. From the both figures it can be noted that for all memory settings Improved CACHEJOIN performs better than existing CACHEJOIN.

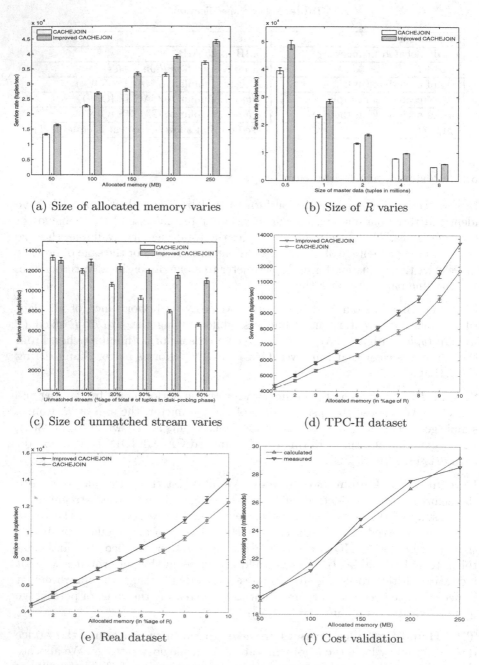

(a) Size of allocated memory varies

(b) Size of R varies

(c) Size of unmatched stream varies

(d) TPC-H dataset

(e) Real dataset

(f) Cost validation

Fig. 3. Performance comparisons

Costs Validation. We also validate the cost model for our new algorithm by comparing the calculated cost with the measured cost. Figure 3(f) presents the comparisons of both costs. The results show that the calculated cost closely resembles the measured cost, which supports the correctness of our cost model.

6   Conclusions

In this paper we propose a revised version of CACHEJOIN with name Improved CACHEJOIN. The new algorithm introduces a hash table that keeps the track of unmatched stream tuples. For every new stream tuple the algorithm first looks it up in this hash table. If match is found the algorithm discard this tuple without loading it into the join module. This saves a significant processing cost and consequently improves the performance. We have provided a cost model and validated it with experiments. We have provided experimental data showing an improvement of new CACHEJOIN over the existing CACHEJOIN algorithm.

References

1. Anderson, C.: The Long Tail: Why the Future of Business Is Selling Less of More. Hyperion (2006)
2. Bornea, M.A., Deligiannakis, A., Kotidis, Y., Vassalos, V.: Semi-streamed index join for near-real time execution of ETL transformations. In: IEEE 27th International Conference on Data Engineering (ICDE 2011), pp. 159–170, April 2011
3. Chakraborty, A., Singh, A.: A partition-based approach to support streaming updates over persistent data in an active datawarehouse. In: IPDPS 2009: Proceedings of the 2009 IEEE International Symposium on Parallel & Distributed Processing, pp. 1–11. IEEE Computer Society, Washington, DC (2009)
4. Karakasidis, A., Vassiliadis, P., Pitoura, E.: ETL queues for active data warehousing. In: IQIS 2005: Proceedings of the 2nd International Workshop on Information Quality in Information Systems, pp. 28–39. ACM, New York (2005)
5. Naeem, M.A., Dobbie, G., Weber, G.: An event-based near real-time data integration architecture. In: EDOCW 2008: Proceedings of the 2008 12th Enterprise Distributed Object Computing Conference Workshops, pp. 401–404. IEEE Computer Society, Washington, DC (2008)
6. Naeem, M.A., Dobbie, G., Weber, G.: A lightweight stream-based join with limited resource consumption. In: Cuzzocrea, A., Dayal, U. (eds.) DaWaK 2012. LNCS, vol. 7448, pp. 431–442. Springer, Heidelberg (2012)
7. Naeem, M.A., Dobbie, G., Weber, G., Alam, S.: R-MESHJOIN for near-real-time data warehousing. In: DOLAP 2010: Proceedings of the ACM 13th International Workshop on Data Warehousing and OLAP. ACM, Toronto (2010)
8. Pandit, S., Chau, D.H., Wang, S., Faloutsos, C.: Netprobe: a fast and scalable system for fraud detection in online auction networks. In: Proceedings of the 16th International Conference on World Wide Web, pp. 201–210. ACM (2007)
9. Polyzotis, N., Skiadopoulos, S., Vassiliadis, P., Simitsis, A., Frantzell, N.E.: Supporting streaming updates in an active data warehouse. In: ICDE 2007: Proceedings of the 23rd International Conference on Data Engineering, Istanbul, Turkey, pp. 476–485 (2007)
10. Polyzotis, N., Skiadopoulos, S., Vassiliadis, P., Simitsis, A., Frantzell, N.: Meshing streaming updates with persistent data in an active data warehouse. IEEE Trans. on Knowl. and Data Eng. 20(7), 976–991 (2008)

Storing and Processing Massive Trajectory Data on SAP HANA

Haozhou Wang[1]($\boxtimes$), Kai Zheng[1], Hoyoung Jeung[2], Shane Bracher[2], Asadul Islam[2], Wasim Sadiq[2], Shazia Sadiq[1], and Xiaofang Zhou[1]

[1] The University of Queensland, Brisbane, Queensland, Australia
{h.wang16,kevinz,shazia,zxf}@uq.edu.au
[2] SAP Innovation Center, Brisbane, Australia
{hoyoung.jeung,shane.bracher,a.islam,wasim.sadiq}@sap.com

Abstract. Owing to the development of cheap RAM-based storage technology, modern computing hardware can afford much larger main memory. Consequently, traditional database systems can be re-designed to store and manage all the data in main memory permanently. Such kind of in-memory database systems (IMDB) have attracted increasing attention from both academia and industry due to its outstanding performance in processing large amount of data. In this work, we will exploit the computational power of SAP HANA, the in-memory column-oriented data analytics platform designed by SAP, to support efficient query processing for moving object trajectories. We have tailored the frame-based data structure designed by our previous SharkDB project and made the trajectory data with variable lengths and sampling rates suitable for relational database model in SAP HANA. Extensive experiments based on large-scale real dataset have demonstrated superior performance of our frame-based design in processing a variant of queries.

1 Introduction

Recently, huge amount of location data that records the motion history of moving objects, known as trajectories, are generated from different sources such as GPS-enabled devices and smartphones. Analyzing trajectory data can help people understand the behavioural pattern of moving objects, and improve the quality of service in applications such as geographical information systems, location-based services, vehicle navigation systems and so on.

Despite the demand for efficient trajectory data processing, non of the existing relational database management systems (RDBMS) has built-in support for trajectory data types and operators/queries. This is mainly caused by the heterogeneity of trajectory data in its length and sampling rate, which makes it difficult to fit into the relational schema with fixed number of columns. Moreover, most traditional RDBMSs adopt disk-based storage with significant I/O overhead when processing large amount of data, which may not meet the real-time requirement in many trajectory related applications such as map services, trip planning and early event detection.

© Springer International Publishing Switzerland 2015
M.A. Sharaf et al. (Eds.): ADC 2015, LNCS 9093, pp. 66–77, 2015.
DOI: 10.1007/978-3-319-19548-3_6

Thanks to the increasing availability of big RAM at lower costs, now it becomes possible and affordable to store and process the entire (or at least a significant portion of) dataset within main memory, which can be orders of magnitudes faster than the traditional disk-based database systems. In our previous work [1], we proposed SharkDB, an in-memory storage system for massive trajectory data, which has achieved promising performance in trajectory query processing. In this work we take a further step to tailor our frame-based data structure into the relational database model and provide built-in support for trajectory query processing in SAP HANA, an in-memory platform developed by SAP for processing high volumes of operational and transactional data in real-time. We are facing two major challenges when migrating our data structure into SAP HANA. The first challenge is how to fit the frame based data structure into relational database model. The second one is how to process the query efficiently using the query language (SQL like) supported by SAP HANA. To address the above challenges, we propose a carefully designed data structure, which is inherited from our previous I/P frame data structure. This new data structure can keep the benefits of the I/P frame data structure, and is compatible with traditional table format perfectly, which can work well in the SAP HANA.

Our key contributions in this work can be summarized as follows.

- We implement the key-value and sample point format, which are commonly used in existing relational database system for storing trajectory data and observe the major drawbacks of these data storages in handling large scale trajectory data.
- We modify our previous I/P frame data structure and fit it into relational database model, which is then used to store trajectory data in SAP HANA.
- We provide efficient built-in data types and operators for trajectory data using the query languages supported by SAP HANA.
- We conduct extensive experiments with large-scale real trajectory datasets. The results demonstrate that our new data storage model can achieve superior performance in ranges of spatial queries compared with traditional trajectory storage structures.

The rest of this paper is organized as follows. A brief introduction of related work is provided in Section 2. In section 3, we propose three data structures on SAP HANA. Section 4 reports on our experimental observations. Section 5 concludes the paper.

2 Relate Work

In this section, we give a brief introduction of SAP HANA database system and review previous works on column-oriented data structure and in-memory data management.

2.1   SAP HANA System

SAP HANA [2,3] is implemented and developed by SAP , which is the newest in-memory based and column-oriented database management system. The goal of SAP HANA is to handle huge amount of data and the real time complex query processing. Currently, the bottleneck of the traditional disk-oriented database system is the I/O cost, since the I/O performance of hard disk is very limited. To remove this bottleneck, SAP HANA moves all of the data into main memory instead of storing them in the hard disk, which means all of the data are accessed and maintained in the main memory directly. The hard disk in SAP HANA will only be used for data backup and maintaining the log files. Moreover, SAP HANA supports both row-oriented data structure and column-oriented data structure, and the column-oriented data structure is recommended by SAP, since the column-oriented data structure can provide better performance for analytic queries.

2.2   Column-Oriented Store Architecture

The column-oriented data structure is stored data column by column, which is a read-optimized data structure. Stonebraker et al. [4] indicate that the column-oriented data structure is a good data structure to store data for analytic queries. A new column-oriented data structure is proposed by Héman et al. [5], called Positional Delta Tree (PDT). The PDT requires less I/O and CPU time for analytic queries. Moreover, Lemke et al. [6] propose a new algorithm to support efficient query process on the compressed column-oriented data structure.

2.3   In-Memory Data Management

Recently, several in-memory RDBMS have been designed and proposed. An earlier version is called IMS/VS Fast Path [7], which is using group commit operation to speed up traditional transactions. Meanwhile, another in-memory database system [8,9] is designed based on IBM's OBE system. DataBlitz Storage Manager is another version of in-memory database system, which is proposed by Baulier et al. [10]. This system provides multi-level APIs for applications. In addition, for in-memory query processing algorithm on traditional transactions, a dictionary-based string compression algorithm is proposed by Bining et al [11] , which is used on the in-memory database systems. Rao and Ross [12] re-desgin the B+-tree as in-memory based indexing structure, called CSB+-trees. Moreover, Manegold et al. [13] propose a cost model for query processing on in-memory database systems. This cost model provides the accurate cost functions for database operations.

3   Implementation Details

In this section, we propose a frame based data structure in SAP HANA. Meanwhile we also implement two common traditional data structures that are used in traditional database systems, which are called key-value format and sample point format.

A key motivation to use SAP HANA in-memory database system to manage trajectory data is to increase the performance of queries. Therefore, we design a set of common queries, which are divided into two categories, basic operations and advanced operations. The basic operations is normal database operations, which include SELECT, INSERT, DELETE and APPEND. All of the basic operations target to single trajectory by trajectory ID. The advanced operations include two analytic queries, which are window query and kNN query. The window query is to find all trajectories passing a given region and active during a given period of time. The kNN query is to find top-k trajectories that are close to a given point and active during a given period of time. We will discuss the details about the implementation of these queries in three data structure in the following.

3.1   Key-Value Format

To store the trajectory data into the traditional data structure, a naive way is to store one trajectory as a single object. Therefore, we can store the whole trajectory dataset into one table with two columns T_{id}, T_{obj}, where T_{id} is the trajectory ID and T_{obj} contains all of simple points that belong to T_{id}. On the other hand, to speed up analytic query processing, for each trajectory object (i.e., each row), we add the auxiliary information that includes the start time, the end time and the MBR of this trajectory. Therefore, in the key-value format, there are five columns, which are $< T_{id}, T_{obj}, T_{st}, T_{et}, T_{mbr} >$.

Basic Operations. Since the key-value format is the traditional row-based data structure, such operations (i.e., SELECT, DELETE, INSERT and APPEND) can be done very straightforward. For INSERT operation, we first encode the trajectory sample points into one object, which is a raw type that supported by SAP HANA. Next we can use normal INSERT SQL command to insert this trajectory into table directly. The SELECT operation is a reverse operation of the INSERT operation, which selects the encoded trajectory object by given trajectory ID and then decodes them directly. For implementing APPEND operation, we combine both SELECT and INSERT operation. Therefore we first select the target trajectory object and encode the trajectory that used for appending as a new raw type object. After this, we add this new object at the end of target trajectory object. Finally, we update the auxiliary information of appended trajectory. The DELETE operation is implemented as the normal DELETE operation, which deletes the trajectory object by trajectory ID.

Advanced Operations. In this subsection, we propose the algorithms of processing advanced operations on key-value format. Given a window query, we first filter out the trajectories, which are out of the query time interval, by using its auxiliary information (i.e., the start and end time of trajectory). After this, we scan the rest of trajectories. If the MBR of a selected trajectory is overlapping the query window, then we decode this trajectory into original sample points and find the actual results. For processing kNN query, we use MINDIST, which is the minimum distance from query point to MBR, as the lower bound. If the MINDIST of a trajectory is larger than the minimum distance of the current k^{th} best trajectory, we can prune this trajectory safety, otherwise we decode this trajectory and calculate actual distance from query to this trajectory.

3.2 Sample Point Format

The major drawback of key-value format is that we need to decode the whole trajectory into sample points when processing advanced operations, which is very time-consuming. Meanwhile, such operations need a larger amount of memory as the buffer to store the sample points. Therefore, to solve this issue, we use another simple schema to store the trajectory in the table, which is called sample point format. In this format, each row only store one trajectory sample point. The format of each row is $< TID, X, Y, T >$, where TID is the trajectory ID of this sample point, X and Y represent the coordinates of this sample point and T is the timestamp of this sample point.

There are two benefits of the sample point format. Firstly, the data in the sample point format structure do not require to decode when dealing with advanced operations, which is an significant advantage of sample point based format. Another one is that we can compress the data of each column in sample point format, since SAP HANA supports to compress the data in column-oriented storage model to save the space, which means the data can be compressed column by column. Moreover, as we discussed before, the sample point format uses simple data type, which can achieve higher compression ratio.

Basic Operations. Both INSERT and APPEND operations on sample point format are the same, we only need to convert the trajectory data into sample point format and insert them into table directly. Due to all of sample points are stored in one table, for APPEND operation, we only need to insert the sample points of new trajectory into table. The SELECT operation is different with key-value format, we need to first select the trajectory sample points by given trajectory ID and then order the sample points by its timestamp to recover the whole trajectory. To delete a trajectory, we can delete the sample points by given trajectory ID directly.

Advanced Operations. For dealing with window query on sample point format, we select the sample points that locate the given spatial window and time interval, then we order them by trajectory ID and timestamp to get the final results. For processing kNN query, we still use the MBR information to speed up the query processing. We first test each MBR with k^{th} minimum distance, if a candidate trajectory is found, we select the sample points of this trajectory within given time interval. After this we calculate the actual distance of each selected sample points and get the final answers.

3.3 Frame Format

As the number of columns is limited in SAP HANA, we have to re-design our original frame based data structure to suit the traditional table structure. Hence, we extract the frame group column ID as a single column and combine with the trajectory ID to identify each frame group. In the table, the information of a single frame group now contains in one row. The information of I-frame point is stored in two columns with its x value and y value. The rest of columns are used for storing P-frame point information, similar with I-frame point, each P-frame point cost two column to store, one for δx and another for δy.

When converting the data from original frame structure to the table based structure, we first select a single frame group from frame data structure, then we get the frame group column id of this frame group and put it into $FGCID$. Next, we extract the trajectory ID and put it into TID. After this, we insert x value and y value of its I-frame point into IFX and IFY respectively. And we insert the P-frame points in the rest of columns. An example of the table structure is shown in Table. 1.

Table 1. Example of Table based Structure

FGCID	TID	IFX	IFY	PF1PX	PF1PY	...
1	1	$p.x$	$p.y$	$\Delta p.x$	$\Delta p.y$	...
1	2	$p.x$	$p.y$	$\Delta p.x$	$\Delta p.y$	...
1	3	$p.x$	$p.y$	$\Delta p.x$	$\Delta p.y$	...
1	4	$p.x$	$p.y$	$\Delta p.x$	$\Delta p.y$	...

Basic Operations. To insert a new trajectory, we first allocate the raw trajectory to frame-based structure, which is proposed in the previous section. Then we split these frame points into frame group and calculate the frame column group ID based on the timestamp of raw trajectory. In the next, we encode each frame group to generate I-frame point and P-frame points. Finally, we insert each frame group as a single raw with trajectory ID and related frame group column ID into the table.

The SELECT operation in this data structure is straightforward. Since the SELECT is based on trajectory ID, so we can select the frame groups by given trajectory ID. In the next step, we order them by frame group column ID. At the end, we decode both I-frame point and P-frame point to the original sample point, the timestamp can be recovered by frame group column ID of each frame group. The DELETE operation is similar as SELECT operation. We remove frame groups by given trajectory ID from table directly.

Different with previous format, to expand a trajectory T_{old} by given a trajectory ID TID and a new trajectory T_{new}, we only need to select the last frame group FG_{last} of T_{old} and decode it to get a trajectory segment. Then, we connect this trajectory segment with T_{new}. After this, we allocate and encode this trajectory to get the new frame groups with related frame group column ID and TID. Finally, we remove that last frame group FG_{last} and insert these new frame groups into database.

Advanced Operations. Given a window query with rectangle query area and a time interval, we first calculate the range of frame group columns (i.e., the range of id of frame group columns) that need to be investigated based on given time interval. Then, we can select all the frame groups inside the given rectangle now. However, the P-frame points in each row have been delta encoded and do not contain actual coordinate information. Hence, for each frame group, we pre-compute its MBR information. If a MBR intersects with given query window, we put the related frame group into the candidate set. After that, for each frame group in the candidate set, we decode the P-frames into sample points. Finally, we filter out the sample points that out of query area and re-construct the segment of trajectory based on these sample points.

For kNN query, we first get the range of frame groups, which is same as window query. Then, for each frame group (i.e., a single row), we first calculate the MINDIST between query point and this frame group by using its MBR information to find the candidate. We keep updating the candidate set if there is a better candidate found. The kNN query requires a min-heap to store the candidate trajectories, hence we create a temporary table during query processing, which contains two columns trajectory ID TID, which is the primary key and the distance $Distance$ from this trajectory to query point. Meanwhile, the elements in this table are ordered by the distance as well. As each trajectory has been encoded as frame groups, $Distance$ is only recording shortest the distance of frame groups that have been investigated with the related trajectory ID. To avoid the different frame groups with same trajectory ID are pushed into the priority queue, we use a table to simulate the min-heap. Because this case can cause such frame groups to be considered as final results, which make the results are incorrect. For example, the distance d_1 of frame group FG_1 (trajectory ID is T_1) is 2, the distance d_2 of frame group FG_2 (trajectory ID is T_1) is 3 and the distance d_3 of frame group FG_3 (trajectory ID is T_2) is 4. Based on our approach, FG_1 and FG_2 are two shortest distance from query point and are pushed priority queue, if $k = 2$, then both of them will consider as the final results. However, they belong to the same trajectory, the actual results will only contain one trajectory, which is not meet the query requirement. The correct answer should return FG_1 and FG_3, since FG_2 belongs to same trajectory as FG_1. Therefore, as we discussed before, using a table and set trajectory ID as primary key can solve this issue, when we find a candidate frame group, we first check the trajectory ID of this frame group in the temporary table. If it exists, we update its distance, otherwise we will insert this ID and distance into table as a new element. Moreover, as this table is simulating the priority queue, we will keep the size of table is equal to k and remove other elements.

4   Experiments

4.1   Experiment Settings

At the present time, there is no any common benchmark or workload for trajectory database test. Therefore, we build a workload test for these table formats. For dataset, we use a very large dataset that contains five months motion history of $20k$ vehicles and the average sampling rate is 30 seconds per sample point. The total number of sample points in this dataset are more than 4 billions and each trajectory contains around $40k$ sample points with one month time duration . All of the data are converted into three table formats respectively, and these tables are all loaded into SAP HANA system and storing in the main memory. The SAP HANA database system contains one TB memory and four Intel 10-cores CPUs. The algorithms are implemented by SQL/L language.

There are three formats of tables used in the experiment test, which includes key-value format, simple point format and frame format. As discussed before, we split the experiments into two main parts in our workload test. The first part is the basic operations experiment that includes four basic operations, SELECT, DELETE, INSERT and APPEND. The second part is the advanced operations experiments includes two trajectory based operations, window query and kNN query. For basic operations experiments, we run each query 1000 times in three tables to get the average running time. For advanced operations experiments, we run each query with different parameter settings. There are two parameters for window query, the spatial search window and the query time interval; and kNN query also has two parameters the number of k as well as the query time interval. The detail of parameter settings are shown in Table 2. In addition, in the frame format, we set the number of n equals 5 and the frame rate is set to 30 seconds, which is same as average sampling rate of the dataset.

Table 2. Parameters Setting

Parameter	Default Value	Range
# results k	5	5-30
area of spatial window AS	3%	1%-11%
length of time interval TI	3h	1h - 11h

4.2   Basic Operations

The basic operation test includes two set of experiments. The first set of experiments includes SELECT by trajectory ID, DELETE by trajectory ID, which are all operated in main memory. Another set of experiments includes INSERT a new trajectory data into table and APPEND a trajectory by trajectory ID, which need transfer the data from other sources (e.g., a PC or a mobile device) to SAP HANA database system via network. In such operations, the main factor that could affect the performance is the length of trajectory (i.e, number of sample points). Hence, in this experiment, we set the trajectory length from $10k$ sample points to $60k$ sample points in a trajectory.

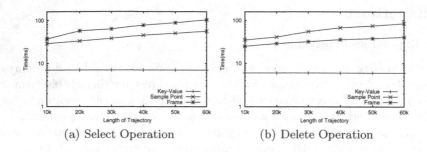

(a) Select Operation   (b) Delete Operation

Fig. 1. Basic Operations(1)

The Fig. 1 shows the results of both SELECT and DELETE operations of three data formats. We can see that the key-value format has the best performance since such basic operations is suitable for row-oriented store, since it can be quick selected by ID and only one row need to be selected in one query request. Therefore, the performance of key-value format is very stable for different trajectory length. For SELECT operation, the frame format needs to decode the data to recover the original trajectory, hence it takes longer time than sample point format. However, there is no need to decode the data in the DELETE operation, the frame format is better than sample point format, since the number of rows that needed to be modified in frame format is less than sample point format. As we can see, even if the frame format structure cannot outperform the traditional row-oriented store database for SELECT and DELETE operations, but the running time is also acceptable due to it is still under one second, which still can be considered as real time processing.

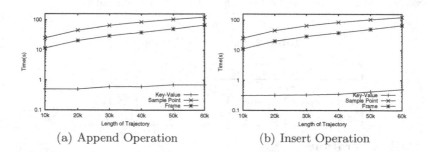

(a) Append Operation   (b) Insert Operation

Fig. 2. Basic Operations(2)

In addition, Fig. 2 shows the result of insert and append operations. As we discussed before, the INSERT operation and APPEND operation are similar, therefore the results between INSERT operation and APPEND operation are also similar. This is no doubt that the key-value format has the best performance, since it only need one INSERT database operation for both INSERT and APPEND operations, which takes less than one second. Both frame format and sample point format need much longer time for these operations, because these two data formats need a large amount of INSERT database operations. On the other hand, the performance of frame format beats the performance of sample point format, even if the frame format needs extra encoding process when dealing with INSERT and APPEND operations. But the number of INSERT database operations and the costs of network transmission of frame format are less than sample point format. Therefore, this experiment also give an evidence that the cost of network transmitting and INSERT database operation are larger than cost of data encoding, which also shows the advantage of the frame format.

4.3   Advanced Operations

Window Query. We first evaluate the efficiency of three formats with different spatial window size with default time interval. In Fig 3(a) show the results, the key-value format cannot handle the analytic query, since it is mainly designed for basic operation. For each query, we can only process the query after decode the trajectory object. In this case, decode the whole trajectory is time consuming and will need amount of memory as the buffer to store the decoded trajectories.

When we increase size of searching areas, the performance of sample point format decreases very quickly, since the sample point format only need to a selection operation with spatial and temporal limitations. Therefore, the larger searching areas are given, the more data need to be retrieved from sample point format. The running time of frame format is slightly increased, when the size of window is increasing. Because the larger spatial window size, the more frame points that needs to be decoded. However, as we can see, this is no doubt that frame format outperforms other two data formats. Such evidences provide that the bandwidth between CPU and memory is now becoming the new bottleneck of in-memory database, and our frame format can reduce the amount of data that needs to transfer to CPU to process.

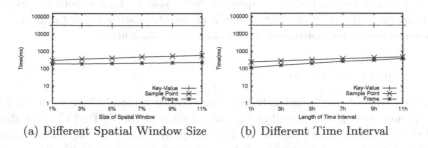

(a) Different Spatial Window Size   (b) Different Time Interval

Fig. 3. Effect of Window Query

We also conduct another set of experiments by using different length of time interval, the results of these experiments are shown in Fig 3(b). Similar with spatial window size tests, when the length of time interval is increasing, the SAP HANA database system needs more time to get the answer of query. Moreover, we can see that the time interval takes less effect than spatial window size, since this dataset is highly density in spatial.

kNN Query. Fig. 4(a) shows the experiment result of kNN query with different query time interval. As we can see, the key-value format cannot process kNN query in the real time as well as window query. The query time of is more than 600 seconds, which is not acceptable for query processing in such advance in-memory database system. Although, both sample point format and frame format can finish the query within one second. The frame format still outperforms the sample point format, this is because the P-frame points can be pruned by our proposed algorithm efficiently, but the sample point format need to calculate the distance of each sample point in the selected area, where the computational cost is very high. Therefore, it provides the frame format is stable in the large trajectory analytic tasks.

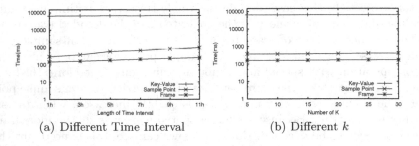

(a) Different Time Interval   (b) Different k

Fig. 4. Effect of kNN Query

We investigate the query performance with regard to the number of k for each query. The results are shown in Fig. 4(b), which is similar with previous experiments. But the different with changing the query time interval, the results of changing number of k of kNN query is quite stable. This is because, we use the temporal table as a priority queue and maintain such small table in SAP HANA is a very small case and can be done very efficiently.

4.4   Compression Ratio

In this subsection, we investigate the compression ratio of three formats. The compression ratio is calculated as the size of data in the SAP HANA database divided by the size of original data. The compression ratio of key-value format is 60.1%, sample point format is 55.2% and frame format is 39.9%. We can see the compression ratio of key-value data structure is not good, since the column of storing trajectory object is hard to compress. As we discussed before, in the frame format, the P-frame points can take less bit for record a simple point and it uses integer data type to instead of double data type, which can increase the compression ratio. Therefore, the frame format achieves the best compression ratio.

5   Conclusion

In this paper, we migrate the frame-based data structure for in-memory trajectory storage into SAP HANA. This new prototype takes advantages of both the power of SAP HANA as a fully functioned database system and the high efficiency of our frame-based structure in trajectory query processing. The extensive experimental results show that our prototype can constantly achieve better performance than the traditional trajectory storage models.

Acknowledgement. This research is partially supported by Australian Research Council (Grants No. LP130100164 and No. DE140100215).

References

1. Wang, H., Zheng, K., Xu, J., Zheng, B., Zhou, X., Sadiq, S.: SharkDB: an in-memory column-oriented trajectory storage. In: CIKM, pp. 1409–1418 (2014)
2. Plattner, H.: A common database approach for OLTP and OLAP using an in-memory column database. In: SIGMOD, pp. 1–2 (2009)
3. Plattner, H.: SanssouciDb: an in-memory database for processing enterprise workloads. In: BTW, vol. 20, pp. 2–21 (2011)
4. Stonebraker, M., Abadi, D.J., Batkin, A., Chen, X., Cherniack, M., Ferreira, M., Lau, E., Lin, A., Madden, S., O'Neil, E., O'Neil, P., Rasin, A., Tran, N., Zdonik, S.: C-store: a column-oriented DBMS. In: VLDB, pp. 553–564 (2005)
5. Héman, S., Zukowski, M., Nes, N.J., Sidirourgos, L., Boncz, P.: Positional update handling in column stores. In: SIGMOD, pp. 543–554 (2010)
6. Lemke, C., Sattler, K.-U., Faerber, F., Zeier, A.: Speeding up queries in column stores. In: Bach Pedersen, T., Mohania, M.K., Tjoa, A.M. (eds.) DAWAK 2010. LNCS, vol. 6263, pp. 117–129. Springer, Heidelberg (2010)
7. Gawlick, D., Kinkade, D.: Varieties of concurrency control in IMS/VS fast path. DEB 8(2), 3–10 (1985)
8. Ammann, A.C., Hanrahan, M., Krishnamurthy, R.: Design of a memory resident DBMS. In: COMPCON, pp. 54–58 (1985)
9. Bitton, D., Hanrahan, M., Turbyfill, C.: Performance of complex queries in main memory database systems. In: ICDE, pp. 72–81 (1987)
10. Baulier, J., Bohannon, P., Gogate, S., Gupta, C., Haldar, S.: DataBlitz storage manager: main-memory database performance for critical applications. In: SIGMOD, pp. 519–520 (1999)
11. Binnig, C., Hildenbrand, S., Färber, F.: Dictionary-based order-preserving string compression for main memory column stores. In: SIGMOD, pp. 283–296 (2009)
12. Rao, J., Ross, K.A.: Making B+- trees cache conscious in main memory. In: SIGMOD, pp. 475–486 (2000)
13. Manegold, S., Boncz, P., Kersten, M.L.: Generic database cost models for hierarchical memory systems. In: PVLDB, pp. 191–202 (2002)

Belief Revision in Uncertain Data Integration

Fereidoon Sadri[✉]

Department of Computer Science, University of North Carolina,
Greensboro, NC, USA
f_sadri@uncg.edu

Abstract. This paper studies the problem of integrating probabilistic uncertain information. Certain constraints are imposed by the semantics of integration, but there is no guarantee that they are satisfied in practical situations. We present a Bayesian-based approach to revise the probability distribution of the information in the sources in a systematic way to remedy this difficulty. The revision step is similar in spirit to tasks like data cleaning and record linkage and should be carried out before integration can be achieved for probabilistic uncertain data.

Keywords: Information integration · Uncertain data · Probabilistic data · Belief revision

1 Introduction

Information integration and modeling and management of uncertain information have been active research areas for decades, with both areas receiving significant renewed interest in recent years [3–6,13,15]. The importance of information integration *with uncertainty*, on the other hand, has been realized more recently. [8,9,11,13–15,17–19,21,24–26]. It has been observed that [15]:

> While in traditional database management managing uncertainty and lineage seems like a nice feature, in data integration it becomes a necessity.

In this paper we study the problem of integrating probabilistic uncertain information. Certain constraints are imposed by the semantics of integration, but there is no guarantee that they are satisfied in practical situations. We present a Bayesian-based approach to revise the probability distribution of the information in the sources in a systematic way to remedy this difficulty. The revision step is similar in spirit to tasks like data cleaning and record linkage and should be carried out before integration can be achieved for probabilistic uncertain data.

This paper is organized as follows: We present the theory of uncertain information integration in Section 2. Probabilistic constraints are discussed in Section 3 and our proposed Bayesian-based approach to revise information sources probabilities is presented in Section 4. Conclusions are presented in Section 5.

© Springer International Publishing Switzerland 2015
M.A. Sharaf et al. (Eds.): ADC 2015, LNCS 9093, pp. 78–90, 2015.
DOI: 10.1007/978-3-319-19548-3_7

2 Preliminaries – Foundations of Information Integration

Foundations of information integration with uncertainty have been discussed in [2,22]. We present a brief summary here. We begin with an example from [22].

Example 1. John and Jane are talking about fellow student Bob. John says "I am taking CS100 and CS101, and Bob is in one of them, but not in both." Jane says "I am taking CS101 and CS102 and Bob is in one of them, but not in both."

Intuitively, if we integrate the information from these two sources (John and Jane), we should infer that Bob is either taking CS101, or he is taking both CS100 and CS102. We present an algorithm for the integration of uncertain information in Section 2.1.   ∎

The model used in [2,22] for the representation of uncertain information is the well-known *possible-worlds* model [1]. We should emphasize that the possible-worlds model is used in the *formalization* of information integration. It is not, in general, efficient for implementation. In Example 1, the information presented by the two sources (John and Jane) is represented by the possible-worlds shown in Figures 1 and 2.

student	course
Bob	CS100

D1

student	course
Bob	CS101

D2

Fig. 1. Possible Worlds of source S1

student	course
Bob	CS101

D3

student	course
Bob	CS102

D4

Fig. 2. Possible Worlds of source S2

We will summarize the integration approach from [22] which uses a simple logic-based technique in Section 2.1). This approach has been shown to be equivalent to the integration approach of [2] which is based on the concept of superset-containment. Interested readers are referred to [22] for details.

First, we should mention that the pure possible world model is not adequate for integration applications. We need additional information, namely, the set of all tuples. The following example demonstrates the possible-worlds with tuple sets model.

Example 2. Andy and Jane are talking about fellow student Bob. Andy says "I am taking CS100, CS101, and CS102 and Bob is in either CS100 or CS101 but not in both." Jane says "I am taking CS101 and CS102 and Bob is in one of them, but not in both."

Intuitively, if we integrate the information from these two sources, we should infer that Bob is taking CS101. The second possibility from Example 1, namely

Bob taking CS100 and CS102, is not valid anymore since Andy's statement rules out the possibility that Bob is taking 102.

However, the possible-worlds representations of these sources (Andy and Jane) are exactly the same as those of Example 1 (Figures 1 and 2). Only when we add the tuple-set to possible worlds of Andy, namely {(Bob, CS100), (Bob, CS101), (Bob, CS102)}, It becomes explicit that Andy's statement eliminates the possibility that Bob is taking CS102.   ■

Hence, we will use the following definition from [2] for uncertain databases that adds tuple sets to the possible-worlds model. To simplify presentation, we assume that possible worlds are sets of tuples in a single relation.

Definition 1. (UNCERTAIN DATABASE). *An uncertain database U consists of a finite set of tuples $T(U)$ and a nonempty set of possible worlds $PW(U) = \{D_1, \ldots, D_m\}$, where each $D_i \subseteq T(U)$ is a certain database.*   ■

This definition adds tuple-set $T(U)$ to the traditional possible-worlds model. In fact, as shown in Example 2, there may be tuples in the tuple set, $t \in T(U)$, that do not appear in any possible world of the uncertain database U. If $T(U)$ is not provided explicitly, then we use the set of all tuples in the possible worlds, *i.e.*, $T(U) = D_1 \cup \cdots \cup D_n$. It is interesting to notice that this model exhibits both closed-world and open-world properties: If a tuple $t \in T(U)$ does not appear in a possible world D_i, then it is assumed to be *false* for D_i (hence, closed-world assumption). In other words, D_i explicitly rules out t. The justification is that the source providing the uncertain information represented by U is aware of (the information represented by) all $t \in T(U)$. If some $t \in T(U)$ is absent from D_i, then the source explicitly rules out t from D_i. On the other hand, all other tuples $t \notin T(U)$ are assumed possible (*unknown*) for possible-worlds D_i (hence, open-world assumption). This distinction is important for integration: Consider integrating D_i, where $t \notin D_i$, with a possible-world D'_j from another source, where $t \in D'j$. For the first case ($t \in T(U)$), D_i and D'_j are not compatible and can not be integrated. This is because D_i explicitly rules out t while D'_j explicitly includes it. On the other hand, for the second case ($t \notin T(U)$), D_i and D'_j can be integrated since D_i can accept t as a valid tuple.

2.1   Integration Using Logical Representation

In this section we review some results from [22]. Given an uncertain database U, we assign a propositional variable x_i to each tuple $t_i \in T(U)$. We define the formula f_j corresponding to a possible world D_j, and the formula f corresponding to the uncertain database U as follows:

Definition 2. (LOGICAL REPRESENTATION OF AN UNCERTAIN DATABASE). *Let D_j be a database in the possible worlds of uncertain Database U. Construct a formula as the conjunction of all variables x_i where the corresponding tuple t_i*

is in D_j, and the conjunction of $\neg x_i$ where the corresponding tuple t_i is not in D_j. That is,

$$f_j = \bigwedge_{t_i \in D_j} x_i \bigwedge_{t_i \notin D_j} \neg x_i \tag{1}$$

The formula corresponding to the uncertain database U is the disjunction of the formulas corresponding to the possible worlds of U. That is,

$$f = \bigvee_{D_j \in PW(U)} f_j \tag{2}$$

Now we can integrate uncertain databases using their logical representations as follows:

Let $S_1, \ldots, S_n$ be sources containing (uncertain) databases $U_1, \ldots, U_n$. Let the propositional formulas corresponding to $U_1, \ldots, U_n$ be $f_1, \ldots, f_n$. We obtain the formula f corresponding to the uncertain database resulting from integrating $U_1, \ldots, U_n$ by conjuncting the formulas of the databases: $f = f_1 \wedge \cdots \wedge f_n$.

Example 3. (INTEGRATION USING LOGICAL REPRESENTATION) Consider Example 1. The uncertain database corresponding to John's statement is represented by $(x_1 \wedge \neg x_2) \vee (\neg x_1 \wedge x_2)$, where x_1, and x_2 correspond to the tuples (Bob, CS100) and (Bob, CS101), respectively. The uncertain database corresponding to Jane's statement is represented by $(x_2 \wedge \neg x_3) \vee (\neg x_2 \wedge x_3)$, where x_2 is as above and x_3 corresponds to the tuple (Bob, CS102). The integration in this case is obtained as

$$((x_1 \wedge \neg x_2) \vee (\neg x_1 \wedge x_2)) \wedge ((x_2 \wedge \neg x_3) \vee (\neg x_2 \wedge x_3))$$
$$= (x_1 \wedge \neg x_2 \wedge x_3) \vee (\neg x_1 \wedge x_2 \wedge \neg x_3)$$

which corresponds to the possible worlds of Figure 3. The result is consistent with our intuition: Based on statements by John and Jane, Bob is taking either CS101 or both CS100 and CS102.

student	course
Bob	CS101

student	course
Bob	CS100
Bob	CS102

Fig. 3. Possible Worlds of the Integration for Example 1

Now consider Example 2. The uncertain database corresponding to Andy's statement is represented by $(x_1 \wedge \neg x_2 \wedge \neg x_3) \vee (\neg x_1 \wedge x_2 \wedge \neg x_3)$, where x_1, x_2, and x_3 represent (Bob, CS100), (Bob, CS101), and (Bob, CS102), respectively. The uncertain database corresponding to Jane's statement is the same as above $(x_2 \wedge \neg x_3) \vee (\neg x_2 \wedge x_3)$. The integration in this case is obtained as

$$((x_1 \wedge \neg x_2 \wedge \neg x_3) \vee (\neg x_1 \wedge x_2 \wedge \neg x_3)) \wedge ((x_2 \wedge \neg x_3) \vee (\neg x_2 \wedge x_3))$$
$$= (\neg x_1 \wedge x_2 \wedge \neg x_3)$$

corresponding to the (in this case, definite) relation consisting only of the tuple (Bob, CS101). Again, this result is consistent with our intuition: Based on statements by Andy and Jane, Bob is taking CS101. ∎

2.2   Probabilistic Uncertain Information

The conceptual model for probabilistic uncertain information is the possible-worlds with tuple-set model with a probability distribution over the set of possible worlds. More formally,

Definition 3. (PROBABILISTIC UNCERTAIN DATABASE). *A probabilistic uncertain database U consists of a finite set of tuples $T(U)$ and a nonempty set of possible worlds $PW(U) = \{D_1, \ldots, D_m\}$. Each $D_i \subseteq T(U)$ is associated with a probability $P(D_i)$ in the [0,1] range, where $\sum_{i=1}^{m} P(D_i) = 1$.*   ■

The integration technique of Section 2.1 can be applied to the probabilistic case to obtain the possible-worlds of the result. We have shown in [22] that very interesting constraints are imposed on the probabilistic structure of information sources in the integration of probabilistic data. We discuss these constraints in Section 3 below. First, we need a few definitions and observations.

Definition 4. (COMPATIBLE POSSIBLE WORLDS). *Let S and S' be sources containing probabilistic uncertain information $\{D_1, \ldots, D_m\}$ and $\{D'_1, \ldots, D'_{m'}\}$, respectively. Let T and T' be the tuple-sets of S and S'. A pair of possible-worlds (D_i, D'_j) from S and S' are said to be compatible if (1) For all tuples $t \in D_i - D'_j$, $t \notin T'$, and (2) For all tuples $t \in D'_j - D_i$, $t \notin T$.*   ■

It is easy to verify that, Given two information sources, only compatible pairs of possible worlds from the two sources can be integrated (combined). Each compatible pair produces a possible world in the answer.

We use a *compatibility* graph G to capture the compatibility relationship defined above. Let S and S' be sources containing probabilistic uncertain information $\{D_1, \ldots, D_m\}$ and $\{D'_1, \ldots, D'_{m'}\}$, respectively. The compatibility graph G for S and S' is a bipartite graph. Nodes of G have a one-to-one correspondence with possible worlds of S and S'. That is, G has nodes $\{N_1, \ldots, N_m, N'_1, \ldots, N'_{m'}\}$, where node N_i, $i = 1, \ldots, m$, corresponds to the world D_i of S, and node N'_j, $j = 1, \ldots, m'$, corresponds to the world D'_j of S'. There is an edge between N_i and N'_j if the pair of possible worlds (D_i, D'_j) are compatible. We sometimes overload the notation and use $\{D_1, \ldots, D_m, D'_1, \ldots, D'_{m'}\}$ for possible-worlds as well as for nodes of the compatibility graph of the two sources.

The following result was proven in [22].

Theorem 1. *Let G be the compatibility graph of sources S and S'. Each connected component of G is a complete bi-partite graph.*   ■

So, if we consider a connected component of G, it has a set of nodes from S (e.g., $\{D_{i1}, \ldots, D_{ik}\} \subseteq \{D_1, \ldots, D_m\}$) and another set of nodes from S' (e.g., $\{D'_{j1}, \ldots, D'_{jk'}\} \subseteq \{D'_1, \ldots, D'_{m'}\}$). Then, by Theorem 1, every node in the first group is connected to every node in the second group. Further, these nodes are not connected to any other nodes. (Note that we are using a symbol D to refer both to a possible-world D and to the node representing D in the compatibility graph.)

3   Probabilistic Constraints

When integrating sources containing probabilistic uncertain information, certain constraints are imposed on the probabilistic distributions of the possible worlds of the sources. The following theorem is from [22]:

Theorem 2. *Let S and S' be sources containing probabilistic uncertain information $\{D_1, \ldots, D_m\}$ and $\{D'_1, \ldots, D'_{m'}\}$, respectively. Let G be their (bipartite) compatibility graph. Let G_1 be a connected component of G, with the set of nodes $\{D_{i1}, \ldots, D_{ik}\} \subseteq \{D_1, \ldots, D_m\}$ and $\{D'_{j1}, \ldots, D'_{jk'}\} \subseteq \{D'_1, \ldots, D'_{m'}\}$. Then the following constraint between the probabilities of the possible-worlds represented by the nodes of the connected component G_1 must hold:*

$$\sum_{D \in \{D_{i1}, \ldots, D_{ik}\}} P(D) = \sum_{D' \in \{D'_{j1}, \ldots, D'_{jk'}\}} P(D') \qquad (3)$$

In other words, each connected component G_1 of the bipartite compatibility graph G of S and S' enforces a constraint that the sum of probabilities of possible-worlds associated with S in the connected component should be equal to the sum of probabilities of possible-worlds associated with S' in the same connected component.

Example 4. The compatibility graph G for the sources of Example 1 (John and Jane) is simple: G has four nodes corresponding to the possible worlds of source 1, D_1 and D_2, and the possible worlds of source 2, D_3 and D_4 (See Figures 1 and 2). The edges of G are (D_1, D_4) and (D_2, D_3). Hence, G has two connected components: $\{D_1, D_4\}$ and $\{D_2, D_3\}$. The probabilistic constraints for this case are $P(D_1) = P(D_4)$ and $P(D_2) = P(D_3)$.

For another example, consider information sources B_1 and B_2 about books and their authors whose possible-worlds are shown in Figures 4 and 5.

book1	Dan
book2	Jen

E1

book1	Dan
book3	Bob

E2

book2	Amy
book3	Bob

E3

Fig. 4. Possible Worlds of source B_1

book1	Dan
book4	Jan

E'1

book1	Dan
book4	Pam

E'2

book4	Jan
book4	Pam

E'3

Fig. 5. Possible Worlds of source B_2

The compatibility graph in this case has two connected components with edges $\{(E_1, E'_1), (E_1, E'_2), (E_2, E'_1), (E_2, E'_2)\}$, and $\{(E_3, E'_3)\}$, respectively (See Figure 6). There are two probabilistic constraints corresponding to the two connected components: $P(E_1) + P(E_2) = P(E'_1) + P(E'_2)$, and $P(E_3) = P(E'_3)$.

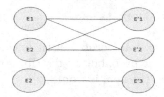

Fig. 6. Compatibility graph for Sources B1 and B2 (Example 4)

These constraints may appear counterintuitive in the first sight. Basically they state that, in general, sources containing probabilistic uncertain information are not independent. Rather, every pair of sources may be correlated. Recent research on *data fusion* (for example, [7,10,12,20,28]) confirms this fact. It has been shown that by taking into account the correlations among sources, significantly better fusion (integration) results can be obtained. Our framework is different from that of data fusion. Nevertheless, the correlation between information sources remains valid.

We will summarize the proof of Theorem 2 below to shed more light on the correlation between sources. Let S and S' be sources containing probabilistic uncertain information $\{D_1, \ldots, D_m\}$ and $\{D'_1, \ldots, D'_{m'}\}$, respectively. Let $\{P(D_1), \ldots, P(D_m)\}$ and $\{P(D'_1), \ldots, P(D'_{m'})\}$ be the probability distributions of the possible-worlds of S and S'. Intuitively, $P(D)$ is the probability of the event that the real world database is D. Note that the probability distribution $\{P(D_1), \ldots, P(D_m)\}$ consists of events that are mutually exclusive and exhaustive. That is, (1) $P(D_i \wedge D_j) = 0$ for $D_i \neq D_j$, in other words, the real world can not be D_i and D_j at the same time, and (2) $\sum_{i=1}^{m} P(D_i) = 1$. Similarly, the probability distribution $\{P(D'_1), \ldots, P(D'_{m'})\}$ is also mutually exclusive and exhaustive. So, we can write

$$P(D_i) = \sum_{j=1}^{m'} P(D_i \wedge D'_j), i = 1, \ldots, m \tag{4}$$

and

$$P(D'_j) = \sum_{i=1}^{m} P(D_i \wedge D'_j), j = 1, \ldots, m' \tag{5}$$

Given a pair of possible-worlds (D_i, D'_j), if D_i and D'_j are not compatible, they contain contradictory information and can not be combined. That is, the events "the real world is D_i" and "the real world is D'_j" are contradictory. Hence, $P(D_i \wedge D'_j) = 0$.

Now consider a connected component G_1 with the set of nodes $\{D_{i1}, \ldots, D_{ik}\} \subseteq \{D_1, \ldots, D_m\}$ and $\{D'_{j1}, \ldots, D'_{jk'}\} \subseteq \{D'_1, \ldots, D'_{m'}\}$. Each possible-world in the first set is compatible with every possible-world in the second set, and vice-versa. Further, these possible worlds are not compatible with any other possible worlds. It follows that $P(D_{iq}) = \sum_{r=1}^{k'} P(D_{iq} \wedge D'_{jr}), q = 1, \ldots, k$, and

$P(D'_{jr}) = \sum_{q=1}^{k} P(D_{iq} \wedge D'_{jr}), r = 1, \ldots, k'$. Then

$$\sum_{q=1}^{k} P(D_{iq}) = \sum_{q=1}^{k} \sum_{r=1}^{k'} P(D_{iq} \wedge D'_{jr})$$

and

$$\sum_{r=1}^{k'} P(D'_{jr}) = \sum_{r=1}^{k'} \sum_{q=1}^{k} P(D_{iq} \wedge D'_{jr})$$

Hence

$$\sum_{q=1}^{k} P(D_{iq}) = \sum_{r=1}^{k'} P(D'_{jr})$$

which is the same as Equation 3.   ∎

We have presented algorithms for the calculation of the probabilities of the possible-worlds of the result of integrating probabilistic uncertain information in the case where probabilistic constraints are satisfied [22,23]. But, in practice, the probabilistic distribution of information sources are provided by the sources themselves or through certain data mining or analytic processing. There is no guarantee that probabilistic constraints are indeed satisfied in practice. The rest of this paper provides approaches for these cases when the probabilistic constraints are not satisfied.

4   Revising Probability Distribution of Sources

In this section we concentrate on the case where we have sources S and S' containing probabilistic uncertain information and one or more of the probabilistic constraints are not satisfied. We use a Bayesian-based approach to revise the probabilistic distributions of the sources such that the revised distributions do satisfy all constraints.

Let the possible-worlds of S and S' be $\{D_1, \ldots, D_m\}$ and $\{D'_1, \ldots, D'_{m'}\}$, respectively. Let us begin by treating S as the original set of events, and S' as the new *evidence* by which the probabilities of the original events, $P(D_i)$'s, are revised. In other words, we want to compute the conditional probabilities

$$P(D_i \mid \text{The evidence provided by } S')$$

which we will simply denote by $P(D_i \mid S')$ henceforth. We will use Q for the *posterior* probability distributions. So, $Q(D_i) = P(D_i \mid \text{The evidence provided by } S')$ is the revised (or posterior) probability of D_i.

This is a case where the evidence itself is probabilistic. Hence, we will use Richard Jeffrey's rule of conditioning [16,27] which is an extension of Bayes' rule to probabilistic evidence.

$$Q(D_i) = \sum_{j=1}^{m'} Q(D'_j) P(D_i \mid D'_j), \quad i = 1, \ldots, m \tag{6}$$

but

$$P(D_i \mid D'_j) = \frac{P(D'_j \mid D_i)P(D_i)}{P(D'_j)} = \frac{P(D'_j \mid D_i)P(D_i)}{\sum_{k=1}^{m} P(D'_j \mid D_k)P(D_k)}$$

Hence, we obtain the following alternative formulation:

$$Q(D_i) = \sum_{j=1}^{m'} Q(D'_j) \left(\frac{P(D'_j \mid D_i)P(D_i)}{\sum_{k=1}^{m} P(D'_j \mid D_k)P(D_k)} \right), \quad i = 1, \ldots, m \qquad (7)$$

As mentioned earlier, Given sources S and S' containing probabilistic uncertain information, probability constraints (Equation 3) may not hold in practice. Next, we prove that the *revised* probability distributions $Q(D_i), i = 1, \ldots, m$ and $Q(D'_j), j = 1, \ldots, m'$, as obtained by Equation 6 (or Equation 7), satisfy all probabilistic constraints.

Theorem 3. *Consider sources S and S' containing probabilistic uncertain information $\{D_1, \ldots, D_m\}$ and $\{D'_1, \ldots, D'_{m'}\}$, respectively. Let G be the compatibility graph of S and S', and $\{P(D_1), \ldots, P(D_m)\}$ and $\{P(D'_1), \ldots, P(D'_{m'})\}$ be their probability distributions. Consider a connected component G_1 of G. Then*

$$\sum_{D_i \in G_1} Q(D_i) = \sum_{D'_j \in G_1} Q(D'_j)$$

where $Q(D_i)$ and $Q(D'_j)$ are revised probability distributions according to Equation 6.

Proof. Consider a node $D_i \in G_1$. Note that $P(D_i \mid D') = 0$ for all nodes $D' \notin G_1$ (D_i and D' are not compatible if they do not belong to the same connected component.) So, we can write by Equation 6,

$$Q(D_i) = \sum_{j=1}^{m'} Q(D'_j)P(D_i \mid D'_j) = \sum_{D'_j \in G_1} Q(D'_j)P(D_i \mid D'_j)$$

Then,

$$\sum_{D_i \in G_1} Q(D_i) = \sum_{D_i \in G_1} \sum_{D'_j \in G_1} Q(D'_j)P(D_i \mid D'_j) = \sum_{D'_j \in G_1} Q(D'_j) \sum_{D_i \in G_1} P(D_i \mid D'_j)$$

and (again, since $P(D \mid D'_j) = 0$ for all nodes D that are not in the same connected component as D'_j – which is G_1):

$$\sum_{D_i \in G_1} Q(D_i) = \sum_{D'_j \in G_1} Q(D'_j) \sum_{i=1}^{m} P(D_i \mid D'_j)$$

But $P(D_i \mid D'_j) = \frac{P(D_i \wedge D'_j)}{P(D'_j)}$. Hence,

$$\sum_{i=1}^{m} P(D_i \mid D'_j) = \sum_{i=1}^{m} \frac{P(D_i \wedge D'_j)}{P(D'_j)} = \frac{\sum_{i=1}^{m} P(D_i \wedge D'_j)}{P(D'_j)} = \frac{P(D'_j)}{P(D'_j)} = 1$$

It follows that

$$\sum_{D_i \in G_1} Q(D_i) = \sum_{D'_j \in G_1} Q(D'_j)$$

Equations 6 and 7 contain two sets of unknowns: Both posterior probability distributions, $Q(D_i), i = 1, \ldots, m$ and $Q(D'_j), j = 1, \ldots, m'$, are unknown. Choosing the values of one set impacts those of the other set. So, our task is to compute these two sets of unknowns. We will discuss below how to use our confidence in the sources to compute these parameters.

4.1 Total Confidence in the Evidence

In some applications, we may have complete confidence in the evidence (*i.e.*, information provided by source S' in our case), and want to revise the probability distribution of the original set of events (information provided by source S) with respect to the evidence. In this case the probability distribution of S' remains unchanged. In other words, we have $Q(D'_j) = P(D'_j), j = 1, \ldots, m$.

We know by Theorem 3 that for every connected component G_1 of the compatibility graph G:

$$\sum_{D_i \in G_1} Q(D_i) = \sum_{D'_j \in G_1} Q(D'_j)$$

So, if we have total confidence in the evidence S':

$$\sum_{D_i \in G_1} Q(D_i) = \sum_{D'_j \in G_1} P(D'_j)$$

In other words, the probability distribution of S is revised in a way that the sum of (revised) probabilities on the S side of a connected component G_1 equals the sum of (original) probabilities on the S' side of G_1.

The "dual" of this situation is when we have total confidence in S, in which case the probability distribution of S' will be revised such that

$$\sum_{D_i \in G_1} P(D_i) = \sum_{D'_j \in G_1} Q(D'_j)$$

4.2 General Case

In general, we will not have total confidence in either of sources. Rather, we may have subjective or analytic confidence measures for the sources. We formalize this situation by introducing *confidence measures* α_i for each source S_i, such that $\sum_{i=1}^{n} \alpha_i = 1$, where n is the number of sources. If there are two sources S

and S', their confidence measures can be denoted by α and $1 - \alpha$. Our approach is to revise the probability distributions of both sources to obtain a weighted probability sum for each connected component G_1 as follows:

$$\sum_{D_i \in G_1} Q(D_i) = \sum_{D'_j \in G_1} Q(D'_j) = \alpha \sum_{D_i \in G_1} P(D_i) + (1 - \alpha) \sum_{D'_j \in G_1} P(D'_j) \qquad (8)$$

The cases for total confidence in S and in S' correspond to $\alpha = 1$ and $\alpha = 0$, respectively.

5   Conclusion

We have studied the problem of integrating probabilistic uncertain information. Certain constraints are imposed by the semantics of integration, but there is no guarantee that they are satisfied in practical situations. We presented a Bayesian-based approach to revise the probability distribution of the information in the sources in a systematic way to remedy this difficulty. The revision step is similar in spirit to tasks like data cleaning and record linkage and should be carried out before integration can be achieved for probabilistic uncertain data.

There is a close relationship between uncertain-data integration and *data fusion*, which refers to the integration of massive amounts of mined data. The process of mining data from sources such as web pages, social media and email messgaes generate large amounts of data with differing degrees of correctness confidence, which can be conveniently modeled by probabilistic uncertain data. In the future, we intend to study the application of our Bayesian probability revision approach to data fusion.

References

1. Abiteboul, S., Kanellakis, P.C., Grahne, G.: On the representation and querying of sets of possible worlds. In: Proceedings of ACM SIGMOD International Conference on Managementof Data, pp. 34–48 (1987)
2. Agrawal, P., Sarma, A.D., Ullman, J.D., Widom, J.: Foundations of uncertain-data integration. Proceedings of the VLDB Endowment **3**(1), 1080–1090 (2010)
3. Antova, L., Jansen, T., Koch, C., Olteanu, D.: Fast and simple relational processing of uncertain data. In: Proceedings of IEEE International Conference on Data Engineering, pp. 983–992 (2008)
4. Antova, L., Koch, C., Olteanu, D.: 10^{10^6} worlds and beyond: Efficient representation and processing of incomplete information. In: Proceedingsof IEEE International Conference on Data Engineering, pp. 606–615 (2007)
5. Chen, D., Chirkova, R., Sadri, F., Salo, T.J.: Query optimization in information integration. Acta Informatica **50**(4), 257–287 (2013)
6. Dalvi, N.N., Ré, C., Suciu, D.: Probabilistic databases: diamonds in the dirt. Communications of the ACM **52**(7), 86–94 (2009)
7. Dong, X.L., Berti-Equille, L., Srivastava, D.: Integrating conflicting data: The role of source dependence. PVLDB **2**(1), 550–561 (2009)

8. Dong, X.L., Halevy, A., Yu, C.: Data integration with uncertainty. In: Proceedings of International Conference on Very Large Databases, pp. 687–698 (2007)
9. Dong, X.L., Halevy, A.Y., Yu, C.: Data integration with uncertainty. The VLDB Journal 18(2), 469–500 (2009)
10. Dong, X.L., Saha, B., Srivastava, D.: Less is more: Selecting sources wisely for integration. Proceedings of the VLDB Endowment 6(2), 37–48 (2012)
11. Eshmawi, A.A., Sadri, F.: Information integration with uncertainty. In: Proceedings of International Database Engineering and Applications, IDEAS, pp. 284–291 (2009)
12. Galland, A., Abiteboul, S., Marian, A., Senellart, P.: Corroborating information from disagreeing views. In: Proceedings of ACM InternationalConference on Web Search and Data Mining, pp. 131–140 (2010)
13. Haas, L.: Beauty and the Beast: The Theory and Practice of Information Integration. In: Schwentick, T., Suciu, D. (eds.) ICDT 2007. LNCS, vol. 4353, pp. 28–43. Springer, Heidelberg (2006)
14. Halevy, A.Y., Ashish, N., Bitton, D., Carey, M.J., Draper, D., Pollock, J., Rosenthal, A., Sikka, V.: Enterprise information integration: successes, challenges and controversies. In: Proceedings of ACM SIGMOD International Conference on Management of Data, pp. 778–787 (2005)
15. Halevy, A.Y., Rajaraman, A., Ordille, J.J.: Data integration: The teenage years. In: Proceedings of International Conference on Very Large Databases, pp. 9–16 (2006)
16. Jeffrey, R.: The Logic of Decision. McGraw-Hill (1965)
17. Magnani, M., Montesi, D.: Uncertainty in data integration: current approaches and open problems. In: Proceedings of VLDB Workshop on Managementof Uncertain Data, pp. 18–32 (2007)
18. Magnani, M., Montesi, D.: A survey on uncertainty management in data integration. ACM Journal of Data and Information Quality 2(1) (2010)
19. Olteanu, D., Huang, J., Koch, C.: SPROUT: Lazy vs. eager query plans for tuple-independent probabilistic databases. In: Proceedings of IEEE International Conference on Data Engineering, pp. 640–651 (2009)
20. Pochampally, R., Sarma, A.D., Dong, X.L., Meliou, A., Srivastava, D.:. Fusing data with correlations. In: Proceedings of ACM SIGMODInternational Conference on Management of Data, pp. 433–444 (2014)
21. Re, C., Dalvi, N.N., Suciu, D.: Efficient top-k query evaluation on probabilistic data. In: Proceedings of IEEE International Conference on DataEngineering, pp. 886–895 (2007)
22. Sadri, F.: On the foundations of probabilistic information integration. In: Proceedings of International Conference on Information and Knowledge Management, pp. 882–891 (2012)
23. Sadri, F., Tallur, G: Integration of probabilistic uncertain information (2014) (manuscript)
24. Sarma, A.D., Benjelloun, O., Halevy, A.Y., Nabar, S.U., Widom, J.: Representing uncertain data: models, properties, and algorithms. The VLDB Journal 18(5), 989–1019 (2009)
25. Sarma, A.D., Benjelloun, O., Halevy, A.Y., Widom, J.: Working models for uncertain data. In: Proceedings of IEEE International Conferenceon Data Engineering, p. 7 (2006)

26. Sen, P., Deshpande, A.: Representing and querying correlated tuples in probabilistic databases. In: Proceedings of IEEE International Conference onData Engineering, pp. 596–605 (2007)
27. Shafer, G.: Jeffrey's rule of conditioning. Philosophy of Science **48**(3), 337–362 (1981)
28. Zhao, B., Rubinstein, B.I.P., Gemmell, J., Han, J.: A bayesian approach to discovering truth from conflicting sources for data integration. Proceedings of the VLDB Endowment **5**(6), 550–561 (2012)

Personal Process Description Graph
for Describing and Querying Personal Processes

Jing Xu[1]([✉]), Hye-young Paik[1], Anne H.H. Ngu[2], and Liming Zhan[1]

[1] The University of New South Wales, Sydney, NSW, Australia
{jxux494,hpaik,zhanl}@cse.unsw.edu.au
[2] Texas State University, Austin, TX, USA
angu@txstate.edu

Abstract. Personal processes are ad-hoc to the point where each process may have a unique structure and is certainly not as strictly defined as a workflow process. In order to describe, share and analyze personal processes more effectively, in this paper, we propose Personal Process Description Graph (PPDG) and a personal process query. The personal process query approach is developed to support different types of graph queries in a personal process graph repository. The approach follows a filtering and refinement framework to speed up the query computation. We conducted experiments to demonstrate the efficiency of our techniques.

1 Introduction

People are often confronted with tasks that are infrequent (even one-off) and not well-described. Examples are applying for jobs, buying a house, or filing a tax return. We refer to these as "personal processes" in which an individual performs a set of logically related tasks to accomplish some personal goals [7,15].

To overcome their lack of familiarity, people typically seek out other people's experiences with the task. It is difficult and time-consuming to capture and share personal processes, although we have many kinds of channels to share variety of information such as photos or video. Furthermore, searching and understanding already shared processes are limited to a simple keyword search via search engines.

- With regards to capturing a personal process, the current model of describing processes, which is based on free texts and bullet points, makes it difficult to (1) understand the process especially if there are many steps involved; (2) compare and contrast different paths/ways to accomplish the goal; and (3) understand the dependencies between data and actions.
- For searching and analyzing a personal process, the current model of keyword search over the textual descriptions is limited and is not able to provide exploratory answers to query. That is, the search always returns a set of Web pages containing the descriptions of the whole processes even if the user just requires a fragment of the process or a specific task in the process.

© Springer International Publishing Switzerland 2015
M.A. Sharaf et al. (Eds.): ADC 2015, LNCS 9093, pp. 91–103, 2015.
DOI: 10.1007/978-3-319-19548-3_8

In order to describe, share and analyze personal processes more effectively, we initiate a novel framework for personal process management named ProcessVidere. Our ultimate aim is to provide a space for users to share their experiences, capture the process knowledge from people, analyze them effectively, and support users to re-use the whole or part of the processes for their own purposes. In this paper, as a concrete step towards building ProcessVidere, we present the following elements as the fundamental components of the system:

- Personal Process Description Graph (PPDG), a graph-based description language to present both control flow and data flow of a personal process.
- A template-based approach to perform structured queries over PPDG, as well as an implementation of the key template queries and preliminary performance evaluation results.

The paper is organized as follows: Section 2 introduces the PPDG language. Section 3 presents the basic design of PPDG query templates. Section 4 describes methods and algorithms for an efficient processing of the query templates. Then we present the experiment results in Section 5. The related work is discussed in Section 6 followed by a conclusion in Section 7.

2   Personal Process Description Graph (PPDG)

Let us consider "attending a graduation ceremony at UNSW". We interviewed six UNSW graduates and asked them to describe the process. A collective summary of the process is as follows:

> *Before the graduation day, a graduand may book a graduation gown online. The booking can be paid for by card (in which case a receipt is issued), or by cash when collected. On the day, the graduand collects the dress, take a photo and register for the ceremony to obtain a seat number. After a briefing session, the graduand attends the ceremony. The dress is returned after the ceremony.*

Not every step is strictly followed, in fact, six different variations of this personal processes are obtained. A typical workflow management or business process modelling (BPM) approaches would attempt to create a single reference model that describes the above process as complete as possible. However, building such a model for a flexible and ad-hoc process is difficult and often makes the model convoluted. As discussed in Section 6, existing workflow or BPM systems are not equipped to deal with ad-hoc processes very well [7,8,15].

In ProcessVidere, instead of developing and managing a reference model, we consider a personal process as process steps experienced and described by a single person. Each description of the process becomes a feasible process, the so-called "model", that leads to the intended goal. PPDG described below is designed to capture the description of an individual experience of a process. Later in the paper, we will explain how queries are written over these descriptions to give a flexible view of the process.

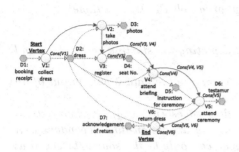
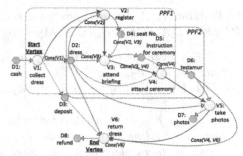

Fig. 1. PPDG1: UNSW graduation ceremony process by Graduand A

Fig. 2. PPDG2: UNSW graduation ceremony process by Graduand B

2.1 PPDG Syntax Overview

PPDG represents a personal process P as a labelled directed graph. It describes the whole process of performing a personal process placing *equal emphasis* on both actions and input/output data related to each action in the process. Figure 1 depicts a PPDG of the graduation ceremony process experienced by Graduand A. An action is noted by a circular node. Actions can be divided into two categories: a simple action and composite action which contains another process. A simple action can be one-off action, repeated action, or duration action. In order to make the visualization of the graph simple, all types of actions are represented using the same notation. The details are stored in the schema associated with the process graph.

Data elements in a process P are represented by hexagonal nodes. A data element can be either basic or composite (i.e., composition of basic data). In PPDG, data could be any artifacts that is available somewhere: physical items such as paper documents, academic dress, or digital items such as digital photos, text messages and so on. A data element can be external (i.e., it is not produced by any action) or processed (i.e., there is at least one action produced it).

The data elements and actions are connected to represent 'action flow' or 'data flow'. Action flows, represented by solid lines, describe temporal sequence of the actions. For example, in Figure 1, 'V_1: *collect dress*' takes places before 'V_2: *take photos*'. Data flows, represented by dotted lines, keep track of data sources and denotes the relationships between the data and actions. For example, 'V_4:attending briefing' takes two data inputs 'D_2: *dress*' and 'D_4: *seat no.*' and produces one data output 'D_5: *instructions for ceremony*'.

PPDG also stores constraints/conditions relating to action, data and the flows. If the constraints concern an action (referred to as 'action constraints'), it may include conditions such as the location or the time the action takes place. If the constraints are on the connecting edges of two actions (referred to as 'transition constraints', it may specify the conditions that should be met for the flow to take place. It can also be used to enforce the sequence of the two actions. We define PPDG more formally as follows.

Definition 1. *A personal process description graph PPDG is a tuple*
$PPDG := (A, D, E_A, E_D, C, \phi, \lambda)$ *where:*

- *A is a finite set of nodes a_0, a_1, a_2,... depicting the starting action (a_0) and actions (a_1, a_2, ...).*
- *D is a finite set of nodes d_0, d_1, d_2,... depicting the data input/output of an action.*
- *E_A is a finite set of directed action-flow edges ea_1, ea_2,..., where $ea_i = (a_j, a_k)$ leading from a_j to a_k ($a_j \neq a_k$) is an action-flow dependency. It reads a_j takes place before a_k. Each node can only be the source/target of at most one action-flow edge : $ea = (a_i, a_j) \in E_A : ea' = (a_k, a_l) \in E_A \setminus ea : a_i \neq a_k$ and $a_j \neq a_l$.*
- *E_D is a finite set of directed data-flow edges ed_1, ed_2,..., where $ed_i = (a_j, d_k)$ leading from a_j to d_k is a data-flow dependency. It reads a_j produces d_k. $ed_l = (d_m, a_n)$ leading from d_m to a_n is a data-flow dependency. It reads a_n takes d_m.*
- *C is a finite set of conditions $c_1, c_2,...$ with $c_i = (< name, descr >, x_j)$ being associated to $x_j \in \{A, D, E_A, E_D\}$ and having name and description of the condition.*
- *ϕ: a function that maps Label $L(A_i)$ to action nodes.*
- *λ: a function that maps Label $L(D_i)$ to data nodes.*

3 Query Templates and Their Basic Constructs

A repository of PPDG is organized by categories and domains. The idea in ProcessVidere is that this repository can be simply keyword searched or browsed, but more importantly, it can be used to perform structured queries for sophisticated analysis. One might pose a query "How to attend a graduation ceremony at UNSW" to understand the whole process. Others might want some answers to particular aspects of the process, such as "What happens after attending the briefing session?", "How do I get a seat no.?", "I have collected the dress, what can I do with it?".

In answering these queries, we will develop various graph-based query processing techniques. In this section, as the first step, we present the basic design of the single match approach. We envisage that this basic set of query constructs will allow us to build more complex query approaches in our immediate future work to implement the similarity (e.g., the semantic mismatch issues in the text labels) or graph composition/aggregation.

For simplicity, we remove C (conditions) from PPDG here on. We propose a template-based approach in which three types of query template constructs are defined: *atomic, path* and *complex* query templates. We also assume that the user can enter one of the query templates directly through ProcessVidere, using an editor not dissimilar to BPMN-Q query editor [12].

Figure 2 introduces another PPDG on graduation ceremony. Although it looks similar to Graduand A process (Figure 1), there are a few differences. For example, 'V_1: collect dress' takes 'D_1: cash' and produces 'D_3: deposit'. 'V_6:

return dress' takes 'D_3: *deposit'* as well as the dress. Also, '*take photo'* appears after '*attend ceremony'* (rather than before '*register'*).

Atomic Query Templates. These will match an action-flow edge ($\in E_A$) or a data-flow edge ($\in E_D$) and the nodes ($\in A, D$) directly connected by it.

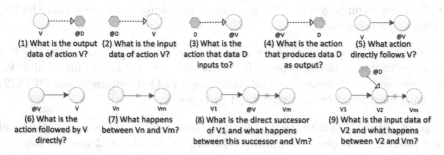

(1) What is the output data of action V?

(2) What is the input data of action V?

(3) What is the action that data D inputs to?

(4) What is the action that produces data D as output?

(5) What action directly follows V?

(6) What is the action followed by V directly?

(7) What happens between Vn and Vm?

(8) What is the direct successor of V1 and what happens between this successor and Vm?

(9) What is the input data of V2 and what happens between V2 and Vm?

Fig. 3. Atomic Query, Path Query and Complex Query Templates over PPDG

Q_1 to Q_6 in Figure 3 represent atomic query templates. The text label shown in each template describes the query contained in it. The prefix symbol "@" in the node label indicates a variable node (i.e., "@D" for a variable data node, "@V" for a variable action node).

Example 1. Let us assume that we issue Q_2 with v as V_1: *collect dress.* The results (processed over PPDG1 and PPDG2) will be $\{PPDG1.ed_1 = (D_1:$ *booking receipt, V_1: collect dress), $PPDG2.cd_1 = (D_1$: cash, V_1: collect dress)*$\}$. So the answer shows that *collect dress* can either take cash or a booking receipt.

Example 2. Similarly, we issue Q_5 with v as V_1: *collect dress.* The results (processed over PPDG1 and PPDG2) will be $\{PPDG1.ea_1 = (V_1:$ *collect dress, V_2: take photos), $PPDG2.ea_1 = (V_1$: collect dress, V_2: register)*$\}$. So the answer shows that *collect dress* can be followed by either *taking photos* or *registering.*

Path Query Templates. Intuitively, a path query template is designed to match a fragment of PPDG contained by two action nodes. Visually, we denote the template as shown in Figure 3 (Q_7). The symbol "$\|$" is used to represent a path query between action nodes specified by V_n and V_m (where n, m are the action node numbers). To explain, we first define a path in PPDG.

Definition 2. *A path of PPDG is a tuple $P_{ath} = (A', D', E'_A, E'_D, \phi, \lambda)$ where:*

- $P_{ath} \subseteq PPDG.$
- $x \in A'$ *iff:*
    - $x = source.$
    - $x = destination.$
    - x *lies on a path from source to destination in ppdg $\in PPDG.$*
- $\forall x, y \in A', e(x, y) \in E_A \rightarrow e(x, y) \in E'_A$
- $d \in D'$ *iff:*

- d *is output data of source action in a path.*
- d *is input data of destination action in a path.*
- d *is input/output data of action x which lies on a path from source to destination in $ppdg \in PPDG$.*
  - $\forall d \in D', e(d, x) \in E_D \rightarrow e(d, x) \in E'_D$ *and* $e(x, d) \in E_D \rightarrow e(x, d) \in E'_D$

A path query template will return a single matching path from each PPDG considered for processing.

Example 3. Let us assume that we set $n = 1$ and $m = 4$ in Q_7, so that Q_7 is a path query to find personal process fragment between 'V_1: *collect dress*' and 'V_4: *attend ceremony*'. Therefore, the returned result of Q_7 (processed over PPDG2) is a path shown as $PPF1$ (dotted box) in Figure 2.

Complex Query Template. A complex query template is a composite of atomic and/or path query templates. Q_8 and Q_9 in Figure 3 are the samples of complex queries that are composed of Q_5 and Q_7 and a constant action, Q_2 and Q_7, respectively. The complex query of PPDG can consist of Q_1 to Q_6 or Q_1 to Q_6 and Q_7 or Q_1 to Q_7 with constant actions/data.

4   Query Processing

Here we formally introduce the definitions of PPDG query and fragment, the schema of PPDG with detailed algorithms to process query templates.

Definition 3 (PPDG Query). *A query graph is a tuple*
$PPDG - Q = (QA, QD, QAF, QADF, QP, T, S)$ *where:*

- *QA is a finite set of action nodes in a query.*
- *QD is a finite set of data nodes in a query.*
- *$QAF \subseteq QA \times QA$ is the action flow relation between action nodes in a query.*
- *$QADF \subseteq QA \times QD$ is the data flow relation between action nodes and data nodes in a query.*
- *QP is the path relation between action nodes which includes data nodes and data edges corresponding to each action node in query.*
- *$T: QA \rightarrow \{CONSTANT\ ACTION, VARIABLE\ ACTION\}$*
- *$S: QD \rightarrow \{CONSTANT\ DATA, VARIABLE\ DATA\}$*

Definition 4 (PPDG Fragment (PPF)). *A connected subgraph (A', D', E'_D), or (A', E'_A), or (A', D', E'_A, E'_D) of a personal process description graph (A, D, E_A, E_D), where $A' \subseteq A, D' \subseteq D, E'_A \subseteq E_A, E'_D \subseteq E_D$, is a fragment of the personal process description graph.*

A query template can return 0, 1 or more PPDG Fragments. Figure 2 shows examples of PPF in $PPF1$ (dotted box) and $PPF2$ (dotted box).

PPDG Data Schema. The schema for our PPDG has the following tables:

- Graph(<u>ID</u>, Name, Description)
- Action(<u>ActionID, GraphID</u>, ActionLabel, Type1, Type2, SubProcess, Description)
- Data(<u>DataID, GraphID</u>, DataLabel, Type1, Type2, SubProcess, Artifacts)
- Edge_Action(<u>sActionID, dActionID, GraphID</u>)
- Edge_Data(<u>DataID, ActionID, GraphID</u>, Direction)

We store the descriptions of graphs, actions, and data in three tables, respectively. The relationships between actions and data are stored in two tables. Specially, the *Direction* item in table *Edge_Data* indicates whether the data is the input or output of an action.

Atomic Queries $(Q_1 \sim Q_6)$. Using the label of an action (data), we can get the *ActionID* (*DataID*) by matching the label in table *Action* (*Data*). Then the *IDs* of related actions and data can be derived from table *Edge_Action* and *Edge_Data*. After that, the label of the related actions and data can be found in table *Action* and *Data*. Each atomic query can be implemented by a single SQL statement. For instance, the results of Q_5 are obtained by the following SQL statement when giving $Q_5(ActionLabel)$.

```
SELECT da.ActionID, da.GraphID, da.ActionLabel
FROM query q, Action sa, Action da, Edge_Action e
WHERE q.ActionLabel=sa.ActionLabel
and e.sActionID=sa.ActionID and e.GraphID=sa.GraphID
and e.dActionID=da.ActionID and e.GraphID=da.GraphID;
```

The other five atomic query templates are straightforwardly implemented with minor modifications of the SQL statement above.

Path Query (Q_7). For the implementation of the path query template, intuitively, we can use the six atomic queries $(Q_1 \sim Q_6)$ above to get the path step by step: (1) First, launch atomic query Q_5 from the first action node sA, and get a set S of action nodes that are direct successors of sA from different processes. (2) Then we perform the same atomic query for each action node in S. (3) The same procedure is executed iteratively until we reach the dA or end of the process. The results are a set $\mathcal{A}$ of action paths. (4) For each action path in $\mathcal{A}$, we launch Q_1 and Q_2 to find the data nodes related to every action nodes. After that, the final results are obtained.

The cost of this initial algorithm is very high due to querying redundant processes which do not contain the end action node dA of the path. Therefore, we filter all the processes to obtain a small set $\mathcal{G}$ of processes which contain both sA and dA nodes before the above steps (1 to 4) are performed.

The Algorithm 1 illustrates the details of path query processing. To enable computing the path in an iterative fashion, we use a tuple T to process the path query. T is employed to maintain the nodes information used for the path computation of a set of action and data nodes in a graph. Particularly, $T.graph$ indicates which graph the path locates in, $T.queue$ is a queue of action nodes

---

Algorithm 1. *Path Query(sA, dA)*

Input : sA, dA: The start and end actions of the path
Output: $\mathcal{R}$: The result set of P_{ath}
1 $\mathcal{G} \leftarrow$ all the processes contain both sA and dA;
2 $\mathcal{T} = \phi$; $\mathcal{T}' = \phi$;
3 **for** *each graph* $\in \mathcal{G}$ **do**
4 | $T.graph{=}graph$; $T.queue.push(sA)$;
5 |___ $\mathcal{T} \leftarrow T$;

6 **while** $\mathcal{T} \neq \emptyset$ **do**
7 | **for** *each* $T \in \mathcal{T}$ **do**
8 | | $r{=}Q_5(T.queue.back())$ on $T.graphID$;
9 | | **if** $r \neq NULL$ **then**
10 | | | $T.queue.push(r)$;
11 | | | **if** $r{=}{=}dA$ **then**
12 | | | |___ $\mathcal{R} \leftarrow T$;
13 | | | **else**
14 | | | |___ $\mathcal{T}' \leftarrow T$;

15 |___ $\mathcal{T}{=}\mathcal{T}'$; $\mathcal{T}' = \emptyset$;
16 **for** *each* $T \in \mathcal{R}$ **do**
17 |___ Find related data nodes of all action nodes in T by Q_1 and $Q_2 \rightarrow T.data$;
18 **return** $\mathcal{R}$;

---

on the path, and $T.data$ is a set of data nodes on the path. First, we filter the graphs by the start and end actions of the path in Line 1. Then, Line 3-5 initialize T for each related graph and store them in a set $\mathcal{T}$. From Line 6 to Line 15, we find the actions on the path using the iterative method. Particularly, Line 8 launches Q_5 to find the next action node r of current action node in the tail of $T.queue$. If there is no result from Q_5, the T is pruned. Otherwise, we push r into $T.queue$ (Line 10). If r is dA, we get one exact result and put it into result set $\mathcal{R}$ (Line 12). If r is not dA, we continue to search for the next action (Line 14, 15). Finally, Line 16-17 launch Q_1 and Q_2 for each actions in $T \in \mathcal{R}$ to fill the data nodes for each path.

Complex Query. We decompose the complex query first, and then utilize methods of atomic and path queries mentioned above to find matched fragments/processes. The filtering and verification methods are used to improve the performance of the algorithm. To begin with, we split the complex query into one *node set* and four *pair sets*: Constant node set S_n, Constant pair set P_{con}, Variable pair set P_v, Path pair set P_{path} and Preprocessing pair set P_{pre}. S_n contains all constant nodes appearing in the query. In any of the four *pair sets*, a *pair* is described as $\{n, n'\}$ - n and n' are nodes linked by one edge or path in a graph. n or n' could be either a constant node or a variable node. We store the pairs with two constant action/data nodes linked by an edge in P_{con}. The pairs which can be processed directly by using atomic queries stored in P_v and

---

Algorithm 2. _Complex Query(Q)_

Input : Q: The complex query
Output: $\mathcal{R}$: The result set of the complex query
1 Decompose Q into nodes set S_n and pair sets $P_{con}, P_v, P_{path}, P_{pre}$;
2 $\mathcal{G} \leftarrow$ all the processes containing the nodes in S_n and matching the pairs in P_{con};
3 $\mathcal{C} = \phi; \mathcal{C}' = \phi$;
4 **for** _each graph_ $\in \mathcal{G}$ **do**
5 $\quad$ $T.graph=graph; T.R \leftarrow$ all nodes in S_n;
6 $\quad$ $\mathcal{C} \leftarrow T$;

7 **for** _each pair_ $\{n, n'\} \in P_v$ **do**
8 $\quad$ $\mathcal{U} \leftarrow$ results of query $Q(n, n')$ on $\mathcal{G}$;
9 $\quad$ Join $\mathcal{U}$ and $\mathcal{C}$ on $graph$ attribute; update $\mathcal{G}$;
10 $\quad$ **for** _each_ $(U, T) \in \mathcal{U} \times \mathcal{C}$ **do**
11 $\quad\quad$ $T.R \leftarrow U.r$;
12 $\quad\quad$ **if** _variable_ $v \in P_{pre}$ _is identified by_ $U.r$ **then**
13 $\quad\quad\quad$ $T.Pair \leftarrow \{u,v'\}$ or $\{v',u\}$;
14 $\quad\quad$ $\mathcal{C}' \leftarrow T$;
15 $\quad$ $\mathcal{C}=\mathcal{C}'; \mathcal{C}'=\phi$;

16 **for** _each pair_ $\{n, n'\} \in P_{path}$ **do**
17 $\quad$ $\mathcal{V} \leftarrow$ results of path query $Q(n, n')$ on $\mathcal{G}$;
18 $\quad$ Join $\mathcal{V}$ and $\mathcal{C}$ on $graph$ attribute; update $\mathcal{G}$;
19 $\quad$ **for** _each_ $(V, T) \in \mathcal{V} \times \mathcal{C}$ **do**
20 $\quad\quad$ $T.R \leftarrow V.R$;
21 $\quad\quad$ $\mathcal{C}' \leftarrow T$;
22 $\quad$ $\mathcal{C}=\mathcal{C}'; \mathcal{C}'=\phi$;

23 **for** _each_ $T \in \mathcal{C}$ **do**
24 $\quad$ Process queries on all pairs in $T.Pair$;
25 $\quad$ $\mathcal{R} \leftarrow T$ if all queries have results on T ;
26 **return** $\mathcal{R}$;

---

those that can be processed by path query stored in P_{path}, respectively. And the pairs which need to be preprocessed before using the methods mentioned above are put into P_{pre}. That is, all the pairs in P_{pre} depend on the results from P_v to identify one node of each pair before using the methods of atomic queries or path query directly. We use S_n and P_{con} to find a candidate set $\mathcal{C}$ of graphs, and then prune and verify $\mathcal{C}$ by P_v, P_{path} and P_{pre}.

The details of the complex query processing is illustrated in Algorithm 2. Similar to Algorithm 1, we also use a tuple T to store the intermediate result. Particularly, $T.graph$ indicates which graph the path locates in, $T.R$ stores the known action/data nodes, and $T.Pair$ stores the pairs that exist in $T.graph$. Line 1 splits query Q into one nodes set and four pair sets. Then, we use S_n and P_{con} to filter all the processes to get a set $\mathcal{G}$ of processes in Line 2. Based on $\mathcal{G}$ and S_n, we get a candidate set $\mathcal{C}$ containing intermediate results T in Line 4-6. From Line 7 to 15, we process atomic queries on each pair in P_v, and then use the results to prune and verify the candidate set. Then, we get the result set $\mathcal{U}$(Line 8) for each query. For each result $U \in \mathcal{U}$, $U.r$ is a result node on graph $U.graph$ for atomic query. Next, $\mathcal{U}$ and $\mathcal{C}$ are joined on their $graph$ attribute(Line 9), so that the unjoined tuples in $\mathcal{C}$ can be pruned safely. If one variable node v of any pair in P_{pre} is identified by $U.r$, we replace the variable node v with $U.r$, and store the corresponding pair in $T.Pair$, because the pair only appears in $T.graph$(Line 10-13). From Line 16 to 22, we process path query on each pair in P_{path}, which can be used for further pruning. Finally, we process atomic query or path query on each pair in $T.Pair$ to obtain the complex query result set $\mathcal{R}$(Line 23-25).

Example 4. To process query Q_8 in Figure 3 (set $m=5$), we split the query as $S_n=\{V_1, V_5\}$, $P_{con}=\{\phi\}$, $P_v=\{\{V_1, @V\}\}$, $P_{path}=\{\phi\}$, and $P_{pre}=\{\{@V, V_5\}\}$. Filtering by S_n and P_{con}, suppose we get the process in Figure 2, then we process the query on P_v by using $Q_5(V_1, @V)$ to get the result $\{V_1, V_2\}$. Note that V_2 matches the node in pair $\{@V, V_5\} \in P_{pre}$, so the new pair $\{V_2, V_5\}$ is stored in $T.pair$. P_{path} is empty and no further processing is necessary. Finally, we process the query on the pair $\in T.pair$ and assemble all results to get the fragment of graph.

5 Performance Evaluations

In this section, we present results of a performance study to evaluate the efficiency and scalability of the proposed techniques in the paper.

Datasets. We have evaluated our query processing techniques on both real and synthetic datasets. We invited 14 volunteers to describe the procedures they went through at Graduation Research School of UNSW, and generated 30 different processes with 295 action nodes and 441 data nodes. The synthetic datasets were generated by randomization techniques. By giving 30 action labels and 30 data labels, we randomly choose n action labels with $n + 2$ data labels to assemble p processes. The number p varies from 100 to 1000 (default value $= 300$). The number n of action nodes in each process varies from 10 to 25 (default value $= 10$). By the default setting, the total number of action nodes and data nodes were 3000 and 3600 respectively in our experiment. We had a total of 10 queries in the experiment. Each query was generated randomly. The average response time and I/O cost of the 10 queries on each dataset represent the performance of our query processing mechanism.

All algorithms have been implemented in Java and compiled on JRE 1.6. We used PostgreSQL to store the data. The experiments were run on a PC with Intel Xeon 2.40GHz dual CPU and 6G memory on Debian Linux.

We have measured the I/O performance of the algorithms by measuring the number of database access. Query response times were recorded to evaluate the efficiency of the algorithms, which contained the CPU time and the I/O latency.

Impact of the Number of Processes (p)**.** We varied the value of p and evaluated the performance of our algorithm against the real datasets where p is

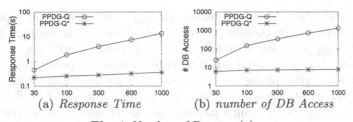

(a) *Response Time*   (b) *number of DB Access*

Fig. 4. Number of Process (p)

30 and the synthetic datasets where p varies from 100 to 1000 in Figure 4. With a larger number of processes, more processes are involved in the computation, thus incurring higher computation cost. From Figure 4(a) and Figure 4(b), we found the result of the response time and number of database access had the same increasing trend as the number of processes increased. That means I/O costs is the main contribution to the query processing cost. Due to the high cost of database access, we can optimize the technique of PPDG query (denoted PPDG-Q here) proposed in Section 4 by storing the actions node of $graph \in \mathcal{G}$ in cache. This optimization is represented as PPDG-Q* in our experiments. As expected, PPDG-Q* significantly outperforms PPDG-Q and is less sensitive to the growth of p.

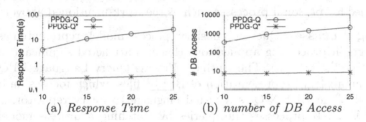

(a) *Response Time*   (b) *number of DB Access*

Fig. 5. Action Nodes of Process (n)

Impact of the Number of Action Nodes in Each Process (n). Figure 5 investigates the impact of the number of action nodes in each process on the algorithms where n grows from 10 to 25. With the growth of n, more action/data nodes are involved in the query computation, thus the response time and the number of database accesses increase. The results show that the scalability of PPDG-Q* is better than that of PPDG-Q regarding the growth of n.

6 Related Work

We discuss related work broadly in two categories: business process modelling languages and process query approaches.

Modelling Business Processes. There are many models developed to represent business processes, Petri Net, Pi-Calculus and recently, Business Process Model and Notation (BPMN) [1,11] to name a few. In all of these models, however, the main focus is in formalizing action flows in a process and data flows are mostly ignored. They also tend to provide highly expressive and sophisticated constructs which make them a complicated tool to use for personal processes. With increasing interests in management of personal workflows, many argue that a process modelling approach that reduces expressive power and simplifying notations for personal processes is necessary [5, 9, 15]. There are some work that takes the simplification view. For example, in [8], a personal workflow management system is proposed using a model called *Metagraph*. It places the data flow at the center of the process. However, doing so makes it difficult to comprehend the temporal execution sequence of the actions. The cooking graph, proposed in CookRecipe [14] system, not only captures individual actions but also the diverse relationship between the actions. However, it is highly customized for summarizing a cooking recipe. [5] proposes a model that simplifies BPMN constructs to only include actions and sequential and parallel action flows. This model does not consider data items in the process.

Our work shares the same view that personal processes do not require complicated expressive power, for example, a parallel execution has no practical meaning when a process is viewed from a single person (i.e., single executor) view point. We also argue that describing data items explicitly is just as important as describing actions to support a useful set of analysis techniques over the personal process model. Therefore our work puts equal emphasis on both. This is a point that is often overlooked in all modelling languages, even the ones that are designed for personal processes such as the work mentioned above.

Querying Processes. Querying business process models represented as graphs is one of graph processing applications that has attracted significant attention (e.g. [4], [2], [3], [13], [6]). The Business Process Query Language BPQL in [4] works on an abstract representation of BPEL[1] files, which focuses on querying actions only. The BPMN-Q is a visual language to query repositories of process models [2]. It processes the queries by matching a process model graph, converted from BPMN, to a query graph [12]. Our PPDG describes personal processes directly as graph (although mapping from/to BPMN is possible) and uses a query paradigm which takes both actions and data into consideration. The approach presented in [3] is based on the notion of partial process models which describe a desired model through a combination of process model fragments and process model queries. Complex query of PPDG has similar structure as partial process models, but the approach in [3] can not be used to perform PPDG queries because PPDG queries need to process both control and data flows. In [10], authors propose the exact subgraph matching approach of assembling graphs to provide answers to a given query graph if no single candidate graph is isomorphic with the query. Another aggregated graph search paper [13] introduces a novel approach for querying and reusing knowledge contained in business process models repositories, which presents the solution for the similar subgraph matching. Due to the structure of PPDG and flexible attribute of personal processes, the above two approaches are not suitable for applying directly to query PPDG graph repositories.

---

[1] http://docs.oasis-open.org/wsbpel/2.0/OS/wsbpel-v2.0-OS.html

7   Conclusion

In this paper, we present Personal Process Description Graphs (PPDG) for describing personally experienced processes. PPDG is the first step towards providing a solution to sharing the process knowledge through a personal process repository, querying and analyzing personal processes and reusing the processes (either as a whole or fragments). Our immediate future work includes improving the query processing algorithms by introducing more punning rules, and utilizing other types of DBMS, *e.g.* graph database, further developing the algorithms to perform similarity and aggregation matches in PPDG repository.

References

1. http://www.bpmn.org/
2. Awad, A.: BPMN-Q: a language to query business processes. In: EMISA, pp. 115–128 (2007)
3. Awad, A., Sakr, S., Kunze, M., Weske, M.: Design by Selection: A Reuse-Based Approach for Business Process Modeling. In: Jeusfeld, M., Delcambre, L., Ling, T.-W. (eds.) ER 2011. LNCS, vol. 6998, pp. 332–345. Springer, Heidelberg (2011)
4. Beeri, C., Eyal, A., Kamenkovich, S., Milo, T.: Querying business processes. In: VLDB, pp. 343–354 (2006)
5. Brambilla, M.: Application and simplification of BPM techniques for personal process management. In: BPM Workshops, pp. 227–233 (2012)
6. Dijkman, R.M., Dumas, M., van Dongen, B.F., Käärik, R., Mendling, J.: Similarity of business process models. Inf. Syst. **36**(2), 498–516 (2011)
7. Hajimirsadeghi, S.A., Paik, H.-Y., Shepherd, J.:Processbook: Towards social network-based personal process management. In: BPM Workshops, pp. 268–279 (2012)
8. Hwang, S.-Y., Chen, Y.-F.: Personal Workflows: Modeling and Management. In: Chen, M.-S., Chrysanthis, P.K., Sloman, M., Zaslavsky, A. (eds.) MDM 2003. LNCS, vol. 2574, pp. 141–152. Springer, Heidelberg (2003)
9. Kim, S.W., Paik, H.-Y., Weber, I.: Automating Form-Based Processes through Annotation. In: Liu, C., Ludwig, H., Toumani, F., Yu, Q. (eds.) Service Oriented Computing. LNCS, vol. 7636, pp. 558–565. Springer, Heidelberg (2012)
10. Le, T.-H., Elghazel, H., Hacid, M.-S.: A Relational-Based Approach for Aggregated Search in Graph Databases. In: Lee, S., Peng, Z., Zhou, X., Moon, Y.-S., Unland, R., Yoo, J. (eds.) DASFAA 2012, Part I. LNCS, vol. 7238, pp. 33–47. Springer, Heidelberg (2012)
11. Lu, R., Sadiq, W.: A Survey of Comparative Business Process Modeling Approaches. In: Abramowicz, W. (ed.) BIS 2007. LNCS, vol. 4439, pp. 82–94. Springer, Heidelberg (2007)
12. Sakr, S., Awad, A.: A framework for querying graph-based business process models. In: WWW, pp. 1297–1300 (2010)
13. Sakr, S., Awad, A., Kunze, M.: Querying process models repositories by aggregated graph search. In: BPM Workshops, pp. 573–585 (2012)
14. Wang, L.: CookRecipe: towards a versatile and fully-fledged recipe analysis and learning system. PhD thesis, City University of Hong Kong (2008)
15. Weber, I., Paik, H., Benatallah, B.: Form-based web service composition for domain experts. TWEB **8**(1), 2 (2013)

Predicting the Spread of a New Tweet in Twitter

Md Musfique Anwar[1]([⊠]), Jianxin Li[2], and Chengfei Liu[1]

[1] Swinburne University of Technology, Melbourne, Australia
{manwar,cliu}@swin.edu.au
[2] RMIT University, Melbourne, Australia
jianxin.li@rmit.edu.au

Abstract. Online social network services serve as vehicles for users to share user-generated contents (e.g. blogs, tweets, videos etc.) with any number of peers. Predicting the spread of such contents is important for obtaining latest information on different topics, viral marketing etc. Existing approaches on spread prediction are mainly focused on content and past behavior of users. However, not enough attention is paid to the structural characteristics of the network. In this paper, we propose topic based approach to predict the spread of a new tweet from a particular user in online social network namely in Twitter based on latent content interests of users and the structural characteristics of the underlying social network. We apply Latent Dirichlet Allocation (LDA) model on users' past tweets of learn the latent topic distribution of the users. Using word-topic distribution from LDA, we next identify top-k topics relevant to the new tweet. Finally, we measure the spread prediction of the new tweet considering its acceptance in the underlying social network by taking into account the possible effect of all the propagation paths between tweet owner and the recipient user. Our experimental results on real dataset show the efficacy of the proposed approach.

Keywords: Online social network · User-generated content · Twitter

1 Introduction

A growing line of recent research has studied the spread prediction of online content in social networks. Predicting how many people a piece of information reaches is very important. For example, successful prediction of the spread of information can improve marketing strategy so that online advertisers can use this information to efficiently target marketing campaigns. Political groups can learn who they should try to propagate their propositions through the "word-of-mouth" message forwarding to as many people as possible. It can also help in identification and ease the adoption of new ideas and trends.

Twitter is a microblogging service that allows users to share information via short messages known as tweets. Information can spread in Twitter in the form of *retweet* which is forwarding a tweet written by another Twitter user. When

© Springer International Publishing Switzerland 2015
M.A. Sharaf et al. (Eds.): ADC 2015, LNCS 9093, pp. 104–116, 2015.
DOI: 10.1007/978-3-319-19548-3_9

a user finds interest to a tweet written by other user, he may share it with his followers (neighbors) by retweeting the tweet. As a result, tweet propagates to users who may not be the direct followers of tweet owner. Thus retweeting is deemed as a key mechanism of information diffusion in Twitter since the original tweet is propagated to a new set of audiences [5]. A number of research have been studied to find out the factors that affect retweetability [3,5,7,11], retweet prediction [4,6,12], and predicting information diffusion by analyzing the properties of tweets and users [8–10].

In this paper, we propose a framework to predict the acceptance and exposure of a new tweet posted by a particular user in Twitter. From a graph topology perspective, the online social network can be viewed as a graph of nodes (users) connected by edges (followers on Twitter, fans in Facebook etc). Most existing studies focused only on possible direct propagation between a user (followee) and his immediate followers along with content features. Our observation is that most of the tweets of a user implicitly exhibit one or more topics that the user is interested in. The intuition is that if the content of a new tweet is related to one or more topics of the past tweets in the social network, then the recipient user interested in those topics will pay attention (accept) on the new tweet and will forward it to his neighbors. In addition, any user may receive a same tweet possibly one or more times from several different paths of different lengths from the tweet owner via different intermediary users. We believe that repeated exposures of a tweet should have significant effect on its acceptance to a recipient user. Our objective is to measure topic based spread prediction of a new tweet from a particular user (twitterer). We assume that there are two aspects that drive the spread of a tweet, latent content interests of users related to the tweet and the network structure i.e. the social relationships among the users in the network. Thereby, along with the content interest, our proposed framework consider the possible effect of all the propagation paths between tweet owner and the recipient user.

In summary, the proposed approach first models the latent topic structure of the network using observed tweets published in the network. Next, we identify top-k topics relevant to the new tweet in order to find recipient user's interest about the new tweet. Finally, to measure the spread prediction of the new tweet, the proposed approach returns a list of expected number of users to whom the new tweet will reach. This work contributes on the following aspects:

- We efficiently list the top-k ranked topics that are most relevant to the content of the new tweet using word-topic distribution of Latent Dirichlet Allocation (LDA) and inverted index data structure technique.
- We propose a novel method to predict the acceptance and exposure of a new tweet in the social network by taking into account the topic distributions of the users and the structural characteristics of the underlying social network.
- We conducted extensive experiments to show the efficacy of our proposed approach using real data sets.

The rest of the paper is organized as follows: Section 2 formally formulates the problem of spread prediction of a new tweet. Section 3 explains topic modeling approach Latent Dirichlet Allocation (LDA). The procedure to find top-k topics relevant to the new tweet is discussed in Section 4. Section 5 explains the technique to efficiently measure the acceptance of the new tweet in the community. Our experimental design is described in Section 6 including a brief discussion. Section 7 reviews related works. Finally, Section 8 concludes the paper.

2 Problem Description and Proposed Framework

First we formally formulate the problem of spread prediction of a new tweet in social network. Then we give an overview of our proposed framework.

2.1 Problem Description

Before defining the problem definition, we introduce several related concepts.

Social Online Relationships: A online relationship refers to user's virtual social relationships in social networks, such as the *following* relationships in Twitter. They can be typically expressed as directed edges between each pair of users in social graph.

Topic: A topic is a distribution over words. For example, *genetics* topic has words like chromosome, DNA, gene, mutation etc. about genetics.

Topic Distribution: Social network users have interests on multiple topics. Formally, each user $u \in U$ is associated with a vector $\theta_u \in \mathbb{R}^T$ of T-dimensional topic distribution ($\sum_z \theta_{uz} = 1$). Each element θ_{uz} is the probability(interest) of the node (user) on topic z.

Whenever a user posts a new tweet m, it will be sent to his followers. We compute a relevance score about m to each recipient user which is defined as,

$$TweetRel(u_i, m) = \sum_{j=1}^{k} TopicRel(t_j, m) \times \theta_{u_i t_j} \qquad (1)$$

where $TopicRel(t_j, m)$, discussed in section 4, is the topical relevance score between the content of m and topic t_j. k in equation 1 represents the number of top-k topics relevant to m. $\theta_{u_i t_j}$ represents user u_i's interest score towards t_j. $TweetRel(u_i, m)$ indicates how relevant the tweet m to user u_i.

In addition, any recipient user may receive a same tweet one or more times either directly from the information source or via the intermediary social contacts which greatly effects one's acceptance decision about the tweet. For each recipient user, we compute his acceptance score about m defined as,

$$AcceptanceScore(u_i, m) = TweetRel(u_i, m) \times \sum_{u_j \in N(u_i)} w_{u_j, u_i} \qquad (2)$$

where $N(u_i)$ represents neighbors of u_i and w_{u_j, u_i} represents weight indicating neighbor u_j's effect (u_j's acceptance score in the proposed framework) on u_i. If the acceptance score of a recipient user is greater than some threshold value, then he will be eligible to perform various actions and behaviors, such as share the tweet with his social contacts. Thus based on the above concepts, we can define the task of predicting the spread m in Twitter.

Problem Definition: We are given a social network in the form of graph $G = (U, E, M)$, where U is the set of users, E is the set of directed edges representing *following* relationship between users and M is the set of all tweets published by all the users. Each tweet $m_i \in M$ implicitly exhibits one or more topics $(t_1, t_2, t_3, \ldots, t_T)$ mentioned in M. The input to this system is then a new tweet m and its owner $u_0 \in U$. The task is then to compute the acceptance scores of m among the users in order to predict its spread in G which is the list of expected number of users to whom m will reach.

2.2   Overview of the Proposed Framework

Our proposed approach first apply statistical topic modeling approach Latent Dirichlet Allocation (LDA) [1] to learn the user interests on different topics. The input to LDA in our framework is all the tweets of all the users and output is the topic distribution of the users and word-topic distribution containing most representative words for each topic. Next, based on the word-topic distribution from LDA, we determine the top-k topics that are relevant and included in m. Finally, we calculate the acceptance score of each recipient user about m based on users topic distribution from LDA and the scores of top-k topics relevant to m. A recipient user having acceptance score above certain threshold value is eligible to forward his acceptance score to his followers. Each follower then compute his own acceptance score using his followee's acceptance score. The final output is the list of expected recipient users.

One of the important steps of the proposed approach is to find the top-k topics that are most relevant to m efficiently and quickly from a number of topics. For doing so, we use inverted index technique along with word-topic distribution from LDA to list the possible related topics for m. Another important task is the calculation of acceptance score of each recipient user by taking into account the different acceptance scores he received from his neighbors through different paths from the tweet owner.

3   Users Topic Distribution

We apply Latent Dirichlet Allocation (LDA) to learn the latent topic distribution of users. The basic idea is that the documents are represented as random mixtures over latent topics, where a topic is characterized by a distribution over words. It assumes a generative process for generating each document as follows:

(i) Randomly choose a distribution over topics
(ii) For each word in the document:
   (a) Randomly choose a topic from the distribution over topics in step (i).

(b) Randomly choose a word from the corresponding distribution over the words associated with the chosen topic.

(iii) The process is repeated for all the words in the document.

This generative process is graphically represented using plate notation in Figure 1. In this figure, shaded and unshaded plates indicate observed and latent variables respectively. An arrow corresponds to conditional dependency between two variables and boxes indicate repeated sampling with the number of repetitions given by the variable in the bottom of the corresponding box.

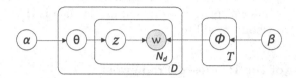

Fig. 1. Graphical Representation of LDA Model

More formally, each of a collection of D documents is associated with a multinominal distribution over T topics, denoted as θ. Each topic is associated with a multinomial distribution over words, denoted as ϕ. θ and ϕ have Dirichlet prior with hyper-parameters α and β respectively. For each word in one document d, a topic z is sampled from the multinomial distribution θ associated with the document, and a word w from the multinomial distribution ϕ associated with topic z is sampled consequently. This generative process is repeated N_d times (N_d is the total number of words in d) to form document d [1,2].

To identify the topics that users are interested in using LDA, we aggregate all the tweets published by a user into a big document d. After running LDA on all the documents in D, we get the following results:

(i) Word-topic distribution set Φ that contains T elements. Each element ϕ_{t_i} in Φ contains most representative words for topic t_i. Each word w_j in ϕ_{t_i} has score denoted as $TermWeight(w_j, t_i)$. Set Φ is used to find top-k topics relevant to the new tweet discussed in Section 4.

(ii) Topic distribution θ_d for each user represented in a $U \times T$ matrix, where U is the number of users and T is the number of topics. Each cell in the matrix represents a specific topic distribution for a particular user. Topic distribution θ is used in Section 5 to measure the spread prediction of m.

4   Top-k Relevant Topics for New Tweet

When a new tweet m is posted, the proposed framework measures the topical relevance between m and the content of each topic ϕ_t in Φ from LDA in order to find the top-k topics that are most relevant to m. Table 1 shows sample word topic distribution in LDA. Each cell in the table contains a word with its relevance score specific to a topic. For example, the word "share" in topic 1 has a score of 0.032740. A same word can appear in more than one topics. For example, the word "time" appeared in both topic 2 and topic 3 having a score of 0.005107 and 0.007483 respectively.

Table 1. Sample Word Topic Distribution in LDA

Topic 1		Topic 2		Topic 3	
share	0.032740	content	0.009946	time	0.007483
reader	0.025515	post	0.007181	work	0.006041
philli	0.010781	googl	0.006835	post	0.005779
todai	0.008514	facebook	0.005971	todai	0.004469
art	0.004406	time	0.005107	open	0.004207

The procedure to find the top-k topics relevant to m is summarized in Algorithm 1. The input is word-topic distribution Φ from LDA and the content (keywords) of new tweet m. We use *inverted index* data structure to associate all the words of all the topics from Φ with their scores to allow fast full text searches. For each keyword w_j in m, the algorithm finds the list of topics that contain w_j along with its score in those topics (line 3 to 5). Then for each relevant topic t_i in the list L, it finds the word in t_i that matches to the keyword w_j of m. The score of w_j denoted as $TermWeight(w_j, t_i)$ is included in ϕ_{t_i} (line 9) and it is added to measure the relevance score $TopicRel(t_i, m)$ of t_i to m (line 7 to 9). The same process continues for all the keywords in m matches to the words in t_i. Finally, the algorithm returns the top-k topics based on their relevance scores.

Algorithm 1. Find Top-k Relevant Topics

Input: Word-topic distribution $\Phi = \{\phi_{t_1}, \phi_{t_2}, \ldots, \phi_{t_T}\}$, keywords Set $V = \{w_1, w_2, \ldots, w_n\}$ in New Tweet m, k
Output: Top-k relevant topics
1: $L \leftarrow Array()$
2: $P \leftarrow PriorityQueue(k)$
3: **for** each $w_j \in m$ **do**
4:     $t_i \leftarrow InvertedList(w_j, T)$
5:     $L.add(t_i)$
6: **for** each $t_i \in L$ **do**
7:     **for** each $w_j \in m$ **do**
8:         **if** $w_j \in t_i$ **then**
9:             $TopicRel(t_i, m) \leftarrow TopicRel(t_i, m) + TermWeight(w_j, t_i)$
10:     $P.add(TopicRel(t_i, m), t_i)$
11: **return** Top-k results from P

For example, if the content of m is (time, art, post, todai), then the relevance score for each topic in table 1 that contain at least one matching words of m is, $TopicRel(Topic\ 1, m) = 0.004406$ (art) $+ 0.008514$ (todai) $= 0.01292$, $TopicRel(Topic\ 2, m) = 0.007181$ (post) $+ 0.005107$ (time) $= 0.01229$ and $TopicRel(Topic\ 3, m) = 0.005779$ (post) $+ 0.007483$ (time) $+ 0.004469$ (todai) $= 0.017731$. With this sample new tweet, we can see that Topic 3 has the highest relevance score followed by Topic 1 and Topic 2, respectively.

5   Acceptance Score Calculation

The final step of the proposed approach is to calculate the acceptance score (according to equation 2) about the new tweet m for each recipient user to estimate the spread prediction of m. If the acceptance score of a recipient user is above certain threshold value ϵ, then he will be *eligible* for passing his acceptance score of m immediately to his followers. The threshold value ϵ is defined as,

$$\epsilon = \lambda \times \left(\sum_{j=1}^{k} avg(\theta_{t_j}) \times TopicRel(t_j, m) \right) \tag{3}$$

where $avg(\theta_{t_j})$ represents the average value of all the users' interests on topic t_j and λ is a control parameter. k in equation 3 represents the number of top-k topics relevant to m. Once a recipient user becomes *eligible*, he can further pass any new acceptance score (computed using acceptance score from other path) later to his followers if it is above another threshold value σ. The value of σ is set as ϵ/η, where η is another control parameter. The purpose of σ is to speed up the process by ignoring insignificant contribution (i.e. too small value of acceptance score) from any path.

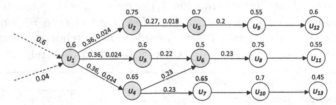

Fig. 2. Sample propagation paths of a tweet with acceptance score

A sample scenario of tweet propagation is depicted in Figure 2. The value on each edge represents the acceptance score of a followee to his follower and any further new acceptance score is separated by comma. The value above a node (user) represents its relevance score about m. Let us consider the value of ϵ and σ as 0.2 and 0.017 respectively. In the figure, user u_1 at first receives acceptance score of 0.6 from a path and then computes its own acceptance score about m which is 0.36 (0.6×0.6). As u_1's acceptance score is above ϵ, so it is *eligible* to forward its acceptance score to its neighbors (u_2, u_3 and u_4). The acceptance scores of u_2, u_3 and u_4 based on their followee u_1's acceptance score of 0.36 are 0.27 (0.36×0.75), 0.22 (0.36×0.6) and 0.23 (0.36×0.65) respectively and all of them become *eligible* to pass their acceptance scores. The acceptance score of u_5 based on its followee u_2's acceptance score of 0.27 is 0.19 (0.27×0.7) which is less than ϵ. The individual acceptance score of u_3 and u_4 is not enough to make u_6 *eligible* but the accumulated acceptance score of 0.45 (0.22+0.23) makes u_6 *eligible* as its acceptance score is 0.23 (0.45×0.5). Upon receiving acceptance score of 0.04 from another path, u_1 then sends its new acceptance score of 0.024 (0.6×0.04) to its neighbors. u_2 then sends its new acceptance score of 0.018 (0.024×0.75) to u_5. Now the accumulated acceptance score (0.27+0.018 = 0.288)

Algorithm 2. Determine the List of Expected Recipient Users

Input: Users Topic Distribution $\theta = \{\theta_{u_0}, \theta_{u_1}, \theta_{u_2}, \ldots, \theta_{u_n}\}$, Tweet Owner u_0, Top-k topics relevant to the new tweet m

Output: List of expected recipient users

```
1:  L ← Array(); initialize an empty queue Q_E ← φ
2:  Set u_0.acceptancescore ← 1; Set u_0.status ← eligible
3:  for each neighbor u_i ∈ u_0 do
4:      L.add(u_i)
5:      Set u_i.status ← process; Set u_i.weight ← u_0.acceptancescore
6:      if u_i.acceptancescore > ε then
7:          Set u_i.status ← eligible
8:          Q_E.add(u_i)
9:  while Q_E is not empty do
10:     NODE-EXPLORE(top of Q_E)
11: return the list of recipient users from L
12: procedure NODE-EXPLORE(u_i)
13:     for each neighbor u_j ∈ u_i do
14:         if u_j.status = eligible then
15:             if u_j.acceptancescore > σ then
16:                 Q_E.add(u_j)
17:         else
18:             if u_j.status = process then
19:                 Update u_j.weight ← u_j.weight + u_i.acceptancescore
20:             else
21:                 Set u_j.status ← process
22:                 L.add(u_j)
23:             if u_j.acceptancescore > ε then
24:                 Set u_j.status ← eligible
25:                 Q_E.add(u_j)
```

---

from u_2 makes u_5 *eligible* as its acceptance score is 0.2 (0.288×0.7). The new acceptance score of both u_3 and u_4 is less than σ and so they are unable to send any further acceptance score to their neighbors. From the figure, we can see that the contribution of a propagation path decreases as the path length increases. The shaded nodes in the figure indicate *eligible* users. The spread prediction of m is the number of recipient users which includes the *eligible* users and their followers. Therefore, in this figure, the list of recipient users are from u_1 to u_9.

Algorithm 2 defines the procedure to measure the spread prediction of a new tweet. We consider several attributes (*status*, *acceptance score* and *weight* indicates neighbor's effect) for each recipient user. The attribute *status* has two possible values: *process* and *eligible*. By *process*, we mean that the user gets one or more acceptance scores from his followee(s) but his own (accumulated) acceptance score is less than ϵ. Once his accumulated acceptance score reaches ϵ, he then becomes *eligible* one. The algorithm begins from tweet owner who passes his acceptance score (initially set as 1 indicated in line 2) to his followers and the eligible ones are kept in a queue Q_E (line 8). Each *eligible* user then passes his acceptance score to his neighbors (through procedure NODE-EXPLORE in line 10). For each *eligible* neighbor, we compute his new acceptance score using his followee's (new) acceptance score. If the status of a neighbor is *process* then his *weight* is updated by adding followee's acceptance score (line 19). If the acceptance score of a neighbor meets the condition to become *eligible*, then we place that neighbor in Q_E (line 23 to 25). For each *eligible* user, the algorithm repeats the above process until Q_E is empty (line 9, 10). Finally, it returns the list of expected recipient users (line 11).

6   Experimental Evaluation

6.1   Data Set

We need complete ground truth dataset of retweeters to measure the accuracy of the proposed framework. To the best of our knowledge, there is no complete annotated ground truth dataset of retweeters in Twitter. We collected two Twitter datasets: a large scale dataset used UDI-TwitterCrawl [13] and a smaller CRAWL dataset [14]. The UDI-TwitterCrawl contains 140 thousand users, 50 million tweets from and 200 million follower links. For each twitter user, this dataset has at most 500 tweets for a period of more than 2 years making it difficult to do the validation of the proposed framework. We used this dataset to show how the proposed framework performs spread prediction of a tweet in large following network. CRAWL dataset contains 9,468 users, 2.5 million follower links and 14.5 million tweets.

We removed stopwords, punctuations, numbers, words with less than 3 characters from users tweets because these words do not help in topic modeling LDA[1]. The remaining words are converted into seed words (stemming words). In both the datasets, we choose users tweets which are posted within June 30, 2011 in order to learn the latent topic distribution of the users using LDA model. Then for each tweet owner, we choose one of his tweets which is posted on or after July 01, 2011 to predict its spread in the network and also measure the accuracy and scalability of the proposed framework. We run all experiments using an Intel(R) Core(TM) i7-4500U 2.4 GHz Windows 7 PC with 16 GB RAM.

6.2   Spread Prediction on Large Scale Dataset

In the experiments, we choose the value of k in Algorithm 1 from 2 to 4. In Table 3, we present the spread prediction result of two different tweets posted by two different tweet owners, Tweet #1 from user 1 (id 14210083) and Tweet #2 from user 2 (id 15958265). Table 2 shows the content of both the tweets. We show the number of both expected recipient users and expected eligible users who are different hops away from tweet owner for both different values of k. All of the cases, the value of λ and η are set as 0.5 and 10 respectively.

The threshold values ϵ and σ vary for different values of k resulting different spread prediction outcome for the same tweet. Table 3 shows that the number

---

[1] LDA is conditioned on Dirichlet hyper-parameters α, β, and topic number T. In this paper, they are set as $\alpha = 50/T$, $\beta = 0.1$ and T from 30 to 60.

Table 2. Example of Tweets

Tweet #	Tweet Content
1	My new social network is an empty pickle jar that you can scream anything you want into
2	POLITICO reporter who covered Sarah Palin quits for Democratic Party job

of expected recipient users increases if we have more eligible users[2]. We also find that for Tweet #2, the number of eligible users increases with increase value of k. Although the content of Tweet #2 is very short compared with Tweet #1, it actually contains more topics like reporter, democratic, political person (Sarah Palin). So the acceptance of Tweet #2 increases with increasing value of k.

Table 3. Spread Prediction Outcomes of Tweets

Tweet #	# of Users in Hops	k	# of Eligible Users	# of Recipient Users	Time(sec.)
1	H1: 30, H2: 4750, H3: 49907, H4: 89961, H5: 2037, H6: 143, H7: 3	2	205 (H1: 5, H2: 82, H3: 90, H4: 28)	23601 (H1: 30, H2: 3183, H3: 4440, H4: 16475, H5: 473)	10
		3	169 (H1: 5, H2: 74, H3: 70, H4: 20)	21150 (H1: 30, H2: 2183, H3: 4259, H4: 14347, H5: 331)	9
		4	143 (H1: 5, H2: 70, H3: 54, H5: 14)	18906 (H1: 30, H2: 2183, H3: 3842, H4: 12638, H5: 213)	9
2	H1: 8, H2: 2907, H3: 38549, H4: 102133, H5: 3049	2	1 (H1: 1)	89 (H1: 8, H2: 81)	6
		3	2 (H1: 1)	245 (H1: 8, H2: 237)	6
		4	6 (H1: 6)	2250 (H1: 8, H2: 2242)	6

6.3   Accuracy and Scalability Result

For each user, CRAWL dataset contains more tweets than UDI-TwitterCrawl dataset. So we performed the accuracy test of our proposed framework on this dataset. Information can be diffused through *retweeters* in Twitter. We randomly choose 30 tweets that were retweeted by at least 20 different retweeters (same as eligible users in the proposed framework). For each tweet, we compare the number of eligible users to the number of retweeters in real scenario. CRAWL dataset does not contain all the tweets for any user. So it is possible that a tweet is retweeted by a user that we do not have information about it. Also the following network is not complete as CRAWL only covers a small sub-network. Both of these will reduce overall accuracy. Again Twitter only shows the most recent tweets of the followees to each follower. So this makes it difficult to perfectly measure the accuracy of the proposed framework since the dataset has no information regarding whether a follower receive or read a tweet posted by his followers. Figure 3 shows the average accuracy result of our proposed framework for different values of λ, T and k.

Figure 3(a) shows the effect of λ on spread prediction of the tweets. Threshold ϵ increases as the value of λ increases. Consequently, the overall accuracy decreases. In Figure 3(b), we show the effect of number of topics T in LDA model. For different values of T, threshold ϵ and σ as well as users' acceptance scores vary resulting different accuracy results. From Figure 3(b), our model achieved best accuracy when the value of T is 30. This is because that CRAWL is small dataset and includes small number of topics. Figure 3(c) shows that the

---

[2] In Table 3, H1: Hop1, H2: Hop2, H3: Hop3, ..., H7: Hop7.

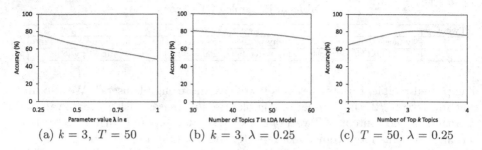

(a) $k = 3$, $T = 50$ (b) $k = 3$, $\lambda = 0.25$ (c) $T = 50$, $\lambda = 0.25$

Fig. 3. Average accuracy result for (a) different values of λ in ϵ, (b) different values of T, (c) different values of k

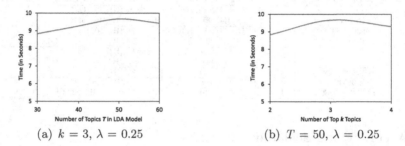

(a) $k = 3$, $\lambda = 0.25$ (b) $T = 50$, $\lambda = 0.25$

Fig. 4. Average run-time for (a) different values of T and (b) different values of k

spread prediction accuracy result varies for different values of k. As each tweet is very short message, it is expected that one tweet may contain only few topics. Our proposed framework achieved best accuracy result when the value of k is 3.

Next, we show the scalability of our approach with respect to different values of T used in LDA model and different values of k. Threshold values ϵ and σ vary with different values of T resulting different number of expected eligible users and hence the computation times vary. Figure 4(a) shows that there is not too much time variation with increase values of T. Figure 4(b) shows the average time taken by our approach to measure the spread prediction for different values of k, where it takes more time when $k = 3$. This coincides with best accuracy result achieved with the value of $k = 3$ as shown in Figure 3(c). In this situation, we have more eligible users (at $k = 3$), thus need more time to calculate the acceptance scores of recipient users.

7   Related Work

There have been many studies on predicting the spread of a tweet considering retweetability in Twitter. Related work can be divided into retweet factor analysis, predicting retweet and analysis of information diffusion in Twitter.

Retweet Factor Analysis. Many researchers studied the factors that affect the retweetability of a tweet. Yang et al. proposed a semi-supervised framework on a factor graph model to predict user's rewteet behaviors considering factors like user, message, time etc [11]. Suh et al. found that content features like URLs and hashtags and contextual features like the number of followees and followers, age of the account affect retweetability [5].

Predicting Retweet. Zaman et al. used a collaborative filtering approach to predict for a pair of users whether a tweet written by one will be retweeted by the other user using feature set like IDs of both users, their number of followers, tweet content [12]. Petrovic et al. used a machine learning approach based on the passive-aggressive algorithm to predict whether a tweet would be retweeted in the future by considering a set of social features and tweet features [6].

Information Diffusion in Twitter. Retweet and spread of URLs, hashtags are treated as information diffusion in Twitter. Yang and Counts constructed networks based on user name mentions to measure three major properties of information diffusion: speed, scale and range [9]. Tsur and Rappoport considered content features as well as temporal and topological features to predict the spread of ideas known as memes within a given time frame [10].

8 Conclusion

In this paper we introduced and evaluated a method to predict the spread of a new tweet from a particular user in Twitter. We took a hybrid approach to predict the spread of a tweet according to its content as well as the topology of the social graph. Different from previous studies is that we consider the possible effect of all the propagation paths between the tweet owner and recipient users. Our experiments demonstrated that multiple propagation paths of a tweet have significant effect on its acceptance in the community.

References

1. Blei, D.M., Ng, A.Y., Jordan, M. I.: Latent dirichlet allocation. Journal of Machine Learning Research **3**, 993–1022 (2003)
2. Griffiths, T.L., Steyvers, M.: Finding scientific topics. Proceedings of the National Academy of Sciences of the United States of America **101**(suppl. 1), 5228–5235 (2004)
3. Boyd, D., Golder, S., Lotan, G.: Tweet, tweet, retweet: Conversational aspects of retweeting on twitter. In: HICSS, pp. 1–10 (2010)
4. Naveed, N., Gottron, T., Kunegis, J., Alhadi, A. C.: Bad news travel fast: A content-based analysis of interestingness on twitter. In: ACM WebSci (2011)
5. Suh, B., Hong, L., Pirolli, P., Chi, E.H.: Want to be retweeted?. Large scale analytics on factors impacting retweet in Twitter network. In: SocialCom (2010)
6. Petrovic, S., Osborne, M., Lavrenko, V.: Rt to win! predicting message propagation in twitter. In: ICWSM, pp. 586–589 (2011)
7. Kwak, H., Lee, C., Park, H., Moon, S.: What is twitter, a social network or a news media? In: WWW, pp. 591–600 (2010)

8. Romero, D.M., Kleinberg, B.M.J.: Differences in the mechanics of information diffusion across topics: Idioms, political hashtags, and complex contagion on twitter. In: WWW, pp. 695–704 (2011)

9. Yang, J., Counts, S.: Predicting the speed, scale, and range of information diffusion in twitter. In: ICWSM, pp. 355–358 (2010)

10. O. Tsur and A. Rappoport. What's in a hashtag? content based prediction of the spread of ideas in microblogging communities. In: WSDM, pp. 643–652 (2012)

11. Yang, Z., Guo, J., Cai, K., Tang, J., Li, J., Zhang, L., Su, Z.: Understanding retweeting behaviors in social networks. In: CIKM, pp. 1633–1636 (2010)

12. Zaman, T.R., Herbrich, R., Gael, J.V., Stern, D.: Predicting information spreading in twitter. In: CSSWC Workshop (2010)

13. Li, R., Wang, S., Deng, H., Wang, R., Chang, K.: Towards social user profiling: unified and discriminative influence model for inferring home locations. In: KDD, pp. 1023–1031 (2012)

14. Bogdanov, P., Busch, M., Moehli, J., Singh, A.K., Szymanski, B.K.: The Social Media Genome: Modeling Individual Topic-Specific Behavior in Social Media. In: ASONAM, pp. 236–242 (2013)

Improvement of Join Algorithms
for Low-Selectivity Joins on MapReduce

Akiyoshi Matono$^{(\boxtimes)}$, Hirotaka Ogawa, and Isao Kojima

National Institute of Advanced Industrial Science and Technology,
1-1-1 Umezono Tsukuba, Ibaraki, Japan
{a.matono,h-ogawa,isao.kojima}@aist.go.jp

Abstract. So far, many studies on join operations on MapReduce have already been proposed. Some of those studies focus on the low-selectivity joins that are frequently used in query processing for datasets among several management domains, such as those used with Linked Open Data. We found there is room for improvement of the state-of-the-art approach for the low-selectivity join on MapReduce called the per-split semi-join[5]. In this paper, we first thus extend the per-split semi-join to improve performance. Our approach can reduce three costs for low-selectivity joins: the amount of network traffic, the number of jobs, and the amount of disk I/O. Moreover, when the number of input relations is large, the selectivity becomes low and thus the effect of our proposed reductions is maximized. Therefore, we also propose an extension of the reduce-side join, which can easily apply three or more inputs, based on the extension of the per-split semi-join. In our experiments, we evaluated two comparisons: the per-split semi-join and our extension of it and the reduce-side join and our extension of it. The first experiment shows that our extension is better than its competitor in any case. In the second experiment, we found that our extension is superior to its competitor when the selectivity is low and the number of inputs is three or more.

1 Introduction

Today, the amount of data is rapidly increasing. Parallel processing frameworks, such as MapReduce[8], increase the importance of such frameworks to analyze such large volumes of data. MapReduce is a parallel programming model on a cluster, where an operation is composed of a set of job pairs of a *map* task and a *reduce* task. In MapReduce, a file is divided into a list of segments, called *splits*, and each task is assigned a split and is run in parallel. In a map task, a split is input, a map method that has been written by a user is performed, and then the output of the map method is partitioned into a set of partitions based on a *partitioner*. In a reduce task, partitions that have been assigned to the reduce task are received from all map tasks, all of the records in all partitions are sorted and merged into a single file, then a reduce method that has been defined by a user reads the file and the output of the reduce method is written into a file.

MapReduce operations work on a distributed file system (DFS). Basically, files that input or output in any operation on MapReduce are put on DFS even

© Springer International Publishing Switzerland 2015
M.A. Sharaf et al. (Eds.): ADC 2015, LNCS 9093, pp. 117–128, 2015.
DOI: 10.1007/978-3-319-19548-3_10

temporary files. So, data transfer between tasks or nodes goes through the DFS. In other words, every data transfer uses both network traffic and disk I/O.

The join operation is one of the most important and the most frequently used operations in query processing in relational databases. There also have been many studies of join operations in a distributed parallel environment. In this paper, we focus especially on the low-selectivity equi-join. An equi-join must determine the intersection between two sets of join keys and then have matching records that have the same key in the intersection. Low-selectivity means that the column is highly selective, in other words, there are a lot of variations in the values in the column. Thus, in general, a low-selectivity equi-join takes very large input relations but the output size is very small.

A low-selectivity equi-join is often used when joining between relations created in different management domains, such as when Linked Open Data [4] is used. In the query processing for Linked Open Data, querying users do not know the structure of the datasets and they also do not know whether a dataset can join to another dataset despite the fact that those datasets are targets of the query processing. Thus, they have to query those datasets to determine the structures or decide whether those datasets can be joined. For example, if you want to make a join between SDBS[1], which is a spectral database for organic compounds, and DBpedia[2], which is a linked dataset that is generated from Wikipedia, then you have to try an equi-join operation using the CAS registry number, which is a unique numerical identifier assigned to every chemical substance. The number of compounds in SDBS is 34,000 and the number of things of DBpedia is 4.58 million[3] but the number of chemical substances assigned a CAS registry number in DBpedia is 1,774[4], and all of them may possibly be contained in SDBS, thus the selectivities are 0.05 and 3.9×10^{-6}, respectively. So the selectivity is expected to be very low.

Several approaches for low-selectivity equi-joins have already been proposed, such as *semi-join*[5] and *Bloom filter join*[14]. All of them are approaches that focus on decreasing the amount of data transfer among nodes by removing unnecessary data before transfer using filters. In MapReduce, the data that is output by a job is written in the distributed file system. Thus a decreasing in the output data of a job leads to decreasing both the amount of network traffic and the amount of disk I/O, and thus directly improves the performance. So in order to reduce the output data, the dangling records that are not contained in the results and will not be used in the following jobs are deleted and are not output in temporary files received by the next job. Thus the more decrease in the number of dangling records sent on to the next job, the more the improvement in the performance. In the processing of a low-selectivity equi-join, there are a lot of dangling records. In other words, there is room for improvement in the low-selectivity equi-join.

---

[1] http://sdbs.db.aist.go.jp/
[2] http://wiki.dbpedia.org/
[3] http://blog.dbpedia.org/2014/09/09/dbpedia-version-2014-released/
[4] http://en.wikipedia.org/wiki/Category:Chemical_pages_needing_a_CAS_Registry_Number

In this paper, we extend the state-of-the-art join algorithms for MapReduce to introduce an index based on position. Per-split semi-join[5] have been proposed for low-selectivity joins, which reads a relation L to extract unique keys, does a semi-join per split of another relation R base on the unique keys, and then finally does an actual join between the results of the semi-join and a split of the relation L. However, we found there are three improvable problems in the per-split semi-join: the number of jobs, the amount of disk I/O, and the amount of network transfer. In particular, because the approach requires three jobs, the I/O cost must be large because the relation L must be read twice, and thus there is room for decreasing of the amount of network transfer because it currently uses keys as they are for the filters' data structure. Therefore, we extend the state-of-the-art in order to address this aspect as well as other factors contributing to the three problems.

Moreover, we considered that if the three or more input relations are joined at one time, then the join selectivity becomes low. However, the per-split semi-join does not support three or more inputs, but the reduce-side joins often used in conventional approaches can be easily extended. Thus we also extend a reduce-side join in order to be able to handle three or more inputs and apply that aspect of the extension of the per-split semi-join.

The remainder of this paper is structured as follows. Section 2 explains several join algorithms as related work. In Section 3, we propose our extensions of the existing algorithms. In Section 4, we evaluate the performance of our extensions through a series of experiments. Finally, we conclude in Section 5.

2   Join Algorithms

So far, many join algorithms using MapReduce have already been proposed [5, 7, 11, 13, 16]. We classify join algorithms using MapReduce into three categories: *reduce-side joins*, *map-side joins*, and *filtering joins*.

2.1   Reduce-Side Join

Reduce-side joins are the most commonly used join strategies using MapReduce; their concept is based on an idea proposed about 30 years ago by [9, 12]. The concept is that records of input relations are partitioned according to the hash values of the join keys, and then a join is performed for each node in parallel.

Standard repartition join[5] is one of the most general methods, and is mentioned in many studies [1, 2, 11, 16, 17], where it may be called a reduce-side join or a different name. Following methods introduced in this section are also have several aliases, but we omit the aliases. Standard repartition join uses both a map task and a reduce task. In the map task, each record of the input relations is read, the name of the input relation is tagged with a value to identify which input the record is from, and then the tagged record is shuffled. In the map task, each record of the input relations is read and then the record is shuffled. In the reduce task, the reduce method receives a key and a list of all records that have the same key, builds buffers to store the records for each relation, and then takes equi-join between the buffers. In this approach, all records with the same key must fit in memory to store them into buffers.

Improved repartition join (IRJ) [5,10,16] was proposed in order to address the problem. In the improved repartition join, values from one relation are guaranteed to occur first in the list of values received by the reduce method, then those from the other relation occur, because this approach sorts values based on composite keys, that is, a comprised of a key and its tag. From this, a buffer for the first relation can be built completely. Then, you can probe the buffer using the values from the other relation.

2.2   Map-Side Join

A Map-side join, which is implemented as a map-only job, is also a well-known join method.

Directed join [3,5,16,17] assumes that input relations are pre-partitioned based on the join keys using the same hash function, and thus all records with the same key exist on a computing node. First, the initialization method in the map task reads a pre-partitioned split of a relation and builds a hash table. Next, the map method scans each record of a pre-partitioned split of another relation, and probes the hash table to match the record.

Fragment replicate join [2,3,10,16] is another map-side join and assumes that the size of one relation must be as small as possible to fit in memory because the relation is broadcast to all nodes and is stored in the memory on the received nodes. Each map task builds a hash table to store the broadcast relation in the initialization method, and then probes the hash table for each record of the split of the larger relation in the map method.

The restriction that the amount of a smaller relation must fit in memory is so difficult to meet that some solutions have been proposed. *Reversed map join*[15] builds a hash table for each split of the larger relation in the map method, and then probes the hash table to find the records of the smaller relation in the close method. *Broadcast join*[5] is a hybrid of both the fragment replicate join and the reversed map join. If the size of the smaller relation is smaller than the split of the larger relation, the fragment replicate join is performed. Otherwise, the reversed map join is done.

2.3   Filtering Join

Filtering joins are used to decrease the amount of records transferred by not sending dangling records in order to minimize communication costs for an equi-join. Almost all filtering joins are composed of a few jobs. The former job(s) generate(s) filtering information to be used to determine whether a record is necessary for the final results or not, and remove(s) unnecessary records using the filtering information. We call the necessary records *joinable* records. The latter job(s) perform(s) the actual join between joinable records in order to obtain the correct results since the filtered joinable records may contain incorrect records.

Filtering joins incur overhead since they contain some additional processes, and thus the filtering joins must be used only the lower the join selectivity for the performance.

Semi-join[5] is a filtering join that requires two given input relations and three jobs. In the first job, the unique join keys of one relation are extracted and a single file containing the unique keys only is generated. In the second job, the unique key file is broadcast, then the dangling records of the other relation are filtered using the unique key data and the files containing only joinable records are generated. In the third job, the final join is executed by broadcasting of the joinable files.

There are similar approaches with the semi-join[14, 16, 18], where a *Bloom filter*[6] is used to decrease the communication costs. Using Bloom filters instead of the unique key file of semi-join, the data after filtering may contain some incorrect records, however the amount of data transfer can be decreased. In addition, the approach[14] lets JobTracker merge Bloom filters into a single filter, that is, the process is operated by the first reduce task in the semi-join. As the result, the approach can reduce the number of jobs.

The semi-join and its similar approaches have a problem due to generating the key data using only one process to construct a single filter. If the input relation is large and there are many splits then the generating process becomes a bottleneck. To cope with this problem, *per-split semi-join (PSSJ)*[5] have been proposed. The first job extracts unique join keys and generates a unique key file for each L's split, but does not merge the unique key files into a single file differently from the semi-join. In the second job, a map task reads an R's split and filters dangling records using all of the unique key files, and thus a reduce method reads R's joinable records and then partitions the records based on L's splits. The third job performs the directed join [5] between each R's partition and L's split as the finally actual join.

3   Proposed Algorithms

3.1   Hash-and Position-Based Per-Split Semi-Join

We found that there is room for improvement in the per-split semi-join. We then extended the per-split semi-join by introducing three reductions: a reduction in the total amount of network traffic, a reduction in the number of jobs, and a reduction of the amount of disk I/O.

In order to reduce the amount of network traffic, we introduce a hash-based approach that sends a set of hash values instead of a set of unique keys. We may be able to say that this approach is a kind of Bloom filter, but we do not use a bit array, because we thought that it is difficult to determine the same number of hash functions and the same length of bit array to be used among the entire set of nodes before reading the input data.

In order to reduce the number of jobs, we execute the actual join in the second reduce task instead of the third map task.

In order to reduce the amount of disk I/O, we introduce a position-based index approach. Per-split semi-join reads the relation L twice, once for extracting unique keys and once for the actual join. For this, the second reading can be skipped using the index that is generated at the first reading.

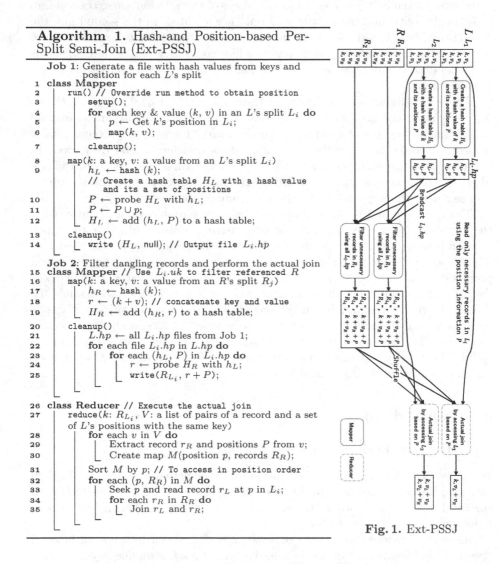

Algorithm 1. Hash-and Position-based Per-Split Semi-Join (Ext-PSSJ)

Job 1: Generate a file with hash values from keys and position for each L's split

```
1  class Mapper
2      run() // Override run method to obtain position
3          setup();
4          for each key & value (k, v) in an L's split L_i do
5              p ← Get k's position in L_i;
6              map(k, v);
7          cleanup();

8      map(k: a key, v: a value from an L's split L_i)
9          h_L ← hash (k);
           // Create a hash table H_L with a hash value
              and its a set of positions
10         P ← probe H_L with h_L;
11         P ← P ∪ p;
12         H_L ← add (h_L, P) to a hash table;
13     cleanup()
14         write (H_L, null); // Output file L_i.hp
```

Job 2: Filter dangling records and perform the actual join

```
15 class Mapper // Use L_i.uk to filter referenced R
16     map(k: a key, v: a value from an R's split R_j)
17         h_R ← hash (k);
18         r ← (k + v); // concatenate key and value
19         II_R ← add (h_R, r) to a hash table;
20     cleanup()
21         L.hp ← all L_i.hp files from Job 1;
22         for each file L_i.hp in L.hp do
23             for each (h_L, P) in L_i.hp do
24                 r ← probe H_R with h_L;
25                 write(R_{L_i}, r + P);

26 class Reducer // Execute the actual join
27     reduce(k: R_{L_i}, V: a list of pairs of a record and a set
          of L's positions with the same key)
28         for each v in V do
29             Extract record r_R and positions P from v;
30             Create map M(position p, records R_R);
31         Sort M by p; // To access in position order
32         for each (p, R_R) in M do
33             Seek p and read record r_L at p in L_i;
34             for each r_R in R_R do
35                 Join r_L and r_R;
```

Fig. 1. Ext-PSSJ

We explain our approach shown at Fig. 1 and Algorithm 1. Our approach is composed of two jobs; the first job generates the filters and indices, and the second job filters dangling records and performs the actual join. At the first map task, we overrode the run method to obtain position information p in a split L_i. The map method calculates the hash value of k and a set of all k's positions P and adds them into hash table H_L, and H_L is output as a file $L_i.hp$ in the cleanup method. In the second job, the map method creates a hash table H_R to store each hash value and its record. The cleanup method filters dangling records using all $L_i.hps$ generated in the first job, and thus outputs R_{L_i} and $r + P$. R_{L_i} is a split name for L, and $r + P$ is the concatenation of a joinable record of R and a set of L_i's positions where the records have the same key with r. First, the data structure must be changed from the list V to a map M with position as the key and its records as values, in order to enable us to sort the map by position order, because from this the second L_i reading can be accessed in the position order for improvement of the performance. Then we seek the position p from which to read the record of L_i. Finally, the actual join is executed.

3.2   Hash-and Position-Based Filtering of a Reduce-Side Join

In filtering joins, the lower the join selectivity is, the more the performance improves. Increasing the number of input relations joined at one join operation can lead to decreasing the join selectivity generally. However, the per-split semi-join and the semi-join do not support three or more input relations. On the other hand, reduce-side joins can easily be extended to handle three or more inputs[5].

Multiple join algorithms in MapReduce have already been studied in [1]. They focus on the multi-way join, that is, relations are joined on different keys. Our focus is on the situation where all the relations are joined on the same key since the purpose of this is to reduce the selectivity.

We extend a reduce-side join to be able to filter dangling records using hash-based filters and position-based indices. Fig. 2 and Algorithm 2 show the algorithm, which is extended based on the improved repartition join[5]. In the first job, the map task overrides the run method in order to obtain position information in the same way as the extension of the per-split semi-join in Algorithm 1. The map method reads a key and a value for each input relation, calculates the hash value h of the key, gets a split name f as a tag, and outputs h as a key and the pair of f and the record's position p as a value. In the reduce task, the reduce method receives a hash value k and a list V of $f + p$ that have the same hash values, extracts a set of all file names F from V, and writes k and V if F contains all input relations. In the second job, the map task also overrides the run method in order to read joinable records and are sorted by the position in the setup method in order to improve the performance of reading disks. The explanation of the following process is omitted because it is similar to that of the improved repartition join[5]. The difference is that our approach is assuming three or more input relations. Thus the actual join performed in the reduce task is different a little. The reduce method inputs the key k and the list of tagged records V, creates buffers $\{B_1, B_2, \ldots\}$ for each relation except the last relation, and finally executes the actual joins using the buffers.

4   Experimental Evaluation

We evaluated the performance of our extended approaches through a series of experiments. In all our experiments, we used m1.medium instances of Amazon

Algorithm 2. Hash-and Position-based Filtering of a Reduce-Side Join (Ext-IRJ)

```
     Job 1: Generate a file with hash values from keys and
            positions for each input
 1   class Mapper
 2       run()// Override run method to obtain position
 3           setup();
 4           for each key & value (k, v) in all relations do
 5               p ← Get current position;
 6               map(k, v);
 7           cleanup();

 8       map(k: a key, v: a value from a split)
 9           h ← hash (k);
10           f ← Get split file name;
11           write(h, f + p);

12   class Reducer // Extract intersection keys
13       reduce(k: a key, V: a list of pairs of a file name and a
            position)
14           F ← Get a set of all file names from V;
15           if F ⊇ {All input relations} then
16               write(k, V)// Output file called U

     Job 2: Filter dangling records and execute actual join
17   class Mapper
18       run()// Random access to joinable records only
19           setup();
20           for position p in U' do // U' is from setup()
21               Seek p;
22               (k, v) ← Extract key & value;
23               map(k, v);
24           cleanup();

25       setup()
26           Read U;
27           U' ← Sort U by position;

28       map(k: a key, v: a value)// joinable records only
29           OMIT: same as IRJ[5]

30   class Partitioner // Partition based on only key
31       OMIT: same as IRJ[5]

32   class Reducer // Execute actual join in a similar way
33       reduce(k: a key, V: a list of tagged records)
34           for v' in V do // V is sorted by tag
35               (t, v) ← Extract a tag and a value from v';
36               Create buffers {B₁, B₂, …} for each
                    relation except the last;
37               if t is the last relation tag then
38                   Execute actual join among v and {B₁,
                        B₂, …};
```

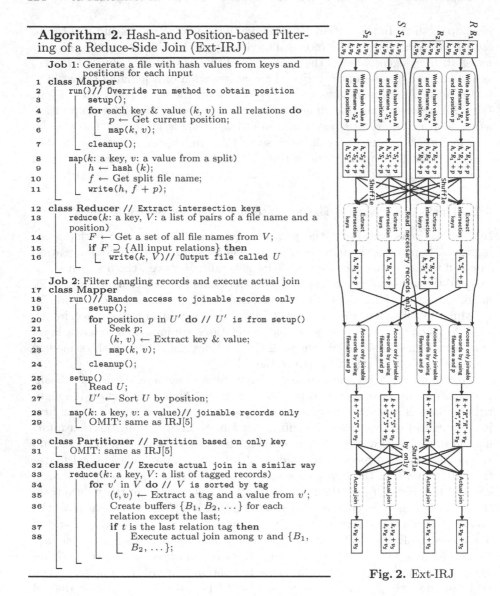

Fig. 2. Ext-IRJ

Web Services (AWS)[5], which has 1 vCPU, 2 EC2 Computing Units (ECU), 3.75 GiB memory, and a 410 GB hard disk storage. Ubuntu 12.04 running Linux 3.2.0 is used as the operating system. The cluster for each experiment is constructed in an availability zone in a region. We used CDH (Cloudera's Distribution including Apache Hadoop) 4.6.0 with Apache Hadoop 2.0.0 running YARN and MR2 and the Oracle Java SE Development Kit 7. We used HDFS as the default configuration, that is, the block size is 128 MB, the number of replications of a block is

---

[5] http://aws.amazon.com/

three and the maximum size of a split is 128 MB. We used 24 or 48 computing nodes running DataNode and NodeManager, 1 node running ResourceManager, and 1 node running NameNode.

We implemented the per-split semi-join and the improved repartition join proposed in [5] as the state-of-the-art baseline join algorithms and their extensions mentioned in Algorithm 1 and Algorithm 2, and compared them respectively. In this section, we call them PSSJ and Ext-PSSJ, and IRJ and Ext-IRJ.

For the experimental dataset, we randomly generated synthetic datasets for all experiments. For the experiment using two relations, we generate three files (e.g., A, B, and S) that are composed of a list of records with a key and a value, whose sets of keys in the three files are disjoint, and merge them into two relations (e.g., $A + S$ and $B + S$) to control the number of intersections, that is, the join selectivity, so that the result of an equi-join between $A + S$ and $B + S$ is S. In our experiment, the number of records in A and B is fixed at 100 million, while the size of S for the intersection varies from 10 to 1 million. In other words, the join selectivity varies from 1×10^{-7} to 0.01. Thus, these selectivities do not depart from the selectivities of the motivational example of DBpedia and SDBS mentioned in Section 1. The size of the key in each record is fixed at 64 bytes, while the size of each record varied: 1024 bytes, 2048 bytes and 4096 bytes. To compare with PSSJ and Ext-PSSJ, we use two relations. In the comparison of IRJ and Ext-IRJ, we use two, three, and four input relations.

Fig. 3 shows the processing times of PSSJ and its extension. The y-axis is the processing times. The x-axis is the size of the intersection, that is, it means the selectivity varies. And the size of record also varies, and the differences are depicted by the shape of the markers of the lines. The circle means that the size of a record is 1 KB, the triangle means that it is 2 KB, and the diamond means that it is 4 KB. Fig. 3a (right) is the result when using 24 computing nodes, and Fig. 3b (left) is that when using 48 nodes. The number of inputs is two, both of whose size are 100 MB plus the size of the intersections. From these figures, we know that our approach improves on PSSJ as the state-of-the-art regardless of the selectivity. And we know that the larger the size of each record is, the bigger the difference in the performance is.

Fig. 4 shows the processing times between IRJ and its extension, Ext-IRJ. The y-axis is the processing times and the x-axis is the size of the intersection. The shape of the markers shows the difference of the size of the record. Fig. 4a show results using 24 computing nodes, and Fig. 4b using 48 nodes. The number of inputs is two, both of whose size is 100 MB plus the size of the intersections. In other words, the conditions are the same as those in Fig. 3. These figures tell us that the difference of the processing times between them is not so large when the selectivity is very low. When the selectivity is high, our approach becomes slow. From only this figure, there is no advantage using our approach over IRJ.

Fig. 5 shows also the processing times between IRJ and Ext-IRJ. The y-axis is the processing times and the x-axis is the size of the intersection. Fig. 4a shows results using 24 computing nodes, and Fig. 4b using 48 nodes. The difference with Fig. 4 is that the size of a record is fixed at 2 KB and the number of input

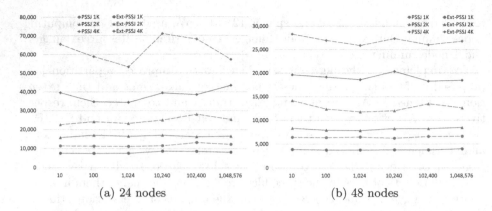

(a) 24 nodes   (b) 48 nodes

Fig. 3. The processing times between PSSJ and Ext-PSSJ

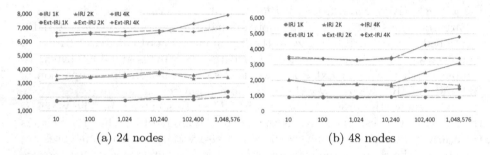

(a) 24 nodes   (b) 48 nodes

Fig. 4. The processing times between IRJ and Ext-IRJ

relations varies in Fig. 5. The shape of the markers shows the difference in the number of input relations. The circle means that the number of inputs is 2, the triangle means that it is 3, and the diamond means that it is 4. The size of every input is 100 MB plus the size of the intersections. From these figures, we know that the more the number of input relation used in an equi-join operation, the better Ext-IRJ is than IRJ.

Thus, as a result, when the number of input relations are two, our proposed approach, Ext-PSSJ, is better than the state-of-the-art approach, PSSJ. And when the number of input relations are 3 or more, if the selectivity is very low, then our approach, Ext-IRJ, must be used instead of IRJ.

5   Conclusions

MapReduce is one of the most powerful programming models for parallel processing. As a matter of course, join operations that need the most expensive processing cost are also performed in MapReduce. Thus several join algorithms on MapReduce have already been proposed.

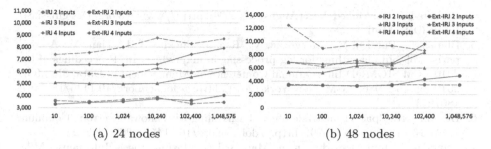

(a) 24 nodes   (b) 48 nodes

Fig. 5. The processing times between IRJ and Ext-IRJ while #inputs varies

In this paper, we focus on the low-selectivity equi-join, which is frequently used in query processing between datasets for which management domains are different, such as those in Linked Open Data. The per-split semi-join proposed in [5] is the state-of-the-art among low-selectivity equi-joins on MapReduce. However, we thought there is still room for improvement in the approach, and we felt we could achieve three reductions: in the amount of the network traffic, in the number of jobs, and in the amount of disk I/O. We thus proposed an extension of the per-split semi-join that achieves those reductions.

In order to decrease the selectivity of an equi-join, if three or more input relations are given and they will be joined by the same key, then you should execute these joins at one time. However, the per-split semi-join does not support joins using three or more input relations. On the other hand, reduce-side joins can support joins performed three or more input relations at one time. Thus we extended a reduce-side join for low-selectivity joins based on our extension of the per-split semi-join.

We evaluated the performance of our proposed approaches through a series of experiments. As a result, we know our extension of the per-split semi-join always surpasses the state-of-the-art approach per-split semi-join. Moreover, our extension of the reduce-side join is better than a reduce-side join when the number of input relations are three or more, and the selectivity is low.

Acknowledgments. An part of this work was supported by JSPS KAKENHI Grant Numbers 24680010, 24240015.

References

1. Afrati, F.N., Ullman, J.D.: Optimizing joins in a map-reduce environment. In: Proceedings of the 13th International Conference on Extending Database Technology, EDBT 2010, pp. 99–110. ACM, New York (2010). http://doi.acm.org/10.1145/1739041.1739056
2. Andreas, C.: Designing a Parallel Query Engine over Map/Reduce. Master's thesis, Informatics MSc, School of Informatics, University of Edinburgh (2010)
3. Atta, F.: Implementation and Analysis of Join Algorithms to handle skew for the Hadoop Map/Reduce Framework. Master's thesis, Informatics MSc, School of Informatics, University of Edinburgh (2010)

4. Berners-Lee, T.: Linked data (August 27, 2006). http://www.w3.org/DesignIssues/LinkedData.html
5. Blanas, S., Patel, J.M., Ercegovac, V., Rao, J., Shekita, E.J., Tian, Y.: A comparison of join algorithms for log processing in mapreduce. In: Proceedings of the 2010 ACM SIGMOD International Conference on Management of Data, SIGMOD 2010, pp. 975–986. ACM, New York (2010). http://doi.acm.org/10.1145/1807167.1807273
6. Bloom, B.H.: Space/time trade-offs in hash coding with allowable errors. Commun. ACM 13(7), 422–426 (1970). http://doi.acm.org/10.1145/362686.362692
7. Chandar, J.: Join Algorithms using Map/Reduce. Master's thesis, Informatics MSc, School of Informatics, University of Edinburgh (2010)
8. Dean, J., Ghemawat, S.: Mapreduce: Simplified data processing on large clusters. In: Proceedings of the 6th Symposium on Operating System Design & Implementation, OSDI 2004, vol. 6, p. 10. USENIX Association, Berkeley (2004). http://dl.acm.org/citation.cfm?id=1251254.1251264
9. DeWitt, D.J., Gerber, R.H.: Multiprocessor hash-based join algorithms. In: Proceedings of the 11th International Conference on Very Large Data Bases, VLDB 1985, vol. 11, pp. 151–164. VLDB Endowment (1985). http://dl.acm.org/citation.cfm?id=1286760.1286774
10. Gates, A.: Programming Pig. O'Reilly, Media (September 2011)
11. Jadhav, V., Aghav, J., Dorwani, S.: Join algorithms using mapreduce: a survey. In: International Conference on Electrical Engineering and Computer Science, ICETACS 2013, Coimbatore, pp. 40–44, April 2013
12. Radu, V.: Application. In: Radu, V. (ed.) Stochastic Modeling of Thermal Fatigue Crack Growth. ACM, vol. 1, pp. 63–70. Springer, Heidelberg (2015)
13. Lee, K.H., Lee, Y.J., Choi, H., Chung, Y.D., Moon, B.: Parallel data processing with mapreduce: A survey. SIGMOD Rec. 40(4), 11–20 (2012). http://doi.acm.org/10.1145/2094114.2094118
14. Lee, T., Kim, K., Kim, H.J.: Join processing using bloom filter in mapreduce. In: Proceedings of the 2012 ACM Research in Applied Computation Symposium, RACS 2012, pp. 100–105. ACM, New York (2012). http://doi.acm.org/10.1145/2401603.2401626
15. Luo, G., Dong, L.: Adaptive join plan generation in hadoop. Tech. rep., Duke University (2010), cPS296.1 Course Project
16. Pigul, A.: Generalized parallel join algorithms and designing cost models. In: Proceedings of the Spring Young Researcher's Colloquium On Database and Information Systems, SYRCoDIS 2012, pp. 29–40 (2012)
17. White, T.: Hadoop: The Definitive Guide. O'Reilly Media / Yahoo Press, 3rd edn. (May 2012)
18. Zhang, C., Wu, L., Li, J.: Efficient processing distributed joins with bloomfilter using mapreduce. International Journal of Grid and Distributed Computing 6(3), 43–58 (2013)

Predicting Users' Purchasing Behaviors Using Their Browsing History

Tieke He[1], Hongzhi Yin[2], Zhenyu Chen [1(✉)], Xiaofang Zhou[2], and Bin Luo[1]

[1] State Key Laboratory for Novel Software Technology, Nanjing University,
Nanjing 210093, China
zychen@software.nju.edu.cn
[2] School of Information Technology and Electrical Engineering,
The University of Queensland, Brisbane St. Lucia, QLD 4072, Australia

Abstract. Some E-commerce giants (e.g., Amazon and Jingdong) with abundant purchasing data achieve highly accurate recommendations, since people have to pay for their choices and their purchasing behaviors are more qualified and valid for capturing users' needs and preferences than other types of users' behavior data (e.g., browsing). However, there is not enough users' purchasing data available for most of small and medium-size E-commerce sites as well as some newly established E-commerce sites. In this paper, we aim to alleviate the sparsity of users' purchasing data by exploiting users' browsing data which is more sufficient. The low validity and reliability of users' browsing data raises great challenge for accurately predicting users' purchasing behaviors since there are many factors leading to users' browsing behaviors. To this end, we propose a novel semi-supervised method to make the most of both high-quality purchasing data and low-quality browsing data to predict users' purchasing behaviors. Specifically, we first use a small amount of purchasing data to supervise the model training of browsing data, and then integrate the results into the item-based collaborative filtering method. We conduct extensive experiments on a real dataset, and the experimental results show the superiority of our method by achieving 25% improvements over traditional collaborative-filtering methods.

Keywords: Recommender systems · Data sparsity · Semi-supervised · Small medium E-commerce

1 Introduction

The rapid development of Internet provides a convenient access to a huge quantity of information. This on one hand helps a lot in our daily life and work, while on the other hand makes us drowned by the information flood. Recommender systems help lead us to the information we are interested in according to some related information, such as demographic information, behaviour data, browsing, purchasing and rating records. E-commerce is the most typical scenario of recommender systems. One of the most famous recommender systems in e-commerce is

M.A. Sharaf et al. (Eds.): ADC 2015, LNCS 9093, pp. 129–141, 2015.
DOI: 10.1007/978-3-319-19548-3_11

Amazon's item-based collaborative filtering recommender system [5]. Item-based collaborative filtering considers the similarities between items, and recommends a list of items which are most similar to items customer used to like. The number of users is usually much larger than that of items on e-commerce sites. It is a challenge to find similar users among a large amount of users. Item-based collaborative filtering method alleviates this problem by exploiting item similarity instead of user similarity. That is why item-based methods are mostly adopted on e-commerce sites.

Almost all the e-commerce sites have more browsing data than purchasing data. When there is no sufficient purchasing data, browsing data is introduced to alleviate the data sparsity problem. The problem is, using this low validity and reliability data may raise the problem of loss of accuracy. Most recommendations are traditionally made merely based on purchasing history and customers' preferences[3]. In the case of lacking purchasing data, it's natural to wonder if we could make use of both data to generate recommendations.

In many other research areas, there also exists the situation where we have a lot of data that is less reliable but another kind of data which is little but much more reliable. One of the these research areas that is most similar to our case is the semi-supervised clustering[11][1]. There are two kinds of data in semi-supervised clustering scenario, one is unlabelled data and the other is labeled data. The amount of unlabelled data is much larger than that of labeled data, and semi-supervised clustering uses limited amounts of labeled data to supervise the clustering procedure of unlabelled data. However, the semi-supervised clustering cannot be applied directly to recommender systems, as the browsing data should not be treated as unlabelled data, and this is one of the challenges of our work which makes our work a non trivial one. Just recall one of your online shopping experiences, you browse items with nearly no cost except time, but when it comes to purchasing you tend to be careful, as you will pay for your choices. Based on this, we can make an assumption that the purchasing data is of high quality, and the browsing data is of low quality. We assume that the high quality data contains some latent labelled information that low quality data doesn't have, and in that way we can adopt the semi-supervised learning method in item-based recommender systems, i.e., we use a small amount of purchasing data to supervise the similarity computation procedure with browsing data.

In this paper, we propose a semi-supervised item-based CF method, which uses purchasing data as supervising information to enhance the validity and reliability of the item similarity computation procedure. The supervising information in our method is in the form of pairwise constraints. A constraint is a pair of items which are bought together by one user, we don't claim that two items in a constraint must be similar, but they shall have some latent connections just as the classic example of *DIAPERS AND BEER*. Browsing data are preprocessed as user-item matrix, which can be used to compute item similarity using classic item-based collaborative filtering [8]. Rather than simply using user-item matrix generated from browsing data to compute item similarity, we use constraints to adjust the user-item matrix through a semi-supervised model [9]. Then the adjusted user-item matrix is used to compute item similarity, and recommendations are generated through item-based collaborative filtering.

Our work is of great importance as guidelines for small and medium e-commerce sites to set up their recommender systems. On one hand, it is hard to gain sufficient high quality data, either for the reason of commercial protection or customer privacy issues. One the other hand, low quality data is easy to get access to, but it results in poor performance and low reliability. One common way of alleviating this problem is to introduce other sources as proof to re-rank the recommendations, but it suffers from information loss. Our method make use of all the available data, both high quality and low quality. The challenges of our work lie in that, first, to access data of different quality in the scenario of e-commerce, second, how to fit these two kinds of data in the semi-supervised model, third, how to split the high quality data to evaluate the experimental results under different supervision ratios.

In summary, our contributions are listed as follows:

1. We are the first to propose the notion of data of different quality in recommender systems.
2. We propose a semi-supervised item-based recommendation method to take advantage of data of both high quality and low quality. This is meaningful for the cases where a lot of emerging small and medium e-commerce sites are lack of purchasing data.
3. Comprehensive experiments are conducted on real world data, and the experimental results indicate that cutting down on low quality data won't affect accuracy too much, while introducing a small amount of high quality data can significantly improve the recommendation accuracy.

The remainder of this paper is organized as follows. Section 2 describes the proposed framework in detail and experimental setting up. Section 3 discusses experimental results and analysis. Some related work on item-based collaborative filtering and semi-supervised learning are discussed in Section 4, followed by conclusion in Section 5.

2 Semi-supervised Item-Based Recommendation

In this section, we describe the framework of our approach and explain each part in detail. Figure 1 illustrates the schematic diagram of our framework. We first preprocess the data to generate browsing matrix B and purchasing matrix P. P will be randomly split into two parts, one is to generate test set T and the other is to generate constraints C. B and C are used as inputs of semi-supervised learning. The semi-supervised learning process adjusts B by constraints to B'. Then B' is used to generate item similarity matrix S based on cosine and pearson similarities. Afterwards, we use matrix B and S to conduct item-based collaborative filtering and generate recommendations. At last, we use test set T to evaluate the effectiveness of our approach.

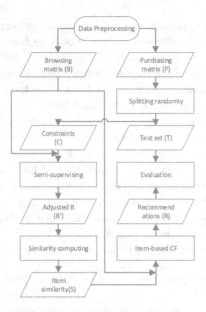

Fig. 1. Framework

2.1 Data Preprocessing

Data preprocessing procedure generates the browsing matrix and purchasing matrix from users' browsing and purchasing records. The left part of Figure 2 shows an example of the browsing matrix B and purchasing matrix P. In matrix B, a number 1 at row i, column j means user U_i browsed item I_j, and 0 means did not. It is the same in matrix P as that in B, 1 means purchased and 0 means not. For the accuracy of similarity computation, we eliminate those users who browsed less than 10 items, and we also drop the corresponding browsing records, because items in test set would appear in browsing matrix in that way. We will discuss in detail in Section 4.4 why we perform such filtering.

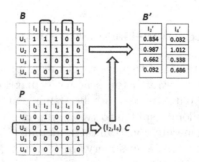

Fig. 2. Example for Semi-supervised Learning

2.2 Constraints

There are two kinds of constraints in semi-supervised clustering, *must-link* constraint and *cannot-link* constraint. Two instances in a *must-link* constraint may not be divided into the same cluster by the original clustering model, but during the semi-supervised procedure, they will be put into the same cluster. A *cannot-link* constraint between two instances means they cannot be in the same cluster after the semi-supervised clustering.

In e-commerce scenario, purchasing data is of more reliability than browsing data or say more reliable, because customers need to pay for their purchasing decisions while not for browsing. Co-purchasing is an indicator that there are some relations among items. In Amazon's item-to-item collaborative filtering recommendation method citeGreg:Amazon, items co-purchased information is a major influential element of item similarity. In our proposed model, we use purchasing data to generate *must-link* constraints. A must-link constraint between two items is defined as they have been frequently purchased together. Figure 2 shows an example of constraint. User U_2 purchased I_2 and I_4, so we get a constraint of (I_2, I_4). There is no *cannot-link* constraint in our case. One way to generate some *cannot-link* constraints is to figure out some really different items artificially. This will be helpful to the semi-supervised learning process but at a great cost, we do not consider *cannot-link* constraints in our method.

2.3 Semi-supervised Learning

There are mainly two functions of the semi-supervised model that we employ [9]. One is to transform the original data into an appropriate new distance space, which complies with the constraints, and in the meanwhile preserves the structure of original data. The other is to reduce the dimensions while not hurting function one too much. As will be described in experiment section, we set a threshold to filter users who browsed less than k items. We mainly make use of the first function of this model. Given a set of browsing records $B=[\vec{b_1}, \vec{b_2}, ...\vec{b_n}]$, in which $\vec{b_i}$ is the ith column of B, and must-link constraints set C, this model generates a weight matrix $W=[\vec{w_1}, \vec{w_2}, ...\vec{w_n}]$ to transform original B into $B'=[\vec{b'_1}, \vec{b'_2}, ...\vec{b'_n}]$, in which the transformed representations $\vec{b'_i} = W^T\vec{b_i}$ can preserve the structure of the browsing records as well as the pairwise constraints C, i.e. instances involved by C should be close in the transformed space. This model uses optimization to get W. The objective function $J(w)$ [9] needs to be maximized while $w^T w = 1$:

$$J(w) = \frac{1}{2n^2} \sum_{i,j} (w^T \vec{b_i} - w^T \vec{b_j})^2$$

$$+ \frac{\alpha}{2n_D} \sum_{(i_i,i_j)\in D} (w^T \vec{b_i} - w^T \vec{b_j})^2$$

$$- \frac{\beta}{2n_C} \sum_{(i_i,i_j)\in C} (w^T \vec{b_i} - w^T \vec{b_j})^2 \qquad (1)$$

The first term expresses the average squared distances of all instances and ensures the model to get an acceptable effectiveness with only a few constraints [9]. The second term is the average squared distances of items in *cannot-link* constraints. The third term is the average squared distances of items in *must-link* constraints. In order to obtain the max value of $J(w)$, the distances in second term need to be expanded such that items in *cannot-link* constraints are adjusted to be more dissimilar, and the distances in third term need to be reduced so that items in *must-link* constraints are modified to be more similar.

In equation 1, α and β are used for leveraging the relative significance of two kinds of constraints. The bigger α is, meaning that we concern more about *cannot-link* constraints, as is β to *must-link* constraints in the opposite way. In our experiments, we do not have *cannot-link* constraints, so we just set α as 0 and β to be 1 to rely on only *must-link* constraints.

In figure 2, we give an example of using semi-supervised learning to adjust B, i_2 and i_4 have been purchased by u_2, so we get a constraint of (i_2, i_4). Then $\vec{b_2}$ and $\vec{b_4}$ are transformed into $\vec{b_2'}$ and $\vec{b_4'}$ in B', $\vec{b_2'}$ and $\vec{b_4'}$ are more similar than $\vec{b_2}$ and $\vec{b_4}$, meanwhile the similarities between other items are mostly remained unchanged.

2.4   Item Similarity

In order to evaluate the effectiveness of our recommender system based on semi-supervised learning, we choose two most widely used similarity computing approaches *cosine* and *pearson* to measure item similarity. Given two item vectors x and y in I', we can compute their similarity as follows:

Cosine similarity:

$$sim(x, y) = cos(\vec{x}, \vec{y}) = \frac{\sum_i x_i y_i}{\sqrt{\sum_m x_m^2}\sqrt{\sum_n y_n^2}} \tag{2}$$

Pearson similarity:

$$sim(x, y) = \frac{\sum_i (x_i - \overline{x})(y_i - \overline{y})}{\sqrt{\sum_m (x_m - \overline{x})^2 \sum_n (y_n - \overline{y})^2}} \tag{3}$$

where x_i is the ith entry in x, $\overline{x}$ is the average value of all entries in x.

2.5   Generate Recommendations

We use item-based collaborative filtering to generate recommendations. The item-based collaborative filtering starts with the items which have been browsed by users, then finds out items most similar to the browsed ones. Because items in our experiments are not too many, we directly compute the similarities between all items and those browsed by users, and then add up the similarities to get the predicted ratings. And then the items with the highest predicted ratings will be recommended.

We predict user i's rating on item j as:

$$R(u_i, v_j) = \sum_k S_{j,k} B_{i,k} \tag{4}$$

where $R(u_i, v_j)$ is the predicted rating of user u_i on item v_j, k will iterate over all items, $S_{j,k}$ is the similarity between item j and item k, $B_{i,k}$ denotes whether user i has browsed item k or not, 1 means yes and 0 means no.

3 Experimental Analysis

3.1 Data Set

In order to evaluate the effectiveness of our approach, we need a dataset that contains both high reliability and low reliability data, which correspond to purchasing and browsing information in our case. We collaborate with GlassesShop[1] to gather users' behaviour information on its web site. GlassesShop is a leading glasses online retailer whose customers mainly come from America and Europe. During the period of collecting data, users have not yet received any recommendation. We eventually gathered 73967 browsing records from 14157 users and 1563 purchases from 806 users between February and March 2012. Among these browsing and purchasing records, there are 701 distinct items in total. There are more browsing records than purchasing records in this dataset, which accounts for our motivation of using purchasing data as supervised information instead of directly using purchasing data to generate recommendations. Table 1 provides some basic statistics about our data set.

Table 1. Statistics of Data Set

average purchases of users	1.9392
# users have both purchasing & browsing data	806
ratio of users purchase more than 1 item	0.0643
# must-links in experiments	168

The browsing data is the original training set and the purchasing data is the original test set. In the semi-supervised learning process, we split the original purchasing data into 10 equal pieces and gradually add one more piece, from 1 to 9, to conduct semi-supervised learning under different ratios.

3.2 Evaluation Method

Recommender systems mainly have two different tasks, one is rating prediction and the other is top-N recommendation. In the prediction scenario, recommender systems try to predict users' ratings for all the items in the data set. *MAE*, *MSE* and *RMSE* are commonly used as error metrics to evaluate predicting accuracy.

---

[1] http://www.glassesshop.com/

However, in the recommendation scenario, we try to recommend a short list of relevant products which best matches users' interest. The most frequently used metrics to measure whether recommendation results are relevant are *Precision*, *Recall* and their related measures.

In our recommender system, we mainly want to navigate users to the products which are most likely to be purchased. It is more of a recommendation task than a prediction one. So we choose top-N *Precision*, *Recall* and F_1 measure to evaluate effectiveness of our recommendation method.

Top-N *Precision* is defined as the ratio of relevant items recommended to the number of all recommended items. In our experiments, relevant items are those items in the test set. These items were eventually purchased by the user after browsing. If our recommender system successfully recommends one relevant item, we call it one *hit*, then the *Precision* can be calculated by following equation:

$$Precision(N) = \frac{hits}{N * |U_T|} \tag{5}$$

where N is the number of items recommended to each user in test set, $hits$ is the number of successfully recommended items, T is the test set, U_T is the set of users in T and U_T is the number of users T contains.

Recall is the ratio of relevant items recommended to the total number of relevant items available. In our experiments, total number of relevant items is the number of items purchased by users. Let $I(u_i)$ be the number of items u_i purchased, U_T means the same as they are in *Precision* calculation equation. Then *Recall* can be defined as follows:

$$Recall = \frac{hits}{\sum_{u_i \in U_T} I(u_i)} \tag{6}$$

We expect both *Precision* and *Recall* to be high. However, they usually conflict with each other, improving one is usually at the expense of the other. Thus, F_1 measure is introduced to combine *Precision* and *Recall*. F_1 measure is calculated as follows:

$$F_1 = \frac{2 \times Precision \times Recall}{Precison + Recall} \tag{7}$$

3.3 Similarity Measure

As discussed in [4], collaborative filtering methods suffer from the large and increasing number of users and items, which makes it difficult to find out neighbour users or items accurately and quickly. In our data set, the number of users is much larger than that of items, and everyday there may be several thousands of more new users coming to this site. But the number of items is relatively small and stable. Therefore, we applied item-based collaborative filtering in our experiment, and we compute item similarity instead of user similarity in our experiments.

With respect to item similarity measurement, we choose two most widely used similarity computation techniques, *Pearson Similarity* and *Cosine Similarity*. We do not adjust the original *Pearson Similarity* and *Cosine Similarity* such as using only co-rated items or adjusting ratings by users' average rating because the item vector contains only 0 and 1.

3.4 Data Filtering

We have to perform data filtering so as to eliminate redundancy of our data set. We eliminate such kind of item browsing records from training set that the customer eventually purchased the item. For example, if user u_1 browsed i_1 and eventually purchased i_1, we delete the record of u_1 browsing i_1 from training set. Otherwise, if the training set contains such browsing record, the recommender system will recommend i_1 to u_1, as i_1 is most similar to itself. Table 2 shows the dramatically different results with and without such records in training set based on *Cosine Similarity*.

Table 2. Results of Data Filtering

	Top-5 Precision
with	0.1846
without	0.0643

Before computing the similarity between items, we filter out the users who browsed less than 10 items. Figure 3 shows the Top-5 *Precision* results when limiting the least items browsed by users to different numbers. *Precision* starts to drop when the limitation of least item browsed exceeds 10. One possible explanation for this phenomenon is that those users who browsed less than 10 items are of low contribution to our approach. They just browse around on the site and do not take a strong interest in the browsed items, so their browsing records are not typical indicator that the co-browsed items have some relations. We eliminate these users' browsing records so as not to affect the accuracy of item similarity. But when limitation increases beyond 10, almost every user's browsing record contributes to item similarity computation, so the more users are eliminated, the worse precision turns out to be. All the experiments conducted hereafter will firstly eliminate users who browsed less than 10 items before computing item similarity.

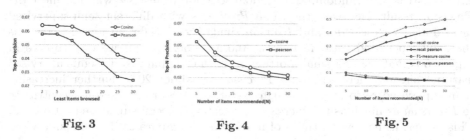

Fig. 3   Fig. 4   Fig. 5

Afterwards, we investigate the recommendation *Precision* when recommending different number of items. We increase the number of recommended items from 5 to 30. Figure 4 shows that top-N *Precision* gradually declines as more items are recommended. From this figure we can see that most of the purchased items in test set have been successfully recommended when providing a recommendation list with only 5 items, although *Recall* increases as more items are recommended, as shown in Figure 5, but the increased *Recall* do not make up for the loss in *precision*, so *F1* measure turns out to be decreasing. Just consider the real case on a web site, presenting a recommendation list of 5 items seems appropriate because recommending too many items may occupy too much space on the web page and bother users with finding out the items they like from the long recommendation list.

3.5 Semi-supervised Results

In order to study the performance of semi-supervised learning with different semi-supervised ratios, we split the original test set which contains the purchasing data, into 10 equal subsets. On each round, we select one set of the 10 pieces of data to do semi-supervised learning and use the rest to evaluate the effectiveness. We gradually add one more piece of purchasing data, from 1 up to 9, to conduct semi-supervised learning, Figure 6 shows the *Precision* changing trend under different semi-supervised ratios. Semi-supervised ratio is the ratio of semi-supervising data to the original purchasing data. In the meanwhile, we also conduct comparison experiments on the same rest part of purchasing data between with and without conducting semi-supervised learning.

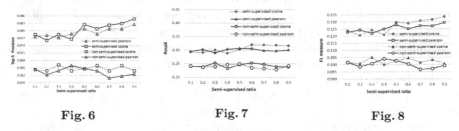

Fig. 6   Fig. 7   Fig. 8

The two lower lines in Figure 6 shows that without semi-supervised learning, top-5 *Precision* fluctuate slightly around 6.4% under both *Cosine* and *Pearson* similarity. This is reasonable because we select supervising data randomly, so the left test set is also random, and random selection of test set will not affect the *Precision* result. However, the top-5 *Precision*s with different semi-supervised ratio under both *Cosine* and *Pearson* similarity do show an upward trend as the semi-supervised ratio increases. Another phenomenon we can observe in Figure 6 is that using a small portion of purchasing data to do semi-supervised learning already increases top-5 *Precision* by about 20%. Further increasing semi-supervised ratio to a very high extent, e.g. 0.8 or 0.9, does not contribute to *Precision* as much as the increased percentage of semi-supervised ratio itself. This means, using several times of more purchasing data to do semi-supervised learning won't increase the *Precision* by several times. A possible explanation for this is that useful item similarity information in must-link constraints concentrates on a relatively small number of items, and applying a relatively small portion of purchasing data can capture most of the must-link information in purchasing data.

Figure 7 provides support for the must-link concentration idea above. *Recall* does not show a clear upward trend, which means when we use purchasing data to do semi-supervised learning, we are mostly likely to strengthen the similarity between a relatively small portion of popular items.

Figure 8 shows that the *F1* measure increases as the semi-supervised ratio increases. The same as *precision*, *F1* measure does not increase sharply even when semi-supervised ratio is very high.

3.6 Comparison with Traditional Methods

We compared our method with the collaborative filtering method directly applied on our dataset.

From Figure 6 we can see that our method improves the top-5 *Precision* by 25% (from 0.064 to 0.08). When employing the *cosine* similarity measurement, as we add up to the ratio of the semi-supervised data, the top-5 *Precision* slightly increases, while the traditional method stays almost the same under different ratios. But when using *pearson* measure the similarity, adding up to the ratio of the purchasing data, the top-5 *Precision* increases, but the traditional method slightly decreases. Figure 7 tells that our method improves the *Recall* by 25% (from 0.24 to 0.3). The same improvement can be seen on *F1* from Figure 8 (from 0.1 to 0.125).

4 Related Work

The main intuitive of collaborative recommendation methods is to exploit information about the past behaviour of an existing user community, and use it to predict which items the current user of the system will most probably like or be interested in. There are two categories of **Collaborative Filtering**, user-based collaborative and item-based collaborative. User-based method first find out several similar users to the active user, and then aggregates items preferred by these similar users to generate a recommendation list [7]. The underlying assumption is that users had similar tastes in the past will have similar tastes in the future, and user preferences are stable over time. Item-based method has been proposed to address the scalability problem of user-based method [8]. The main idea is to compute predictions using item similarity other than the similarity between users. Many adjustments to original collaborative filtering have been proposed. Resnick et al. [7] computed user similarity based on ratings on items which have been co-rated by both users. Chee et al. [2] used the average ratings of a small group of users as default ratings.

Researchers made some attempts of introducing semi-supervised learning into recommendation. Zhou et al. [10] developed a framework using semi-supervised learning to recommend graphs. Ro et al. [6] applied semi-supervised learning in relational classification for personalized tag recommendation. As far as we know, this is the first time of combining browsing data and purchasing data in e-commerce site with semi-supervised learning.

5   Conclusions

In this paper, we proposed a semi-supervised item-based recommendation method to evaluate the effectiveness of combining purchasing data as supervising information with browsing data to improve item-based collaborative filtering. We extract must-link constraints from purchasing data and use semi-supervised learning model to adjust browsing data using these constraints. Experimental results show that using a small portion of purchasing data can improve recommendation accuracy remarkably. Our method outperforms the traditional collaborative filtering method that directly applied on the browsing and purchasing data as much as 25%. For small medium e-commerce sites and newly established comprehensive ones, our method can help them overcome the dilemma of lacking sufficient purchasing data for accurate recommendations. In the future, we will explore proof of cannot-link constraints so as to enhance the performance of our proposed method. Also, more experiments will be done on data from e-commerce sites selling different kinds of goods in future work.

Acknowledgments. This is supported in part by the National Basic Research Program of China (973 Program 2014CB340702), the National Natural Science Foundation of China (Grant No. 11171148).

References

1. Basu, S.: Semi-supervised learning. In Encyclopedia of Database Systems, pp. 2613–2615 (2009)
2. Chee, S.H.S., Han, J., Wang, K.: RecTree: An Efficient Collaborative Filtering Method. In: Kambayashi, Y., Winiwarter, W., Arikawa, M. (eds.) DaWaK 2001. LNCS, vol. 2114, pp. 141–151. Springer, Heidelberg (2001)
3. Chen, L.-S., Hsu, F.-H., Chen, M.-C., Hsu, Y.-C.: Developing recommender systems with the consideration of product profitability for sellers. Information Sciences **178**(4), 1032–1048 (2008)
4. Deshpande, M., Karypis, G.: Item-based top-n recommendation algorithms. ACM TOIS **22**(1), 143–177 (2004)
5. Linden, G., Smith, B., York, J.: Amazon. com recommendations: Item-to-item collaborative filtering. IEEE Internet Computing **7**(1), 76–80 (2003)
6. Marinho, L.B., Preisach, C., Schmidt-Thieme, L.: Relational classification for personalized tag recommendation (2009)

7. Resnick, P., Iacovou, N., Suchak, M., Bergstrom, P., Riedl, J.: Grouplens: an open architecture for collaborative filtering of netnews. In: Proceedings of the 1994 ACM Conference on Computer Supported Cooperative Work, pp. 175–186. ACM (1994)
8. Sarwar, B., Karypis, G., Konstan, J., Riedl, J.: Item-based collaborative filtering recommendation algorithms. In: WWW, pp. 285–295. ACM (2001)
9. Zhang, D., Zhou, Z.-H., Chen, S.: Semi-supervised dimensionality reduction. In: SDM (2007)
10. Zhou, D., Zhu, S., Yu, K., Song, X., Tseng, B.L., Zha, H., Giles, C.L.: Learning multiple graphs for document recommendations. In: WWW, pp. 141–150. ACM (2008)
11. Zhu, X., Ghahramani, Z., Lafferty, J., et al.: Semi-supervised learning using gaussian fields and harmonic functions. ICML 3, 912–919 (2003)

Bus Arrival Time Prediction Using a Modified Amalgamation of Fuzzy Clustering and Neural Network on Spatio-Temporal Data

Sonia Khetarpaul[1]([✉]), S.K. Gupta[1],
Shikhar Malhotra[2], and L. Venkata Subramaniam[3]

[1] Department of Computer Science and Engineering,
Indian Institute of Technology, Delhi, India
{sonia,skg}@cse.iitd.ernet.in
[2] Department of Computer Science and Engineering, Thapar University,
Patiala, India
shikharmalhotra1@gmail.com
[3] IBM Research Laboratory, Delhi, India
lvsubram@in.ibm.com

Abstract. This paper presents a dynamic model that can provide prediction for the estimated arrival time of a bus at a given bus stop using Global Positioning System (GPS) data. The proposed model is a hybrid intelligent system combining Fuzzy Logic and Neural Networks. While Neural Networks are good at recognizing patterns and predicting, they are not good at explaining how they decide their input parameters. Fuzzy Logic systems, on the other hand, can reason with imprecise information, but require linguistic rules to explain their fuzzy outputs. Thus combining both helps in countering each other's limitations and a reliable and effective prediction system can be developed. Experiments are performed on a real-world dataset and show that our method is effective in stated conditions. The accuracy of result is 86.293% obtained

Keywords: Spatio-temporal data · GPS data · Exponential smoothing · Data clustering · Neural networks

1 Introduction

Growing traffic congestion has caused a lot of problems in many countries. Congestion not only leads to a time loss, but also causes grave inconvenience to the commuters and is a major source of air pollution. In developed countries, most of the people still use private vehicles. A lot of effort has been put in to improve the quality of travel experience in public vehicles. One of them is providing travelers with reliable travel information through the help of Advanced Public Transport System (APTS). Availability of Vehicle Location Information (VLI), as a part of Advanced Travel Information (ATIS) helps to increase accuracy of prediction.

© Springer International Publishing Switzerland 2015
M.A. Sharaf et al. (Eds.): ADC 2015, LNCS 9093, pp. 142–154, 2015.
DOI: 10.1007/978-3-319-19548-3_12

APTS applications include pre-trip schedule, real-time information systems, automatic vehicle tracking, timed transfers, bus arrival notification systems, and systems providing priority of passage to buses at signalized inter-sections.

Global Positioning Systems, enable collection of vast data, for example, about arrival of a transit vehicle. However, the random nature of urban traffic poses challenges for processing of this vast data including Prediction. Delays at inter-sections, unfortunate events, weather changes, and stoppage of buses is challenging to track and process.

Studying the bus travel pattern and analyzing it can lead to other benefits as well. Stoppage time on busy stations can be increased, a new flyover can be proposed where congestion is maximum, or the number of buses plying a route could be increased. Using public transport helps in reducing pollution levels and is an easy alternative to personal transportation. Building faith of commuters is important and therefore a suitable intelligent prediction system should be placed to provide accurate Estimated Arrival Time.

There is need for an algorithm that can predict travel time with reasonable accuracy to address these problems.

We develop a hybrid intelligent system, consisting of three stages - (1) Exponential Smoothing, (2) Fuzzy Clustering, (3) Back-Propagation Neural Network with a learning rate adapted according to cluster membership of the data record.

1.1   Literature Review

A variety of prediction models for forecasting traffic states such as travel time and traffic flow have been developed. The five most widely used models include: **Historical Data Based Models** [8,9], These models assumes that historical data patterns remain the same and therefore maybe inaccurate in real situations. These models are reliable only when the spatial-temporal factors in the area of study are relatively unwavering. **Time Series models** [6,16], assume that all external factors on the system either are accounted for or are constant. Variation in the historical data patterns or a dissimilarity in real time data and historical data can lead to results being wrong and erroneous. **Regression models** [9,14], determine a dependent variable with a complex mathematical function formed by a set of independent variables. In general, the applicability of the regression models is limited because variables in transportation systems are highly inter-correlated. **Kalman Filtering model** [1,7,11], have ability to adequately accommodate traffic fluctuations with their time-dependent parameters. **Machine Learning models** [9,13], use Artificial Neural Network (ANN) and Support Vector Regression methods, are able to deal with complex relationships in the data set, can easily process complex data and can predict in the presence of noise as well. Recently, hybrid models e.g. a combination of Kalman Filtering and Neural Network [7,10], a combination of Time Series and Kalman Filtering [16], also showed good results. There are no experimental results to compare these models in a meaningful way.

Our proposed model is a hybrid intelligent system combining Fuzzy Logic and Neural Networks. This results in a reliable and effective prediction system.

2   Bus Arrival Time Prediction Model

The architecture of proposed model is given in figure 1 consisting of three modules: (1) Exponential Smoothing is applied to the data set and a base prediction is done. Also we use the output of this phase as inputs to clustering algorithm along with our previously selected inputs (2) All records are then normalized and fed into Fuzzy Clustering and membership of each record to each of the clusters is found (3) Finally all records are fed into a modified Back Propagation Neural Network with its learning rate adapted according to level of membership of each record to each cluster.

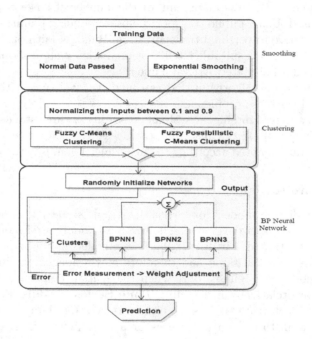

Fig. 1. The Architecture of Proposed Model

2.1   Exponential Smoothing

Smoothing is applied to reduce the variations that occur in data so that an effective system could be developed. A number of varieties of smoothing are available like simple average, weighted moving average and many others. We use Exponential Smoothing. This type of smoothing takes into consideration that the recent observations of bus arrival times are more relevant and important than the past observations. Thus it assigns weights to them in an exponentially decreasing magnitude.

Smoothing begins by setting a smoothed value S_2 to an original observation Y_1. The subscripts refers to the time periods in which observations are spaced. The first smoothed value, alternatively, can be set as an average of the first few observations or any other approximation. For the nth time period, the smoothed value S_n is calculated (described in [17]) by:

$S_n = \alpha Y_{n-1} + (1 - \alpha)S_{n-1}$, where $0 < \alpha \leq 1$ and $n \geq 3$, (α is a smoothing constant)

Substituting for S_{n-1} in the basic equation, we get:

$$S_n = \alpha Y_{n-1} + \alpha(1 - \alpha)Y_{n-2} + (1 - \alpha)^2 S_{n-2}$$

The expanded equation can be written as (by substituting further):

$$S_n = \alpha \sum_{i=1}^{n-2} (1 - \alpha)^{i-1} Y_{n-1} + (1 - \alpha)^{n-2} S_2, \text{where } n \geq 2$$

The weights, $\alpha(1 - \alpha)^n$ decrease geometrically, and adds to unity:

$$\alpha \sum_{i=0}^{n-2} (1 - \alpha)^i = \alpha[1 - (1 - \alpha)^t/(1 - \alpha)] = 1 - (1 - \alpha)^t$$

A smaller value of α, the more important is the selection of the initial estimation. α in our model is calculated by minimizing the error between target and output values by reducing the mean squared error.

2.2 Fuzzy Clustering

Clustering of data is important to determine records which follow similar trend. The records in which buses take nearly similar amount of time to reach at a given stop are clustered together.

Fuzzy C-Means Clustering (FCM), described in [2], poses a lot of advantages over crisp clustering as each data point can have a particular membership with each cluster centre rather than the whole belonging to one entity only. But it has some disadvantages as well. It does not work well in the presence of outlier points.

Possibilistic C Means Clustering (PCM), described in [5], was developed to counter it. Optimization of its objective function sometimes helps to identify outliers (noise points). This algorithm may lead to overlapped cluster formations. Also typicalities may be very sensitive to the additional parameters needed at the time of initialization.

Thus the need for both possibilities (i.e. typicality and membership) values is justified, and a model is proposed to optimize the algorithm. Fuzzy Possibilistic C-Means Clustering (FPCM) normalizes the possibility values, so that the sum of possibilities of all data points in a clusters is 1. FPCM is less prone to the problems of both FCM and PCM. First step in the fuzzy clustering is the data normalization.

Normalization. In data normalization, all the data values of records in the data set need to have their values between 0.1 to 0.9 or -1 to +1 so that the contribution of a specific field may not overshadow the contribution of other. The normalization equation to be followed is:

$$X_i = 0.1 + 0.8 * (K_i - min(K_i))/(max(K_i) - min(K_i))$$

where X_i is normalized value corresponding to $(K_i)^{th}$ record set value which needs to be normalized. This transforms all the values between 0.1 to 0.9.

Fuzzy Possibilistic C-Means Clustering (FPCM). In FPCM clustering, typicalities come into play along with membership values. The centroids of clusters may be shifted in presence of outliers. Typicalities (t_{ij}) are to correct and to alleviate the undesirable effects of erroneous observations and outliers. To classify a data point (x_i), the cluster centroid needs to be closest to the data point which is given by membership matrix (u_{ij}). u_{ij} is a function of x_i and all c centroids, whereas t_{ij} is a function of x_i and c_j alone. FPCM is based on minimization of function as described in [12,15].

$$J(Z; U, C) = \sum_{j=1}^{c} \sum_{i=1}^{N} (u_{ij}^m + t_{ij}^\eta) \|x_i - c_j\|^2$$

$$\text{With constraints, } \sum_{j=1}^{c} u_{ij} = 1, \text{for all } i$$

$$\sum_{i=1}^{N} t_{ij} = 1, \text{for all } j$$

and η is the typicality weight, lies in the range [3,5].

Typicality of a data point to a cluster, will be normalized with respect to the distance of all N data points from that cluster. Additionally, membership values will sum to 1 for each data point from each cluster. Before using FPCM algorithm, the parameters must be specified: the fuzziness exponent, m, and the termination tolerance, ϵ.

Fuzziness Exponent- m, significantly influences the fuzziness of the resulting partition. As m approaches to one, the partition becomes hard i.e. $(\mu_{ij} \in \{0, 1\})$ and c_j are cluster centres. As $m \to \infty$, the partition becomes completely fuzzy $(\mu_{ij} = 1/c)$. Usually, m = 2 is initially chosen.

Termination Criterion- The FPCM algorithm stops iterating when the norm of the difference between U in two successive iterations is smaller than the termination parameter ϵ. For the maximum norm $max^{ik}(|\mu_{ij}^{k+1} - \mu_{ij}^k|)$, the usual choice is $\epsilon = 0.001$, even though $\epsilon = 0.01$ works well in most cases, while drastically reducing the computing times.

Algorithm 1 describes Fuzzy Possibilistic C-Means Clustering, Where X represents Normalized Matrix (Delay, Time since start of journey, Average speed, Time slot of bus journey and Arrival Time of all bus stations).

Membership Level. In this stage, we use the sigmoid function to improve the precision of results and to accelerate the training process of neural networks.

Algorithm 1. Fuzzy Possibilistic C-Means Clustering Algorithm

---

Data: Normalized Matrix X number of cluster centres, and standard values
(Weighting exponent(m), Termination Threshold(ϵ), typicality weight(η)
where η lies in[3,5], and maximum number of epochs)

Result: Clustered centres(matrix V), Cluster Membership level of each record
(matrix U), typicalities of each cluster centre(T) and Error after each
iteration(E)

begin

    1. Randomly select cluster centres

    2. Initialize $U = [\mu_{ij}]$ matrix, $U^{(0)}$
Calculate the μ_{ij} using:
$$u_{ij} = \frac{1}{\sum_{k=1}^{c}[\frac{\|x_i - c_j\|}{\|x_i - c_k\|}]^{\frac{2}{m-1}}} \text{ for all } i, j$$

    3. Calculate the typicality T=$[t_{ij}]$ matrix, $T^{(0)}$
Calculate t_j using:
$$t_{ij} = \frac{1}{\sum_{k=1}^{N}[\frac{\|x_i - c_j\|}{\|x_k - c_j\|}]^{\frac{2}{(\eta-1)}}} \text{ for all } i, j$$

    4. At kth-step: calculate the centre vectors $C^{(k)} = [c_j]$ with $U^{(k)}$ and $T^{(k)}$
$$c_j = \frac{\sum_{i=1}^{N}((u_{ij}^m + t_{ij}^\eta) * x_i)}{(\sum_{i=1}^{N}(u_{ij}^m + t_{ij}^\eta))} \text{ for all } i$$

    5. Update $U_{(k)}$, $U_{(k+1)}$, $T(k)$ and $T(k+1)$
$$u_{ij} = \frac{1}{\sum_{k=1}^{c}[\frac{\|x_i - c_j\|}{\|x_i - c_k\|}]^{\frac{2}{m-1}}} \text{ and } t_{ij} = \frac{1}{(\sum_{k=1}^{N}[\frac{\|x_i - c_j\|}{\|x_k - c_j\|}]^{\frac{2}{(\eta-1)}}} \text{ for all } i, j$$

    6. If $((\|U(k+1) - U(k)\| < \epsilon$ and $\|T(k+1) - T(k)\| < \epsilon)$ or the minimum j
is achieved, then STOP; otherwise return to step 2.

    Return(clusters)

The advanced fuzzy distance (AFD_k) between records data (X_i) and a cluster center (c_j) is presented as follow:

$$AFD_k(X_i) = sigmf(\|c_j - x_i\|, [n, m])$$
$$sigmf(x, [a, c]) = 1/(1 + \exp^{-(a(x-c))})$$

Where k varies from 1 to number of clusters(c). Typically the values are taken as n=50 and m=0.5 for sigmoid function. Now to calculate the membership level we look at the AFD of each record and use the formula:

$$CML_k(X_i) = 1/AFD_k(X_i)$$

where CML denotes the Cluster Membership Level. Thus, we construct a new training sample $(X_i, CML_1(X_i), CML_2(X_i), CML_3(X_i), \ldots, CML_c(X_i))$ and fed into the Neural Network.

2.3 Back Propagation Neural Network

Back Propagation Neural Networks as described in [19], consists of an input layer, hidden layer(s), as well as an output layer. At this stage, we propose an adaptive neural networks evaluating system which consists of number of neural networks as there are cluster centres chosen in the Fuzzy Clustering stage. Each cluster K is associated with the K_i^{th} Back Propagation (BP) network. For each cluster, the training sample is fed into a parallel Back Propagation networks with a learning rate adapted according to the level of clusters membership (CML_k) of each record of training data set.

The Back-Propagation algorithm is based on *Widrow-Hoff delta learning rule* [4] in which the weight adjustment is done through mean square error of the output response to the sample input. The set of these sample patterns are repeatedly presented to the network until the error value is minimized.

The BPNN multilayer network with M layers. N_j represents the number of neurons in j^{th} layer. Here, the network is presented the pth pattern of training sample set with N_o-dimensional input $X_{p1}, X_{p2}, \ldots, X_{pN_o}$ and N_M dimensional known output response $T_{p1}, T_{p2}, \ldots, T_{pN_M}$. The actual response to the input pattern by the network is represented as $O_{p1}, O_{p2}, \ldots, O_{pN_M}$. Let Y_{ji} be the output from the ith neuron in layer j for p^{th} pattern; W_{jik} be the connection weight from k^{th} neuron in layer (j-1) to i_{th} neuron in layer j; and δ_{ji} be the error value associated with the i_{th} neuron in layer j.

Algorithm for Back-Propagation Neural Network is described and explained in [19]. The algorithm is repeated for every training sample pattern p, and until the root mean square (RMS) of output errors is minimized.

The RMS of the errors in the output layer are defined as:

$$E_p = 1/N_M \sqrt{\sum_{j=1}^{N_M} (T_{pj} - O_{pj})^2}$$

for the pth sample pattern.

In the *generalized delta rule* the error value δ_{ji} associated with the ith neuron in layer j is the rate of change in the RMS error E_p respect to the sum-of-product of the neuron:

$$\delta_{ji} = -\partial(E_p)/(\partial net_{ji})$$

Back-Propagation procedure has "Local Minima" problem associated with it. This occurs because the algorithm always changes the weights in such a way as to cause the error to fall. But the error might briefly have to rise as part of a more general fall. If this is the case, the algorithm will "get stuck" (because it can't go uphill) and the error will not decrease further. Solution is to add "momentum" to the weight change. This means that the weight change of this iteration depends not just on the current error, but also on previous changes.

Modified Neural Network with Adaptive Learning Rate. For every record X_i, the output from all the BPNNs corresponding to each cluster are combined according to cluster membership level(CML). Let us say, we have c clusters, then:

$$O_p = \frac{\sum_{k=1}^{c} CML_k(X_i) * O_p(X_i))}{(\sum_{k=1}^{c} CML_k(X_i))}$$

where O_p is the combined outputs from all the BPNNs, i.e., our predicted arrival time. Our goal is to minimize E so that the weight in each link is accordingly adjusted and the final output can match the desired output. For the iteration n and for BPN_k (associated to k^{th} cluster), the adaptive learning rate in BPN_k and the variation of weights Δw_k can be expressed as:

$$n_k = (CML_k(X_i) * n) / \sum_{k=1}^{c} CML_k(X_i))$$
$$\Delta w_k(n + 1) = n_k * \Delta w_k(n) + \alpha * (\delta E / (\delta w_k(n)))$$

If the value of membership level (CML_k) of data record X_j to k_i^{th} cluster is close to zero then the changes in BPN_k network weights are very minimal.

3   Experiments

3.1   Data Collection and Selection

Data were collected using Automatic Vehicle Location (AVL) systems. In these systems, GPS receivers are usually interfaced with a GSM modems and placed in the buses. They record point locations in latitude-longitude pairs, delay, vehicle ID, trip number, date, time, etc. The data was collected by [18] for the month of January, 2013 for different buses on different routes in Dublin, Ireland.

The entire dataset contains a total of 1,015,234 GPS probes for the entire month spaced at an interval of roughly 30 seconds. Out of which bus route 46A (Phoenix Park to Dun Laoghaire in Dublin, Ireland) has been chosen for the case study. This route was selected because it is one of the longest (approximately 19km) routes in the Dublin Bus network. This makes it an excellent candidate for a long term arrival time prediction study. It is also a very busy route, According to the static timetable there is a bus approximately every 6-7 minutes from 06:30 until 22:30, Monday to Friday, with the frequency decreasing on weekends. Route 46A contains a total number of 59 bus stops. Figure 2 is an image of 59 stations on bus route 46A.

3.2   Data Cleansing

In a real life environment, data suffers from outliers, mismeasurements and mis-recording. Typical scenarios include poor satellite visibility, signal interference and bus's labeling GPS probes with an incorrect route id and some rare events like accidents, etc. Data needs to clean for further processing to address these issues. Cleansing steps involve filtering outliers, wrong measurements and mis-recorded data. It is very important to detect these observations and deal with them as soon as possible.

In our chosen routes 46A, there are some cases, where data may be incomplete, for example, some bus journeys didn't have the complete records or for some the GPS device was left tuned on even after trip completion. Some journeys didn't start from the appropriate station depicting a bus journey of different route with some of its route overlapping 46A. Some other constraints which were imposed in order to clean the data are as follows -

1. A vehicle journey must start and end at the terminus bus station only or at a location slightly deviated.
2. We remove trips which contain a large gap in space or in time between any two sequential probes. These gaps happen occasionally when a bus's GPS unit fails to record a large portion of its trips data.
3. The final check we do concerns the special case where a bus waits at its first or last bus stop. Buses do this to realign with their timetables, these points need to be removed because they don't add any information.

3.3   1st Phase- Exponential Smoothing

Pre-processing the Data Set. Out of the 28 days data available with us, 21 days is used for training our model while the rest 7 days are used for testing purposes. On examination of our data set deeply, we found that Winter's Exponential Smoothing (for seasonal and trend patterns) would not work well as we segregated our data station wise. So, to consider the seasonal pattern in our data, we apply the Exponential Smoothing according to vehicle journey's time slot for different days of week. We propose that any events like delays, congestion or traffic patterns should occur for every weekday at that specific time slot only e.g. School rush, office timings, local market opening, etc.

After dividing 3 weeks training data according to weekdays (keeping in mind the similarity of patterns on same weekdays), we apply single exponential smoothing in order to smooth the arrival time values for each station. Our purpose to apply this smoothing is twofold - firstly we can input the smoothed values into our next phase for Fuzzy Clustering (as this method won't work well in the presence of outlier points); secondly we get an initial prediction on our data set (i.e, when bus is not started). For example, After applying Exponential Smoothing for station 808 (2nd station out of 59 stations) for all Wednesdays, prediction was made for the fourth Wednesday i.e 23rd January, 2013, assuming real time data being received. Graph is shown in figure 3.

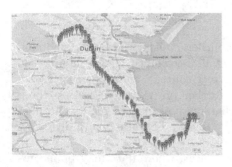

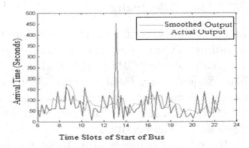

Fig. 2. Route 46A   **Fig. 3.** Predicted Smoothed Output Vs. Actual Output for 23rd January

3.4 2nd Phase - Fuzzy Clustering

In this phase we propose to apply Possibilistic C-Means Clustering(FPCM) to group similar journeys (in fact GPS points) so that the input to the next phase can be made refined. FPCM alleviates the undesirable effects of outliers in determining the cluster centres as well as keeping the data points close to centroids.

We first normalize our dataset to transform the values between 0.1 and 0.9 and then pass on the datasets day wise so that fuzzy clustering could be applied. We performed experiments with different number of clusters. After experimenting, we found that choosing number of clusters as 3 gives the best possible results. A plausible explanation for the same can be that there are three situations which may arise - (1) The bus is exactly on time and no deviation in the schedule has occurred (2) Due to bottleneck situations, the bus is lagging behind in its schedule (3) There is no visible traffic clogging and the bus delays at intersections and traffic lights are minimum, resulting in the bus being ahead of its schedule.

Finally our 63 dimensional data i.e. arrival time at all 59 stations, time slot of day, average speed, time since start of journey and delay, is passed through the clustering model. The outputs of the dataset are recorded and the Advanced Fuzzy Distance(AFD) is calculated, using which finally the Cluster Membership Level(CML) is determined.

3.5 3rd Phase - Back-Propagation Neural Network

The Back Propagation architecture employed in our model consists of three layers - input layer, hidden layer and the output layer. The number of hidden layers were varies from 1 to 3. However, the best results were observed on taking a single hidden layer. The layers were tested by minimizing the prediction error.

Five values (i.e. 30, 40, 50, 70, 100) were taken for the number of neurons in the hidden layer and tested for best possible results. The best number was selected for each model. In other words the number of hidden neurons in the Neural Network for some midway station can be different from the number of hidden neurons for predicting the arrival time at the immediate next station. Similarly for weekday period and weekend period the optimal number of hidden neurons may vary. Although on a collective average, the prediction results from taking different neurons did not vary much.

Out of the 28 days data available, with 21 days data, the model was trained and the outputs were checked for the remaining 7 days of real time data. For training the Back Propagation Neural Network, the Levenberg - Marquadt training algorithm [20] has been employed with the learning rate modified.

For prediction of the arrival time of next 20 stations from a particular station on a specific day, the previous data was captured. Say, we are currently at 22nd station, then the arrival time of 21 stations along with other input parameters(time slot of day, average speed, time since start of journey and delay), and the cluster membership level of each record to the 3 cluster centres were passed to the Neural Network corresponding to each cluster centre. Each of the three networks then predicted the output for the remaining 37 stations with the learning rate adapted according to the membership level of each record. The outputs of the 3 Neural Networks were then combined to give the final output.

We have compared results of our hybrid prediction model with baseline approach and actual measurements. Figures 4 and 5 show some example journey results for 28th January and 31 January, 2013 respectively. These predictions are made by giving as input first 2 and 15 bus stops real time data. Figure 6 shows the comparison of prediction results computed using baseline algorithm (BPNN) and our proposed model for a day. This graph shows that by combining fuzzy clustering with BPNN gives much better results than using only BPNN. Figure 7 shows the comparison of prediction average results computed using baseline algorithm (BPNN) and proposed model with actual measurement for randomly selected bus stops.

Mean absolute percentage error E_a calculated using equation:

$$E_a = 1/N \sum_{j=1}^{N} (T_j - O_j)/T_j * 100\%$$

where, T_j= Target Value (i.e. Measured Arrival Time at given Transit Stop)
O_j=Predicted Value (i.e. Predicted Arrival Time at given Transit Stop)
N = Number of predictions
Prediction Accuracy, A_p, is: $A_p = 100$ - E_a.

The complete model was applied for 2930 observations. On comparing them with the measured actual outputs the Mean absolute percentage error is 13.70%, hence accuracy of our prediction model came out to be 86.293% with the average root mean square error of 0.0334. On the other hand, accuracy of baseline algorithm (BPNN) came out to be 74.229 % .

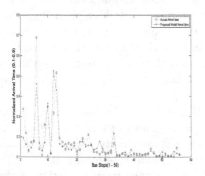
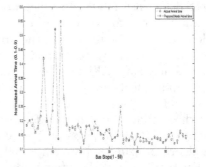

Fig. 4. Predicted Vs Actual Arrival Time for 28 January 2013

Fig. 5. Predicted Vs Actual Arrival Time for 31 January 2013

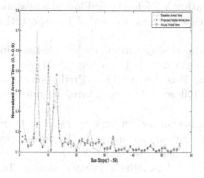

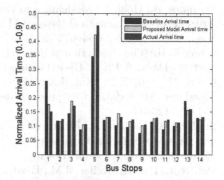

Fig. 6. Baseline, Actual and Proposed model results

Fig. 7. Baseline, Proposed model and actual arrival time values for randomly selected bus stops

4 Conclusions

This paper presents a modified Fuzzy Neural system designed to evaluate probable traffic conditions based on identified factors (input variables). The model is checked for real data(bus route number 46A in Dublin, Ireland). Historical data is obtained and is used to train the prediction system. Recent data are given as input to the prediction system. With fuzzy based system, membership functions are calculated and Back-Propagation Neural Network is then employed to predict bus arrival times. Amalgamation of the Fuzzy and Neural Network techniques counters each other's shortcomings and presents a reliable real time model for prediction. The results are entrusting and promising based on the flexibility and ease of adaptability.

References

1. Kalman, R.E.: A New Approach to Linear Filtering and Prediction Problems. Transactions of the ASME Journal of Basic Engineering **82**(D), 35–45 (1960)
2. Dunn, J.C.: A Fuzzy Relative of the ISODATA Process and Its Use in Detecting Compact Well-Separated Clusters. Journal of Cybernetics **3**, 32–57 (1973)
3. Bezdek, J.C.: Pattern Recognition with Fuzzy Objective Function Algoritms. Plenum Press, New York (1981)
4. Beale, R., Jackson, T.: Neural Computing: An Introduction. Hilger, Philadelphia (1991)
5. Krishnapuram, R., Keller, J.M.: The possibilistic C-means algorithm: insights and recommendations. IEEE Transactions on Fuzzy Systems **4**(3), 385–393 (1996)
6. Al-Deek, H., D'Angelo, M., Wang, M.: Travel Time Prediction with Non-Linear Time Series. In: Proceedings of the ASCE 1998 5th International Conference on Applications of Advanced Technologies in Transportation, pp. 317–324 (1998)
7. Chien, S.I.J., Ding, Y., Wei, C.: Dynamic Bus Arrival Time Prediction with Artificial Neural Networks. Journal of Transportation Engineering 128(5) (2002)

8. Williams, B., Hoel, L.: Modeling and Forescating Vehicle Traffic Flow as a Seasonal Arima Process: Theoretical Basis and Empirical Results. Journal of Transportation Engineering **129**(6), 664–672 (2003)
9. Jeong, R.H.: The Prediction of Bus Arrival time Using Automatic Vehicle Location Systems Data. A Ph.D. Dissertation at Texas A and M University (2004)
10. Chen, M., Liu, X.B., Xia, J.X.: A Dynamic Bus Arrival Time Prediction model based on APC data. Computer Aided Civil and Infrastructure, Engineering, 364–376 (2004)
11. Shalaby, A., Farhan, A.: Bus travel time prediction for dynamic operations control and passenger information systems. In: 82nd Annual Meeting of the Transportation Research Board. National Research Council, Washington D.C (2004)
12. Pal, N.R., Pal, K., Keller, J.M., Bezdek, J.C.: A Possibilistic Fuzzy c-Means Clustering Algorithm. IEEE Transactions of Fuzzy Systems **13**(4), August 2005
13. Bin, Y., Zhong-Zhen, Y., Baozhen, Y.: Bus Arrival Time Prediction using Support Vector Machines. Journal of Intelligent Transportation Systems **10**(4), 151–158 (2006)
14. Ramakrishna, Y., Ramakrishna, P., Sivanandan, R.: Bus Travel Time Prediction Using GPS Data. In: proceedings of Map India, New Delhi (2006)
15. Saad, M.F., Alimi, A.M.: Modified Fuzzy Possibilistic C-means. In: IMECS 2009, Hong Kong, March 18–20, 2009
16. Thomas, T., Weijermars, W.A.M., Van Berkum, E.C.: Predictions of Urban Volumes in Single Time Series. IEEE Transactiuons on Intelligent Transportation Systems **11**(1), 71–80 (2010)
17. Exponential Smoothing, October 17, 2014. http://en.wikipedia.org/wiki/Exponential_smoothing
18. Dublin Dataset, October 17, 2014. http://dublinked.com/datastore/datasets/dataset-304.php
19. BPNN, October 17, 2014. "http://wwwold.ece.utep.edu/research/webfuzzy/docs/kk-thesis/kk-thesis-html/node22.html
20. Levenberg-Marquardt algorithm, October 17, 2014. http://en.wikipedia.org/wiki/Levenberg-Marquardt_algorithm

A Domain Independent Approach for Extracting Terms from Research Papers

Birong Jiang[1], Endong Xun[1], and Jianzhong Qi[2]([✉])

[1] Beijing Language and Culture University, Beijing, China
{jiangbirong,edxun}@blcu.edu.cn
[2] University of Melbourne, Melbourne, Australia
jianzhong.qi@unimelb.edu.au

Abstract. We study the problem of extracting terms from research papers, which is an important step towards building knowledge graphs in research domain. Existing terminology extraction approaches are mostly domain dependent. They use domain specific linguistic rules, supervised machine learning techniques or a combination of the two to extract the terms. Using domain knowledge requires much human effort, e.g., manually composing a set of linguistic rules or labeling a large corpus, and hence limits the applicability of the existing approaches. To overcome this limitation, we propose a new terminology extraction approach that makes use of no knowledge from any specific domain. In particular, we use the title words and the keywords in research papers as the seeding terms and word2vec to identify similar terms from an open-domain corpus as the candidate terms, which are then filtered by checking their occurrence in the research papers. We repeat this process using the newly found terms until no new candidate term can be found. We conduct extensive experiments on the proposed approach. The results show that our approach can extract the terms effectively, while being domain independent.

Keywords: Terminology extraction · Word2vec · Statistical approach

1 Introduction

We study the problem of *terminology extraction* from a corpus of research papers, i.e., we extract the terms from the research papers. Here, a *"term"* refers to a word or compound words representing a concept of a specific *domain*, e.g., in chemistry, "alcohol" is a term that refers to an organic compound in which the hydroxyl functional group is bound to a saturated carbon atom[1]. Terminology extraction has various applications in text mining such as semantic analysis and machine translation. For instance, terminology extraction is a first step of building a knowledge graph from a large corpus, which has gained much

---

Birong Jiang—This work is done when Birong is a visiting student at the University of Melbourne.

[1] http://goldbook.iupac.org/A00204.html

© Springer International Publishing Switzerland 2015
M.A. Sharaf et al. (Eds.): ADC 2015, LNCS 9093, pp. 155–166, 2015.
DOI: 10.1007/978-3-319-19548-3_13

popularity in recent years in both the research community (e.g., YAGO2 [8]) and the industry (e.g., Google Knowledge Graph[2]).

Existing terminology extraction approaches [5,7,9,16,18] are using rule based techniques, supervised machine learning techniques, or a combination of these two types of techniques, all of which rely on some domain knowledge, e.g., manually composed linguistic rules or labeled corpuses. Acquiring such domain knowledge requires much human effort (e.g., manually labeling a large corpus) and limits the applicability of the existing approaches.

To overcome this limitation, we propose a novel approach that does not require any domain knowledge. This approach extracts terms from research papers (which can be of any domain) based on the following observations:

1. The title and the keywords of a research paper usually contains the key terms in the paper.
2. Related terms of the same domain tend to appear in similar context.
3. While a large manually labeled corpus of a particular domain is difficult to obtain, a massive open-domain corpus is relatively simple to acquire (e.g., Wikipedia provides free downloads of its full content[3]).

Following these observations, we extract terms as follows. We start with forming a set of seeding terms consisting of the title words and the keywords (after removing the stop words) of the research papers. We then identify words with similar context to that of the seeding terms from a massive open-domain corpus as the candidate terms. These candidate terms will be filtered by checking their occurrence in the research papers, the remaining of which will be added to the set of seeding terms. We repeat this process until no more candidate terms can be found and the set of seeding terms will be returned.

In our approach, we make use of *word2vec* [13,14] to identify the words with similar context to that of the seeding terms from a massive open-domain corpus. Here, word2vec is an open-sourced tool released by Google for computing vector representations of words. It takes a text corpus as input and outputs a numeric vector for each word. Using these word vectors, given a word w, word2vec can compute and produce a list of words that are similar to w (e.g., having similar context) and their cosine similarity values. For example, when "france" is input, "spain" would be one of the similar words in the produced list, and the cosine similarity may be 0.678515. We train a word2vec model with a massive open-domain corpus (detailed in the experimental section), feed it with the seeding terms, and take the produced lists of similar words as our candidate terms. We observe that, even though word2vec is not trained on the corpus of research papers, it can still produce most of the terms appearing in the research papers. This is because of the massiveness of the open-domain training corpus that covers many of the concepts in various domains.

---

[2] http://www.google.com.au/insidesearch/features/search/knowledge.html
[3] http://en.wikipedia.org/wiki/Wikipedia:Database_download

To summarize, we make the following contributions in this paper:

- We analyze the key characteristics of term occurrence in research papers that may allow us to overcome the limitation of the domain dependent terminology extraction algorithms.
- Based on the analysis we propose a novel domain independent term extraction algorithm. This algorithm uses the title words and the keywords in research papers as the seeding terms and word2vec to identify similar terms from an open-domain corpus as the candidate terms. It filters the candidate terms by checking their occurrence in the research papers. The surviving candidate terms are added to the seeding terms and the above process repeats until there is no new candidate terms.
- We conduct extensive experiments on the proposed algorithm, and the results confirm the effectiveness of the proposed algorithm in extracting terms from research papers.

The rest of the paper is organized as follows. In Section 2 we review related studies. In Section 3 we present the proposed algorithm. In Section 4 we discuss the experimental results and in Section 5 we conclude the paper.

2 Related Work

Existing terminology extraction approaches [5,7,9,16,18] can be categorized into three groups: rule-based, statistical, and a combination of the two. We briefly review each group of the studies. A more detailed survey of the studies on terminology extraction can be found in [15].

Rule-based approaches use linguistic features to identify candidate terms. Various linguistic rules [4,5,7] have been devised for term extraction. For example, a simple rule is that terms are mostly nouns or noun phrases in the form of *[noun, noun]* or *[adjective, noun]*. This rule has been applied in systems such as the PHRASE system [4]. Evans et al. [5] combine the simple noun phrases and compose more complex rules for term extraction. Daille et al. [2] identify the most common linguistic rules from human produced terminological data banks. Their study confirms that terms are mainly short noun phrases in the form of *[noun, noun]* and *[adjective, noun]*. They call terms in such form the *base-terms*. Using morphological or syntactic variations, more complex and longer terms can be formed from the base-terms. Other than word level structural rules for terms, sentence level structural rules have also been proposed. For example, Xun and Li [19] use rules such as *"[noun] is ..."* or *"[noun] refers to ..."* to identify the terms and their definitions at the same time.

Using the linguistic rules, a terminology extraction algorithm generally works in four steps [15]:

- Perform Part-Of-Speech (POS) tagging on the corpus for terminology extraction.
- Identify and extract candidate terms from the corpus using the basic linguistic rules.

- Collapse variations of the same term into one unique form.
- Apply more sophisticated linguistic rules to further filter the candidate terms.

After the four steps a list of candidate terms will be produced. Depending on the quality of the linguistic rules and the domain of the corpus, the probability of these candidate terms being true terms varies. A further refinement step is required to identify the true terms, which is usually done by human experts manually.

Statistics-based approaches use statistical features to determine the possibility (i.e., the *termhood* [15]) of a word (or compound words) being a term. For example, Salton et al. [16] extract all 2-word combinations as candidate terms, and use *tf-idf* to measure their termhood. Jones et al. [9] extract all N-word combinations as candidate terms, and use term *length* and *frequency* to measure the termhood. Krenn [11] also uses frequency as the termhood measure and empirically shows that it is a reliable measure. More sophisticated measures have been used, such as the Z-score [3], the C-value [6], and the Multiword Expression Distance [1]. The main disadvantage of statistics-based approaches is that they need a large training data set to learn the parameters for determining whether a candidate term is indeed a term, which is usually manually labeled and requires much human effort.

Hybrid approaches combine rule-based techniques and statistical techniques. A common practice is to use linguistic rules to first identify a set of candidate terms and then filter them by statistical termhood measures. For example, Earl [4] first extracts noun phrases as the candidate terms and then filters them by the frequency of the noun elements. Justeson and Katz [10] also first extract the candidate terms using linguistic rules, and then rank the candidates by frequency. Semantic and contextual information is also used in the hybrid approaches to help rank the candidate terms. For example, Maynard and Ananiadou [12] use *context-factor* to incorporate the semantic and contextual information. Velardi et al. [18] use *domain relevance* and *domain consensus* to achieve similar purpose.

Word2vec [13,14] is an open-sourced tool released by Google for computing vector representations of words using deep learning techniques. It takes a text corpus as input and computes a numeric vector for each word. Using these vectors, given a word w, word2vec can compute and produce a list of words that are similar to w (e.g., having similar context) and their cosine similarity values. For example, given "france" as the input, a list as shown in Table 1 may be produced.[4] If the training corpus is labeled properly, word2vec can also handle phrases, i.e., computing phrase vectors as well as similar phrases. In this study, we train a word2vec model with a massive open-domain corpus (detailed in the experimental section), feed it with the seeding terms, and take the produced lists of similar words as our candidate terms. We observe that, even though word2vec is not trained on the corpus of research papers, it can still produce most of the

---

[4] https://code.google.com/p/word2vec/

Table 1. Similar Word List

Word	Cosine similarity
spain	0.678515
belgium	0.665923
netherlands	0.652428
italy	0.633130
switzerland	0.622323
luxembourg	0.610033
portugal	0.577154
russia	0.571507
germany	0.563291
catalonia	0.534176

terms appearing in the research papers. This is because of the massiveness of the open-domain training corpus that covers many of the concepts in various fields. We will detail how word2vec is used in the following section.

3   The Term Extraction Algorithm

Given a set of research papers P of the *same domain*, we aim to extract all *terms* in the research papers. Here, the research papers need to be of the same domain, but we do not have any assumption on of which specific domain the research paper should be; a "term" refers to a word or compound words representing a concept of the domain.

To gain insight to the problem of extracting terms from research papers we conducted an empirical study on a corpus of 200 medical research papers. We have the following observations.

- *60.13% of the nouns in the title are terms; 83.61% of the noun phrases in the title are terms (after removing the stop words).*
- *96.72% of the keywords are terms.*
- *64.44% of the terms not in either the titles or keywords appear in similar context (in the abstract or body) to that of the terms in the titles or keywords.*

These observations motivate us to use the nouns and noun phrases in the title and the keywords as a set of seeding terms, find the words or phrases in the abstract and body parts of the research papers that are similar to the seeding terms, and return them together with the seeding terms as the answer set. To facilitate the process of finding the words or phrases similar to the seeding terms, we make use of *word2vec* [13,14].

We propose to extract terms from a set of research papers P as follows.

1. We use a POS tagging tool[5] to identify the noun phrases from the titles of the research papers in P.

---

[5] There are many existing POS tagging tools, e.g., the Stanford POS Tagger [17].

Algorithm 1. Term Extraction(P, k, α)

---

1 $N \leftarrow$ the noun phrases in the titles of P;
2 $K \leftarrow$ the keywords in the titles of P;
3 $W \leftarrow N \cup K$;
4 **foreach** $w \in W$ **do**
5 $\quad\lfloor$ Compute $ranking\_score(w)$;

6 Sort W by the ranking scores;
7 $T \leftarrow$ top-k words/phrases in W;
8 $has\_new\_term \leftarrow true$;
9 **while** has_new_term **do**
10 $\quad$ $has\_new\_term \leftarrow false$;
11 $\quad$ $L \leftarrow \emptyset$;
12 $\quad$ $C \leftarrow \emptyset$;
13 $\quad$ **foreach** $t \in T$ **do**
14 $\quad\quad$ $L_t \leftarrow$ words/phrases similar to t returned by word2vec;
15 $\quad\quad\lfloor$ $L \leftarrow L \cup \{L_t\}$;
16 $\quad$ **foreach** $L_t \in L$ **do**
17 $\quad\quad$ **foreach** $w \in L_t$ **do**
18 $\quad\quad\quad$ **if** w *appears in at least α lists in* L **then**
19 $\quad\quad\quad\quad\lfloor$ $C \leftarrow C \cup \{w\}$;

20 $\quad$ **foreach** $c \in C$ **do**
21 $\quad\quad$ **if** c *appears in* P **then**
22 $\quad\quad\quad$ $T \leftarrow T \cup \{c\}$;
23 $\quad\quad\quad$ $has\_new\_term \leftarrow true$;

$\quad$ **return:** T

---

2. Together with the keywords of the research papers in P, we add the noun phrases to a set W.

3. We use the top-k words/phrases in W as a set of seeding terms T, where k is an algorithm parameter that will be chosen empirically. To choose the top-k words/phrases, we rank the words/phrases in W by their length (number of words contained by a phrase) and frequency in P. As suggested by existing studies such as [9,11], these two measures have been reliable in identifying the true terms. Given a word/phrase $w \in W$, we use the following empirical equation to compute its ranking score:

$$ranking\_score(w) = \log(f_w) \times l_w$$

Here, f_w denotes the frequency of w and l_w denotes the length of w. Intuitively, words/phrases of larger frequency or length have higher probability of being true terms.

4. For each seeding term $t \in T$, we use word2vec to compute a list of words/phrases that are similar to t.

5. If a word/phrase appears in at least α lists returned by word2vec, we add it to the set of candidate terms C. Here, α is an algorithm parameter that will be chosen empirically.
6. For each candidate term $c \in C$, we scan the research papers in P and see if it can be found in the papers. If yes, we add it to the set of seeding terms T. Otherwise, we simply drop c.
7. We repeat Step 4 through Step 6 until no more terms can be found. Then we return T as the answer set.

Discussion. Algorithm 1 summarizes the procedure above. In the algorithm, word2vec is trained on an open-domain corpus which does not require any labeling effort. Meanwhile, the algorithm does not require any domain specific knowledge and is fully automatic. Therefore, we achieve an algorithm that avoids the drawbacks of the existing approaches. Note that a potential limitation is that when word2vec is trained on an open-domain corpus, it may not contain all the terms that appear in P and thus cannot help extract those terms. This limitation can be easily overcome by adding the documents in P to the corpus used to train word2vec.

4   Experimental Study

We implement the proposed algorithm in Perl and conduct experiments on a Desktop computer with a 3.0GHz CPU and 3GB memory.

We test the algorithm on a set of 200 medical research papers, where a total of 1704 unique terms with 9780 occurrences have been labeled by domain experts as the gold standard. We measure the term extraction accuracy and recall.

$$Accuracy = \frac{number\ of\ unique\ true\ terms\ extracted}{total\ number\ of\ unique\ terms\ extracted} \times 100\%$$

$$Recall = \frac{number\ of\ unique\ true\ terms\ extracted}{total\ number\ of\ unique\ true\ terms} \times 100\%$$

To the best of our knowledge, there is no existing domain independent terminology extraction algorithms. Hence, no baseline algorithm is used in the experiments.

We train a word2vec model with an open-domain corpus containing 500,000 documents from science and technology periodicals. The corpus has approximately 20 billion words.

The proposed algorithm takes in the 200 medical research papers, extracts the nouns and noun phrases from the titles and the keywords as the seeding terms, and use word2vec to identify candidate terms. We only use the words/phrases that appear in at least 2 word2vec similar word list as the candidate terms. We filter the candidate terms by checking their respective number of appearance in the research papers and only return a candidate term as a term if it appears for at least once.

4.1   The Effect of the Number of Initial Seeding Terms

We first verify the effect of the number of initial seeding terms k. We expect higher recall and lower accuracy as more initial seeding terms are used. Table 2 shows the result where we vary k from 10 to 40. We use the top 15 similar words returned by word2vec for each seeding term as the candidate terms in this set of experiments. As expected, the more initial seeding terms are used, the higher the recall will be. This is natural because more initial seeding terms will bring more candidate terms and eventually more terms. Also, the recall increases as more iterations of the algorithm are performed since more candidate terms will be find by word2vec (for conciseness we only show the result of the first two iterations). Meanwhile, the accuracy drops as k increases because more nouns and noun phrases that are less likely to be true terms are used as the seeding terms, and the candidate terms similar to them are less likely to be the true terms.

Table 2. Varying Initial Seeding Term Set Size

k	10	20	30	40
Recall (iteration 1)	56.69%	66.90%	71.83%	74.30%
Recall (iteration 2)	62.68%	72.89%	76.06%	77.82%
Accuracy (iteration 1)	40.21%	36.54%	32.64%	30.35%
Accuracy (iteration 2)	36.73%	33.50%	30.07%	28.14%

4.2   The Effect of the Size of the Word2vec Similar Word Lists

Given an input word w, word2vec can return a list all words/phrases appearing in the corpus sorted by their similarity to w. The words/phrases appearing at the later part of the list are not very similar to w. Therefore, we limit the size of the word2vec similar word lists and only use the words/phrases that are most similar to the seeding terms as the candidate terms.

Let the size of the word2vec similar word list be l. Table 3 shows the accuracy and recall where we vary l from 15 to 240. We use 40 initial seeding terms in this set of experiments. We can see that, the more words returned by word2vec are used as the candidate terms, the higher the recall will be. Also, the recall increases as more iterations of the algorithm are run. After 2 iterations, using 240 similar words for each seeding terms will achieve a recall of 90.14%. This is very close to the recall upper bound 91.20% (We scan the open-domain corpus where word2vec is trained on and find that 1554 among the 1704 unique terms are contained in the open-domain corpus. This means that using word2vec, the recall upper bound is $\frac{1554}{1704} \times 100\% = 91.20\%$). Adding more iterations will not significantly increase the recall and thus we terminate the algorithm after 2 iterations. Meanwhile, the accuracy drops as the recall increases, which is also expected because currently the proposed algorithm has a very simple candidate term checking procedure and has a relatively small pruning power in filtering the candidate terms. When more similar words are considered as the candidate terms, more will pass the filtering stage and be returned as terms. How to achieve a better accuracy while maintaining a good recall value is our next step of study.

Table 3. Varying the Size of the Word2vec Similar Word Lists

l	15	30	60	120	240
Recall (iteration 1)	74.30%	76.41%	80.63%	83.45%	83.08%
Recall (iteration 2)	77.82%	81.69%	86.27%	88.03%	90.14%
Accuracy (iteration 1)	30.35%	28.48%	26.83%	25.49%	25.92%
Accuracy (iteration 2)	28.14%	25.58%	23.45%	21.62%	21.94%

4.3 The Effect of the Pruning Threshold of the Candidate Terms

We explore the following two ways to increase the term extraction accuracy.

Table 4. Varying the Number of Word2vec Similar Word Lists that a Candidate Term Needs to Appear in

α	2	3	4	5	6
Recall (iteration 2)	77.82%	76.41%	76.06%	75.35%	75.35%
Accuracy (iteration 2)	28.14%	29.51%	30.02%	30.19%	30.36%

(1) Requiring the words/phrases to appear in more than 2 word2vec similar word lists so that they can be added to the candidate terms. In this set of experiments we set the word2vec similar word list size at 15 and the number of initial seeding terms at 40. We vary the number of word2vec similar word lists a word/phrase needs to appear in, i.e., α, so as to be treated as a candidate term from 2 to 6. Table 4 shows the accuracy and recall values. Intuitively, as α increases, the accuracy increases while the recall decreases. We notice that the increase and the decrease are at a similar rate. This can guide us to choose an appropriate value of α given a recall or accuracy value we want to achieve.

Table 5. Varying the Number of Research Papers that a Candidate Term Needs to Appear in

p	1	2	3	4	5
Recall (iteration 2)	62.68%	49.30%	39.44%	30.99%	26.41%
Accuracy (iteration 2)	36.73%	40.84%	42%	40.03%	40%

(2) Requiring the candidate terms to appear in more than one research papers.
In this set of experiments we set the word2vec similar word list size at 15 and
the number of initial seeding terms at 10. We vary the number of research papers
a candidate terms needs to appear in (denoted by p) so as to be returned as a
term from 1 to 5. Table 5 shows the accuracy and recall values. It shows that,
as p increases, the accuracy increases while the recall decreases. We notice that
the recall decreases much faster than the accuracy increases. Further, accuracy
starts to also decrease when the number p is larger than or equal to 4. This
suggests that the value of p should be kept at a small value (e.g., 1 or 2).

4.4   Error Case Study

We conduct an error case analysis on the experimental results and make the
following observations.

1. As shown in Section 4.2, our algorithm is able to achieve a recall of 90.14%.
   Among the terms that are not extracted, i.e., the false negatives, 89% are not
   contained in the open-domain corpus used to train the word2vec model. For
   example, "OVX rat" (a female rat whose ovaries have been removed) is term
   that is not contained in the training corpus and is missed by the algorithm.
   These false negatives can be recovered by adding the test corpus into the
   training corpus. The rest 10.3% of the false negatives (18 out of a total of
   1704 terms labeled in the test corpus) are missed by the proposed algorithm
   because they have extremely low number of occurrences in the word2vec
   training corpus. As described in the experiments, we constrain the number
   of similar words returned by word2vec to be used as the candidate terms.
   If a word has a low occurrence and is not ranked as the top similar word
   to the seeding terms, then it may not be found the proposed algorithm. For
   example, "2,3-dimethyl-2,3-butanediol" is such a term.
2. As discussed in the subsections above, the accuracy of the proposed algorithm is relatively low due to a relatively simple strategy used to filter the
   candidate terms. We find that the false positives are mainly the words used
   as terms in the open-domain corpus while not being viewed as terms in the
   research papers, such as "portable transformer rectifier", "animal migration", and "average life expectancy". To filter such false positives, more
   sophisticated techniques such as the transfer learning techniques are to be
   explored to integrate the knowledge in the open-domain corpus into our
   target domain more systematically.

5   Conclusions and Future Work

We proposed a novel approach for extracting terms from research papers. We
use the title words and the keywords in research papers as the seeding terms and
word2vec to identify similar terms from an open-domain corpus as the candidate terms, which are then filtered by checking their occurrence in the research

papers. We repeat this process using the newly found terms until no new candidates terms can be found. Compared with the existing approaches, our approach has the advantage of not requiring any domain knowledge, i.e., being domain independent, which is an important advantage considering the amount of effort needed to acquire and prepare the domain knowledge. As shown in the experimental study, our approach can extract the terms effectively. We analyze the false positives and false negatives in our experimental results and observe that: (i) the false positives are mainly the words used as terms in the open-domain corpus while not being viewed as terms in the research papers; (ii) the false negatives are the terms that have extremely low number of occurrences in the open-domain corpus. These observations are expected as the open-domain corpus used to train the word2vec model is very different from the corpus of research papers. Our future study is to build upon the current results and design algorithms to work together with word2vec to reduce the false negatives as well as algorithms to filter the false positives.

Acknowledgments. This work is supported by the Natural Science Foundation of China under Projects No. 61300081, No. 61170162 and No. 2012BAH16F00, and Melbourne School of Engineering Early Career Researcher Grant under Project Reference Number 4180-E55.

References

1. Bu, F., Zhu, X., Li, M.: Measuring the non-compositionality of multiword expressions. In: Proceedings of the 23rd International Conference on Computational Linguistics (COLING), pp. 116–124 (2010)
2. Daille, B., Hubert, B., Jacquemin, C., Royauté, J.: Empirical Observation of Term Variations and Principles for their Description. Terminology **3**(2), 197–258 (1996)
3. Dennis, S.F.: The construction of a thesaurus automatically from a sample of text. In: Proceedings of the Symposium on Statistical Association Methods for Mechanized Documentation, pp. 61–148 (1965)
4. Earl, L.L.: Experiments in automatic extracting and indexing. Information Storage and Retrieval **6**(4), 313–330 (1970)
5. Evans, D.A., Zhai, C.: Noun-phrase analysis in unrestricted text for information retrieval. In: Proceedings of the 34th Annual Meeting on Association for Computational Linguistics (ACL), pp. 17–24 (1996)
6. Frantzi, K.T., Ananiadou, S.: Extracting nested collocations. In: Proceedings of the 16th Conference on Computational Linguistics (COLING), pp. 41–46 (1996)
7. Gianluca, R.B., Rossi, G.D., Pazienza, M.T.: Inducing terminology for lexical acquisition. In: Proceedings of the Conference on Empirical Methods in Natural Language Processing (EMNLP) (1997)
8. Hoffart, J., Suchanek, F.M., Berberich, K., Lewis-Kelham, E., de Melo, G., Weikum, G.: Yago2: exploring and querying world knowledge in time, space, context, and many languages. In: Proceedings of the 20th International Conference Companion on World Wide Web (WWW), pp. 229–232 (2011)
9. Jones, L.P., Gassie Jr., E.W., Radhakrishnan, S.: Index: The statistical basis for an automatic conceptual phrase-indexing system. Journal of the American Society for Information Science **41**(2), 87–97 (1990)

10. Justeson, J.S., Katz, S.M.: Technical terminology: some linguistic properties and an algorithm for identification in text. Natural Language Engineering **1**, 9–27 (1995)
11. Krenn, B.: Empirical implications on lexical association measures. In: Proceedings of the 9th EURALEX International Congress (2000)
12. Maynard, D., Ananiadou, S.: Identifying contextual information for multi-word term extraction. In: 5th International Congress on Terminology and Knowledge Engineering, pp. 212–221 (1999)
13. Mikolov, T., Sutskever, I., Chen, K., Corrado, G.S., Dean, J.: Distributed representations of words and phrases and their compositionality. In: Proceedings of the 27th Annual Conference on Neural Information Processing Systems (NIPS), pp. 3111–3119 (2013)
14. Mikolov, T., Yih, W., Zweig, G.: Linguistic regularities in continuous space word representations. In: Proceedings of the Conference of the North American Chapter of the Association of Computational Linguistics on Human Language Technologies (HLT-NAACL), pp. 746–751 (2013)
15. Pazienza, M., Pennacchiotti, M., Zanzotto, F.: Terminology extraction: an analysis of linguistic and statistical approaches. In: Sirmakessis, S. (ed.) Knowledge Mining, Studies in Fuzziness and Soft Computing, vol. 185, pp. 255–279. Springer, Heidelberg (2005)
16. Salton, G., Yang, C.S., Yu, C.T.: A theory of term importance in automatic text analysis. Journal of the American Society for Information Science **26**(1), 33–44 (1975)
17. Toutanova, K., Klein, D., Manning, C.D., Singer, Y.: Feature-rich part-of-speech tagging with a cyclic dependency network. In: Proceedings of the Conference of the North American Chapter of the Association for Computational Linguistics on Human Language Technology (HLT-NAACL), pp. 173–180 (2003)
18. Velardi, P., Missikoff, M., Basili, R.: Identification of relevant terms to support the construction of domain ontologies. In: Proceedings of the ACL Workshop on Human Language Technology and Knowledge Management, pp. 5:1–5:8 (2001)
19. Xun, E., Li, C.: Applying terminology definition pattern and multiple features to identify technical new term and its definition. Journal of Computer Research and Development **46**(1), 62–68 (2009)

TK-SK: Textual-Restricted K Spatial Keyword Query on Road Networks

Xiaopeng Kuang[1], Pengpeng Zhao[1(✉)], Victor S. Sheng[2], Jian Wu[1], Zhixu Li[1], Guanfeng Liu[1], and Zhiming Cui[1]

[1] School of Computer Science and Technology, Soochow University, Suzhou, China
20145227006@stu.suda.edu.cn,
{ppzhao,jianwu,zhixuli,szzmcui,szzmcui}@suda.edu.cn
[2] Computer Science Department, University of Central Arkansas, Conway, USA
ssheng@uca.edu

Abstract. With the rapid development of GPS-enabled devices, spatial keyword query, considering both spatial proximity to a query location and the textual relevance to the query keywords, is applied to many real-world applications. In this context, we study a specific type of spatial keyword query Textual-restricted K Spatial Keyword query (TK-SK query), which returns the nearest k points of interest (POIs) whose textual description is not less than a specified textual relevance threshold and whose location is close to the query location. We further propose a baseline approach and two advanced approaches (a separated index approach and a hybrid index approach) with different indexing strategies to solve this problem. Our comprehensive experiments conducted on real spatial datasets clearly demonstrate the efficiency of our two advanced approaches.

Keywords: Spatial keyword query · Hybrid index · Road networks

1 Introduction

The rapid development of techniques for both geo-positioning and mobile communication has made location awareness query necessary in many applications. For instance, Google and Baidu maps provide target location information with a short textual description (points of interest, POIs). People are able to find the POIs that are closely related to their input keywords. To the best of our knowledge, there are researches dealing with boolean spatial keyword query, range spatial keyword query, and top-k spatial keyword query. However, POIs returned from these queries, sometimes, do not satisfy users' intents when they just need the top k POIs whose textual description is strongly related with their query keyword and whose location is close to the query location. For example, boolean spatial keyword query may not return any POI when no POI meets query requirements, or returns too many POIs for users to choose if many POIs meets query requirements. Top-k spatial keyword query could recommend POIs having low textual relevance but high spatial scores.

© Springer International Publishing Switzerland 2015
M.A. Sharaf et al. (Eds.): ADC 2015, LNCS 9093, pp. 167–179, 2015.
DOI: 10.1007/978-3-319-19548-3_14

For example, a user searches for the nearest POI, which is strongly related to his query keywords (e.g., keywords are *Chinese*, *Music* and *Hotel*), as shown in Figure 1. Given a user query $Q(q.l, q.d = \{Chinese, music, Hotel\}, q.K=1)$, where $q.l$ represents query location, $q.K$ means the max number of POIs to be returned, and the query keywords (represented as $q.d$) are {*Chinese, Music, Hotel*}. Top-k spatial query returns P7 as the top one POI (spatial weight $\alpha = 0.7$). However, P7 is not very consistent with the user's query intent on the dimension of text. Actually, P3 is more accordant to the intent of the user, because P3 possesses higher textual relevance and is also close to $q.l$. P7 could be excluded by adjusting the spatial weight α. However, a small α could cause side-effects. A remote POI possessing higher textual relevance but far from the query location may be returned. What's more, finding an appropriate spatial weight α is also not an easy task. Boolean spatial keyword query will could not return any POI, because there does not exist any POI that contains all the query keywords.

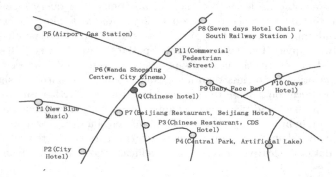

Fig. 1. Spatial keyword query on road networks

In view of this, we propose a new type of spatial keyword query on road networks, namely, Textual-restricted K Spatial Keyword (TK-SK) query. TK-SK query specifies a spatial location and a set of query keywords. it is to find top k POIs whose textual relevance to the query keywords is not less than a given threshold and whose location is closer than the rest POIs satisfied in textual relevance. Now, we redefine the previous query Q as $Q_N(q.l, q.d = \{Chinese\ music\ Hotel\}, q.K=1, q.T=0.6)$, where $q.T$ represents the textual relevance threshold. Q_N searches the nearest K POIs whose textual relevance is not less than 0.6, Then P3 will be returned as the best choice. To formalize the new TK-SK query, we first propose a baseline for processing TK-SK query on road networks, which utilizes the Dijkstra's algorithm to explore road networks. To improve its efficiency, we develop a separate index approach. This approach first obtains the candidate POIs whose textual reference is not less than the threshold T using a CA algorithm [1], and then gets k POIs from the candidates POIs employing the G-tree road network index [2]. To further improve the efficiency of processing these queries, we develop a hybrid index approach which considers text and spatial information synchronously.

The rest of this paper is organized as follows. First, we describe the related work in Section 2. We have the problem statement in Section 3. Then, we describe the baseline approach in Section 4,the separated index approach in Section 5, and hybrid index approach in Section 6. Finally, we present experiment evaluation in Section 7, and conclude the paper in Section 8.

2   Related Work

To the best of our knowledge, there is no existing work on the query problem of Textual-restricted K Spatial Keyword (TK-SK) query on road networks. In this section, we will introduce three existing spatial keyword queries, which are somehow related to our research.

Top-k Spatial Keyword Query. The top-k spatial keyword query aims to find a set of objects with geographical marks whose rankings are based on their spatial proximities and query keywords similarities. Many efficient indexing techniques have been proposed such as IR [3], RCA[4]. All of them are based on Euclidean distance. To the best our knowledge, the state-of-art approach processing top-k spatial keyword query on road networks was proposed by Roach Junior [5], which employs an overlay on the top of actual road networks to improve the query processing performance.

Range Spatial Keyword Query. Range spatial keyword query is applied to search objects whose location is intersecting with or contained in a query region [6,7]. The state-of-art approach processing spatial keyword on road networks was proposed by W.Li et al. [8], which partitions an entire road networks into a group of interconnected subnets and organizes them in a hierarchy structure.

Boolean Spatial Keyword Query. Boolean spatial keyword query requires the returned objects meeting all the query requirements, and these objects are ranked by their distances to the query location. Felipe et.al. [9] considered AND semantics. They proposed a hybrid indexing structure called IR^2-Tree (Information Retrieval R-Tree) which combines an R-Tree with superimposed text signatures. Different from our approach, [9] is restricted to boolean keyword queries and Euclidean distance.

3   Problem Defination

As we described before, TK-SK query on road networks retrieves k POIs of a road networks for a given query Q_N, which are the closest to the location specified in the query and their textual relevance θ is not less than a given threshold T.

Formally, a query Q_N has four components, denoted as $< q.l, q.d, q.K, q.T >$, where $q.l$ is a query spatial location(including its latitude and longitude), $q.d$ is the set of query keywords, $q.K$ is the maximal number of results to return, and $q.T$ is the textual relevance threshold. Each POI (denoted as p), as previously mentioned, lies on a vertex of a road networks. Given a query Q_N, a road network G, and a set of POIs P, Q_N returns k POIs in a descending order of a function score τ. The function score τ is proportional to the textual relevance between the description of a candidate POI $(p.d)$ in the road network and the query keywords $(q.d)$, but inversely proportional to the distance cost of the location of a candidate POI $(p.l)$ to a query location $(q.l)$, defined as :

$$\tau = \frac{\lceil \theta(q.d, p.d) - q.T \rceil}{\delta(q.l, p.l)} \tag{1}$$

where the function f is a specific boolean satisfaction function. We will explain it further after we explain the notations of its parameters. $\theta(p.d, q.d)$ represents the textual relevance of $p.d$ according to the query keywords $q.d$, between 0 and 1. T is a textual relevance threshold. This threshold is used as a textual restriction to decide whether a POI p meets the query textual requirements. For a POI p, if $\theta(q.d, p.d) \geq T$, we think p satisfies the query q in the text dimensions. In other words, $f(\theta(q.d, p.d) - T)$ is to determine where the query q is satisfied. If the query q is satisfied, $f(\theta(q.d, p.d) - T) = 1$. Otherwise, $f(\theta(q.d, p.d) - T) = 0$. $\delta(p.l, q.l)$ represents the network proximity between the query location$(q.l)$ and the POI location $(p.l)$, i.e, the shortest distance cost of the location of a POI $(p.l)$ to a query location $(q.l)$, defined as:

$$\delta(q.l, p.l) = \|q.l, p.l\| \tag{2}$$

4   A Baseline Approach

In this section, we present the baseline approach based on Dijkstra's algorithm to support TK-SK queries on road networks.

4.1   A Baseline Index Structure

Our baseline index structure includes a spatial component and a spatial-textual component. It preserves the locations and connectivity of road networks, and can be employed to process TK-SK queries on the road networks. Its spatial component including a road network R-tree and an adjacency file [10], stores the spatial structure of road networks and permits starting a query at any location of road networks. The spatial-textual component aims to store the textual information of POIs. Given the vertexes closer to Q_N, the spatial-textual component computes out the POIs whose textual relevance scores are not less than T. As Figure 2 shows, the textual information of a POI P_i can be accessed according to the indexes of both P_i and its related vertex V_k.

4.2   Query Processing

The baseline query processing is an expansion-based approach employing the Dijkstra's algorithm. First, road network R-tree locates the vertex v where query location lies and seeks for candidates from the POIs belonging to v. Then explore the road network employing the Dijkstra's algorithm to find more candidates until the size of candidates arrives $q.k$ or the entire road network has been accessed. The candidates found are the final results.

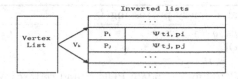

Fig. 2. A baseline index structure: spatial-textual component

5 Separated Index Approach

While the baseline approach presented in the previous section could be easily implemented, it has a performance drawback that each vertex accessed will cost one I/O to find candidate POIs. In this section, a separated index approach is proposed to reduce the I/O caused by accessing vertexes.

5.1 Separated Index Structure

Our approach uses separated indexes, which include a road network index for searching by location and an inverted file index for searching by keyword.

Textual Inverted File. The inverted file we proposed is constituted by posting lists identified by a keyword. Each posting list stores POIs that have a term t in their description. For each POI p, the posting list stores:1)the vertex in which p lies on a road networks, and 2) the importance of term t_i in the description of the POI p in the descending order.

G-tree Road Network Index. G-Tree is an efficient tree-based road network index for kNN search on road networks, which preservers the spatial structure information in an effective way [2]. We employ it to find k nearest candidate POIs located in the vertexes of a road network. The distance matrices keep the shortest-path distance between a border and a border/vertex(non-leaf/leaf). The distance matrix of a non-leaf node keeps the shortest-path distance between its child borders while the distance matrix of a leaf node keeps the shortest-path distance between vertexes in the node and borders in the node.

5.2 Querying Processing

Processing TK-SK queries on the separated index can be performed using Algorithm 1 as follows.

Algorithm 1 call function $findCandidatePOI$ to find candidate POIs C [4] and related vertexes N with CA algorithm (line 2), After getting the candidate POIs C and the related vertexes N, nearest candidate vertexes will be found from N with G-tree[2] road network structure (line 4). After that, the candidate POIs C is reevaluated with constraint R. Each candidate p belonging to C is removed from C when R does not include the vertex on which p lies. Finally, the k best POIs are be returned.

Algorithm 1. Separated query processing(Q_N)

Input: $Q_N = (q.l, q.d, q.k, q.T)$
Output: top k best POI
  1: $R \leftarrow \phi, N \leftarrow \phi, C \leftarrow \phi, H \leftarrow \phi$
  2: Call *findCandidatePOI*() for candidate POIs C and vertexes N, update C and N
  3: $v_0 \leftarrow$ *vertex* in which $q.l$ lies
  4: Call *Gtree_nearestKNeighbor*(N, v_0, k) to find nearest k candidate vertexes R
  5: **for** each POI $p \in C$ **do**
  6:     $n \leftarrow$ node p lies
  7:     **if** R contains n **then**
  8:        update distance of p in C
  9:     **else**
10:        delete p from C
11:     **end if**
12: **end for**
13: sort C in the increasing order of distance
14: **return** C

6　Hybrid Index Approach

While the separated index approach performs better than the baseline approach, the separated index approach takes time to calculate the occurrence list for each query. In this section, we propose a hybrid index approach, which only calculates the occurrence list once for all queries.

6.1　Hybrid Index Structure

We propose a hierarchic indexing IG-tree (inverted G-tree), which integrates the inverted file and G-tree road network, and organizes the textual description by hanging in the G-tree nodes. An IG-tree is built in a bottom-up way. Specifically, the inverted file of a node in IG-tree is built before the inverted file of its parent node is complete. All nodes in IG-tree except the root node are associated with an inverted file. The vertexes of a road network are the leaf nodes in IG-tree, and the inverted files connected with leaf nodes preserve the detail textual information of POIs. Each non-leaf node presenting a partition of a graph G connects with an inverted file, which only keeps the max textual score for each keyword.

6.2　Querying Processing

The hybrid index query processing algorithm takes advantage of the combination of the G-Tree to find the nearest region node and inverted file to prune nodes which would never meet the querying requirements.

Algorithm 2 presents the hybrid index query processing algorithm. The algorithm receives a query Q_N as its input parameter and returns the k best POIs in the decreasing order of the function score $\tau(p)$. The beginning of the algorithm is the same with beginning part of Algorithm 1. if n is a leaf node (i.e., n is a

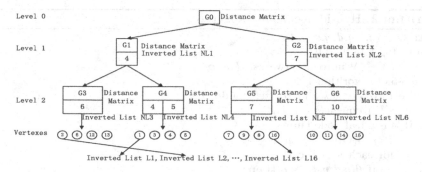

Fig. 3. Hybrid Index Structure

vertex of a road network), its corresponding POI p with $\theta(p.d, q.d) \geq T$ lying on the node n will be inserted into R. If n is a non-leaf node, each of its children n' will be explored. After calculating the relevance of each child to the query, if the textual relevance of a child $\theta(n'.d, q.d)$ is not less than the textual relevance threshold T, this child is a vertex candidate. This is the node n' will be added to the minHeap H (line 18 or 21), and the distance to the query location would be updated at the same time(line 18 or 20). This procedure terminates when the entire road networks is accessed, or there is the k best POIs found.

In order to help understand this algorithm better, we use an example as case study to show the procedure of finding the top 1 POI for a query as follows.

Assume a query Q_N whose $q.l$ is near to vertex v_7 as shown in Figure 3, $q.d = \{Chinese, music, Hotel\}$, $q.k = 1$ and $q.T = 0.6$. First, we construct the occurrence list, and then compute the top 1 answers as follows:

Step 1: insert the root node $< G0, 0 >$ into the minHeap H queue;

Step 2: poll$< G0, 0 >$ from H. Find the children of G0, i.e. G1 and G2. Calculate the textual relevance of its children. Get $\theta < G1.d, q.d > = 1$ and $\theta < G2.d, q.d > = 1$. Both of them are higher than the threshold T. Then, calculate the distance to q.l for both G1 and G2. Since $\delta(G2.l, q.l) = 0$ and $\delta(G1.l, q.l) = 3$, G2 is queued before G1.Thus, queue H has $< G2, 0 >, < G1, 3 >$

Step 3: poll$< G2, 0 >$ from H. Find the children of G2, i.e., G5 and G6. Calculate the textual relevance for two children. Get $\theta < G5.d, q.d > = 0.87$ and $\theta < G6.d, q.d > = 0.81$. Then calculate the distance to q.l. Since $\delta(G5.l, q.l) = 0$, G5 is inserted into the queue H before $< G1, 3 >$. Since $\delta(G6.l, q.l) = 8$, G6 is inserted after $< G1, 3 >$. Thus, queue H has $< G5, 0 >, < G1, 3 >, < G6, 8 >$

Step 4: poll$< G5, 0 >$ from H. Again, find the children of G5. After calculating the textual relevance of each child,we find that there is no leaf node n' that its textual relevance is not less than the textual threshold. The same story occurs for $< G1, 3 >$. Thus, queue H has $< G6, 8 >$ left

Step 5: poll $< G6, 8 >$ from H. Again, find the children of G6, and calculate the textual relevance and the distance for each child. We find that the leaf node v10 (vertex 10 of the road networks) satisfies. $< v10, 8 >$ is queued. Thus, queue H hash $< v10, 8 >$

Algorithm 2. Hybrid query processing(Q_N)

Input: $Q_N=(q.l, q.d, q.k, q.T)$
Output: top k best POI
 1: $R \leftarrow \phi, N \leftarrow$ V in road networks, $H \leftarrow \phi$
 2: line 3-5 of Algorithm 1
 3: **while** R.size()$< q.k$ and H.size()>0 **do**
 4:     $n \leftarrow H$.Dequeue()
 5:     **if** n is a leaf node **then**
 6:         $P' \leftarrow$ POIs lying on n
 7:         **for** each $p \in P'$ **do**
 8:             **if** $\theta(q.d, p.d) \geq T$ **then**
 9:                 insert p into R
10:             **end if**
11:         **end for**
12:     **else if** n is a non-leaf node **then**
13:         **for** child node $n' \in O(n)$ **do**
14:             **if** $\theta(q.d, n'd) \geq T$ **then**
15:                 continue;
16:             **else**
17:                 **if** v_0 is in n' **then**
18:                     H.Enqueue($< n', SPDist(v_0, n') = 0 >$)
19:                 **else**
20:                     MINDIST-OUTSIDE-NONLEAF(v_0, n')
21:                     H.Enqueue($< n', SPDist(v_0, n') >$)
22:                 **end if**
23:             **end if**
24:         **end for**
25:     **end if**
26: **end while**
27: **return** at most k POIs from R

Step 6: poll $< v10, 8 >$ from H. Explore the POI lying on vertex 10. P3 is returned as , and it is the first one found. Since top 1 answer is generated. The algorithm terminates.

7   Performance evaluation

In this section, we will investigate the performance of the three proposed approaches, i.e., the baseline approach, the separated index approach, and the hybrid index approach. All approaches are implemented in Java. We conduct the our experiments on a PC with 8G memory and I5-3470CPU@3.20GH, running Windows 7 (64bits).

7.1   Datasets

We use several different scale POIs datasets [5] for scalability evaluation and use London, Florida(FLA) and central USA(CTR) as the three road network in our experiments. The characteristics of the three road networks are shown in Table 1. The statistics of the POI datasets are shown in Table 2. the parameters setting in the experiments are shown in Table 3. Without special indications, bold values in Table 3 are default values used in our experiments.

Table 1. Characteristics of the road networks

Attributes	London	FLA	CTR
Total number of vertexes	203,382	1,070,376	14,081,816
Total number of edges	274,946	2,712,798	34,292,495

Table 2. The POIs datasets

datasets	Number of Distinct terms	Average Number of terms
50K	26,314	3.44536
100K	42,450	3.24614
150K	55,548	3.1491866
300K	93,930	3.1063807

Table 3. Parameters evaluated in experiments

Parameters	Values
Number of results(k)	**10**, 20, 30, 40, 50
Number of keywords	1, 2, **3**, 4, 5
Textual threshold(T)	0.2, **0.4**, 0.6, 0.8
Road networks	***London***, FLA, CTR
Datasets	50K, **100K**, 150K, 300K

7.2 Experimental Results

In this section, we evaluate the query processing performance of the three approaches on scalable POI datasets and different road networks

Varying the Number of Results(k). In this experiment, we investigate the performance of three approaches under different numbers of results k. We increase the number of results k from 10 to 50. From Figure 4 It is obvious that the baseline approach needs significant more time to process the query than the separated index approach and the hybrid index approach. this is because the baseline approach needs exploring more vertexes or edges than the other two approaches. Exploring vertexes and edges is time consuming. On the contrary, the separated index approach filtrates most POIs using the CA algorithm and gets top k nearest POI from the candidates with the help of G-tree road network structure. Figure 4 also shows that the hybrid index approach does further improve the performance of the separated index approach. This is because the hybrid index approach does not use extra time to filtrate vertexes. Its Inverted lists hung on nodes in IG-tree play an important role in pruning road networks. We also notice that the response time of the baseline approach grows a little with the increment of k. However, it seems that there is no obvious growth with the increment of k for the separated index approach and the hybrid index approach.

Varying the Number of Keywords. Figure 5 presents the response time of the three approaches while varying the number of keywords in queries from 1 to 5. From Figure 5, we can see clearly that the response time of the baseline approach is significantly higher than that of the separated index approach and of the hybrid index approach. We also can see that the hybrid index approach does perform better than the separated index approach. Besides, Figure 5 also shows that the response time of both the baseline approach and the separated index approach is going down with the increment of the number of keywords. This is because there are more candidates in each querying iteration with more query keywords.

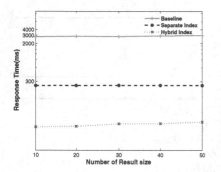
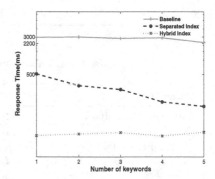

Fig. 4. The response time varying the number of results

Fig. 5. The response time varying the number of keywords

Varying the Textual Threshold T. In this experiment, we adjust the textual relevance T to evaluate its impact on the performance of three approaches. The threshold varies from 0.2 to 0.8 with a step length 0.2. The experimental results are shown in Figure 6. It is obvious that the response time of the baseline app-roach is much higher than that of the other two approaches. Between these two approaches, the response time of the hybrid index approach is lower than that of the separated index approach. Figure 6 also shows that the response time of the baseline approach increases when increasing the threshold. This is because when the threshold is set to be larger, the baseline approach takes more time to find POIs that meet the query requirements. The separated index approach would spend more time on seeking k POIs on its G-tree. However, when the threshold is larger, there are only a few candidates available, even non candidates available, there is no need to explore the G-tree for k POIs (e.g., $T=0.8$). That is why we can see the response time of the separated index approach goes down when the threshold is higher than 0.6. We also notice that the response time of the hybrid index approach is invariant with the increment of the threshold from Figure 6. This is due to its pruning functionality.

Varying POI Datasets Size. In this experiment, we study the response time of the three approaches under different sizes of POIs shown in Table 2. The experimental results are shown in Figure 7. Again, from Figure 7, we can see that the response time of the baseline approach is much higher than that of the other two approaches. Between these two approaches, the response time of the hybrid index approach is lower than that of the separated index approach. Figure 7 also shows that the response time of the baseline approach increases when increasing the POI dataset size. However, the response time of the separated index approach and the hybrid index approach does not increase when increasing the POI dataset size. This is because the baseline approach starts from the query location to explore a road network, using dijkstras algorithm. Its I/Os for data access, including vertexes, edges and POIs, is much larger than the other two approaches. The separated index approach just spends I/Os on loading inverted lists related to query keywords at the beginning of the query, so that the influence of the POI dataset sizes can be neglected. The hybrid index approach prunes many nodes, so that it can find and return results quickly.

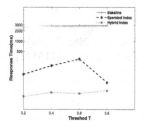 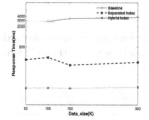 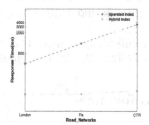

Fig. 6. The response time while varying the textual threshold

Fig. 7. The response time with different POI datasets sizes

Fig. 8. The response time under different road networks with different sizes

Varying Road Networks with Different Sizes. In this experiment, we study the response time of the three approaches on three different road networks with different size, i.e., London, FLA and CTR shown in Table 1. Our experimental results are shown in Figure 8. From Figure 8, we can see that the response time of the hybrid index approach is significantly lower than that of the separated index approach under different road networks. Figure 8 shows that the response time of the separated index approach becomes larger along with the bigger scale of road networks. This is because the separated index approach calculates the occurrence list for each query. With the expansion of road network, its computation takes significantly more time. However, the inverted list on each node used in the hybrid index approach is already in memory, and we just calculate the occurrence list once when procedure starts for the hybrid index approach. Thus we can see that the response time of the hybrid index approach does not impact by the sizes of different road networks. Therefore, it can be applied to different scale road networks. Finally, we want to mention that we have no results of the baseline approach in Figure 8, because it does not scale well for larger scale road network,such as CTR.

8   Conclusions

In this paper, we introduce a new query type Textual-restricted K Spatial Keyword query (TK-SK) on road networks. Given a spatial location and a query keyword, TK-SK queries return the k nearest POIs, which are ranked in terms of the function scores τ. We first presented a baseline approach to process these queries. Then, we presented a separated index approach that deals with the textual description, and the spatial information dividedly to find satisfied POIs quickly. We further proposed a hybrid index approach with a new index named IG-tree, which permits pruning road networks, so that we can reduce the cost of traveling the road networks and POIs. Our experimental results under all different settings show that the hybrid index approach performs the best under all settings, followed by the separated index approach. Both the hybrid index approach and the separated index approach perform much better than the baseline approach under all settings.

Acknowledgments. This work was partially supported by Chinese NSFC project (61170020, 61402311, 61440053), and the US National Science Foundation (IIS-1115417).

References

1. Fagin, R., Lotem, A., Naor, M.: Optimal aggregation algorithms for middleware. Journal of Computer and System Sciences **66**(4), 614–656 (2003)
2. Zhong, R., Li, G., Tan, K.L., Zhou, L.: G-tree: an efficient index for knn search on road networks. In: Proceedings of the 22nd ACM international conference on Conference on information & knowledge management, pp. 39–48. ACM (2013)
3. Cong, G., Jensen, C.S., Wu, D.: Efficient retrieval of the top-k most relevant spatial web objects. Proceedings of the VLDB Endowment **2**(1), 337–348 (2009)
4. Zhang, D., Chan, C.Y., Tan, K.L.: Processing spatial keyword query as a top-k aggregation query. In: Proceedings of the 37th international ACM SIGIR conference on Research & development in information retrieval, pp. 355–364. ACM (2014)
5. Rocha-Junior, J.B., Nørvåg, K.: Top-k spatial keyword queries on road networks. In: Proceedings of the 15th international conference on extending database technology, pp. 168–179. ACM (2012)

6. Khodaei, A., Shahabi, C., Li, C.: Hybrid indexing and seamless ranking of spatial and textual features of web documents. In: Bringas, P.G., Hameurlain, A., Quirchmayr, G. (eds.) DEXA 2010, Part I. LNCS, vol. 6261, pp. 450–466. Springer, Heidelberg (2010)
7. Li, Z., Lee, K.C., Zheng, B., Lee, W.C., Lee, D.L., Wang, X.: Ir-tree: An efficient index for geographic document search. IEEE Transactions on Knowledge and Data Engineering **23**(4), 585–599 (2011)
8. Li, W., Guan, J., Zhou, S.: Efficiently evaluating range-constrained spatial keyword query on road networks. In: Han, W.-S., Lee, M.L., Muliantara, A., Sanjaya, N.A., Thalheim, B., Zhou, S. (eds.) DASFAA 2014. LNCS, vol. 8505, pp. 283–295. Springer, Heidelberg (2014)
9. De Felipe, I., Hristidis, V., Rishe, N.: Keyword search on spatial databases. In: IEEE 24th International Conference on Data Engineering, ICDE 2008, 656–665. IEEE (2008)
10. Rocha-Junior, J.B., Gkorgkas, O., Jonassen, S., Nørvåg, K.: Efficient processing of top-k spatial keyword queries. In: Pfoser, D., Tao, Y., Mouratidis, K., Nascimento, M.A., Mokbel, M., Shekhar, S., Huang, Y. (eds.) SSTD 2011. LNCS, vol. 6849, pp. 205–222. Springer, Heidelberg (2011)

Handling Query Skew in Large Indexes:
A View Based Approach

Weihuang Huang(✉), Jeffrey Xu Yu, and Zechao Shang

The Chinese University of Hong Kong, Hong Kong, China
{whhuang,yu,zcshang}@se.cuhk.edu.hk

Abstract. Indexing is one of the most important techniques to facilitate query processing over a multi-dimensional dataset. A commonly used strategy for such indexing is to keep the tree-structured index balanced. This strategy implies that all queries are uniformly issued, which is partially because the query distribution is not possibly known and will change over time in practice. A key issue we study in this work is whether it is the best to fully rely on a balanced tree-structured index in particular when datasets become larger and larger. This means that, when a dataset becomes very large, it becomes unreasonable to assume that all data in any subspace are equally important and are uniformly accessed by all queries at the index level. Given the existence of query skew, in this paper, we study how to handle such query skew at the index level without sacrifice of supporting any possible queries in a well-balanced tree index and without a high overhead. To tackle the issue, we propose index-view at the index level, where an index-view is a short-cut in a balanced tree-structured index to access objects in the subspace that are more frequently accessed, and propose a new index-view-centric framework for query processing using index-views in a bottom-up manner. We study index-views selection problem, and we confirm the effectiveness of our approach using large real and synthetic datasets.

1 Introduction

A large number of real applications need to deal with a multi-dimensional dataset which is a set of points in a d-dimensional space for $d > 1$. As a multi-dimensional dataset, spatial data is extensively studied for geographical positioning and wireless communication technologies. LBS (Location Based Service) is a fundamental application based on spatial data, where an object can be considered as a POI (Point Of Interest), which is composed of the location of the instance, such as restaurant checkin, and some detailed information. LBS enables users to find interesting results according to their location.

Indexing is one of the most important techniques to facilitate query processing over a multi-dimensional dataset. Many index techniques such as R-TREE [9], QUADTREE [7], and KD-TREE [3] are studied in the literature. In [16], Samet gives a comprehensive introduction and thorough analysis of numerous widely used techniques for multi-dimensional data processing. A commonly used strategy for such indexes is to keep the tree-structured index balanced. This strategy

© Springer International Publishing Switzerland 2015
M.A. Sharaf et al. (Eds.): ADC 2015, LNCS 9093, pp. 180–193, 2015.
DOI: 10.1007/978-3-319-19548-3_15

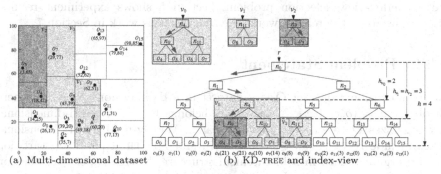

(a) Multi-dimensional dataset   (b) KD-TREE and index-view

Fig. 1. Multi-dimensional dataset & Its index

reduces query processing cost in the worst case, and can handle all different queries equally well. In other words, this strategy implies that all queries are uniformly issued, which is partially because the query distribution is not possibly known and will change over time in practice [6] [9].

A key issue we study in this work is whether it is the best to fully rely on a balanced tree-structured index in particular when datasets become larger and larger. This means that it becomes unreasonable to assume that all data in any subspace are equally important and are uniformly accessed by all queries at the index level. It is important to note that Pareto principle or 80-20 rule is well known such that a small number of data will be frequently accessed. In this paper, we study how to handle such query skew at the index level. The support of such query skew at the index level should not offer sacrifice of supporting any possible queries in a well-balanced tree index and should not come with a high overhead. The main contribution of this work is summarized:

- To tackle the issue, we propose *index-view* at the index level, where an index-view is a short-cut in a balanced tree-structured index to access objects in subspaces that are more frequently accessed. This strategy is effective when a tree-structured index is built in memory over a large dataset, such as [11] [2].
- We give a class of queries that can be proposed using index-view, and show that almost all existing queries can be proposed using index-view. In addition, we propose an index-view-centric framework for query processing using index-views, which is different from the widely used filter-and-refinement framework adopted for querying over an index. There is no need to access the global scope.
- We study how to select top-K index-views over a tree-structured index as size constraint index-view selection. We propose an algorithm for selecting top-K index-views from a set of candidate index-views based on a given query distribution at a specific time.
- We conduct extensive experimental studies based on 4 real dataset and synthetic datasets to confirm our finding.

The paper is organized as follows. Section 2 gives the problem statement. Section 3 discusses index-view based query processing. Section 4 proves and discusses index-view selection problem. Section 5 shows experiment results. Section 6 discusses the related work. We conclude this work in Section 7.

2   The Problem Statement

A multi-dimensional dataset O is a dataset in a d-dimension space $\mathbb{R}^d$, where each object $o = (o.x_1, o.x_2, ..., o.x_d) \in O$ belongs to $\mathbb{R}^d$. A multi-dimensional index organizes O to support efficient search. Widely-used multi-dimensional indexes include R-TREE [9], QUADTREE [7], and KD-TREE [3]. They partition data and organize data in a hierarchical structure. In this paper, we focus on KD-TREE. Our approach can be easily extended to support other tree structured indexes where applicable.

KD-TREE is a binary search tree. It recursively divides a multi-dimensional space into two parts with a hyperplane perpendicular to one of the d-dimensions. The dimension (x_i) used to partition is called a *discriminator*. A node n in KD-TREE with value v_i is called x_i-discriminator, where all nodes with a smaller x_i value than v_i are in the left child of n and all nodes with a value equal to or greater than the x_i value, i.e. v_i, are in the right child. It cycles through d-dimension to choose the discriminator to partition. For instance, the discriminator of nodes in a level h is $(h \bmod d)$, where mod gets the remainder after $h \div d$. In the original version of KD-TREE [3], v_i is chosen as $o.x_i$, where $o \in O$. But this leads to difficulty for deletion. The partition method proposed in [8] partitions data in a way that each partition does not occur on data itself, which we will use in this work. We denote the root node of a KD-TREE by r. Fig. 1 shows a dataset O containing 16 spatial objects in a KD-TREE.

It is well known that the indexes built in a hierarchical structure are usually balanced [6], because they are supposed to support any possible queries efficiently without the knowledge on how queries are issued. Queries are assumed uniformly distributed. However, in practice, queries are usually not uniformly issued. There are more queries in certain subspaces in the multi-dimensional space and a few queries in other places. In other words, a query distribution can be highly skewed. A key issue is whether it is possible to support skewed query distributions at the index level.

In this paper, we assume that we know the query distribution. Here, a query distribution can be denoted by L which is composed of $|O|$ query records. Each query record e is in the format of (o, t_o), which indicates that o is accessed up to t_o times. Fig. 1 shows a skewed query distribution for the objects in the KD-TREE. Here, o_i at the bottom of the tree indicates the object o_i in the multi-dimensional space, and t_{o_i} is shown in $(\cdot)$. L contains all objects. For example, in L, o_4 and o_5 are most frequently queried objects with $t_{o_4} = t_{o_5} = 21$, whereas o_2 and o_{12} have never been visited.

To support such query skew at the index level, we have to consider two cases. One case is when queries are skewed in certain subspaces. The other case is when queries are issued in any subspace. Obviously, we need to support both cases well. This requires us to adapt a well-balanced index to support any possible queries, and in addition, to support queries when they are skewed on top of the well-balanced index.

Problem Statement: The problem to be studied in this paper is how to support queries, that are highly skewed, at the index level, without sacrifice of the ability of supporting any possible queries in a well-balanced tree index and without high overhead.

3   Index-View Based Query Processing

To address the issue, in this paper, we propose index-view. An index-view v of a tree-structured index $\mathcal{I}$ is a subtree of $\mathcal{I}$ rooted at n_v. We denote the subspace represented by v as s_v which covers object o if $o \in s_v$. In Fig. 1(b), it shows 3 index-views v_0, v_1, and v_2, rooted at n_4, n_{11}, and n_9 respectively. We denote an index-view set by $\mathcal{V}$ over the index $\mathcal{I}$. In the example, $\mathcal{V} = \{v_0, v_1, v_2\}$.

The concept of index-view looks similar to the concept of view defined over relations using SQL, but is different. First, we deal with views at the index level. Second, the index-view set is not defined by users but automatically constructed by the query-distribution. We will discuss how to select index-views in later section.

In this section, we focus on how to conduct query processing for given $\mathcal{V}$. It is worth noting that there are many different kinds of queries proposed. For multi-dimensional data, typical queries include point query, range query and kNN query [16] and their approximate variants [2]. For spatial data, with the proliferation of LBS, a large number of innovative types of query arise, such as: aggregate query [14], top-k spatial keyword query [5], direction-aware query [12]. Intuitively, an index-view v can be used to answer a query q if q can be answered by s_v, the subspace covered by v. However, it is not always the case. Consider a kNN query $q = (62, 20)$ located in the 2-dimensional space in Fig. 1(a). Its top-2 nearest neighbors are o_8 and o_{11}. This kNN query q cannot be answered by the index-view v_1 even though it is located in the subspace covered by v_1, because o_{11} is not covered by v_1. Here, the questions are what kind of queries can be supported by index-view and how such queries can be processed given index-view.

In the following, we define a category of queries that can be answered by index-view, and propose a new index-view-centric framework to process the category of queries using index-view. The two things are interrelated. It is important to mention that most of these queries cannot be answered by index-view directly, because they are supposed to be processed in the commonly-used filter-and-refinement framework in a top-down manner. The filter-and-refinement framework starts from the dataset which is represented by the root of tree-structured index, and filters subspaces which cannot contain any result by narrowing down in the index. The search space decreases until certain conditions are satisfied. The query execution is performed in the search space, which is small. Different from the filter-and-refinement framework, we propose the index-view-centric framework, which is to start from an index-view instead of the root of the $\mathcal{I}$, and expand the index-view to the search space needed to find the answers.

3.1   Index-View Friendly Query

We discuss the category of queries that can be answered by index-view in our index-view-centric framework which we call index-view friendly query.

Definition 1. *(Index-view Friendly Query) Let the result of a query q over a dataset O be denoted as $\mathcal{S} = \Pi(q, O)$, where $\mathcal{S} \subset O$. q is index-view friendly if it satisfies following conditions:*

1. $|\mathcal{S}| \ll |O|$, where $|\cdot|$ is the size of set.
2. q is decomposable [17], i.e. if O_1 and O_2 form a partition of O $(O_1 \cup O_2 = O$ and $O_1 \cap O_2 = \emptyset)$, $\mathcal{S}_1 = \Pi(q, O_1)$ and $\mathcal{S}_2 = \Pi(q, O_2)$, $\mathcal{S}$ can be computed from $\mathcal{S}_1$ and $\mathcal{S}_2$ efficiently.
3. $d(\cdot, \cdot)$ is the distance metric defined on O, such that: for $o, o_i, o_j \in O$, $d(o_i, o_j) \geq 0$, $d(o_i, o_j) = 0$ iff $o_i = o_j$, $d(o_i, o_j) = d(o_j, o_i)$ and $d(o_i, o_j) \leq d(o, o_i) + d(o, o_j)$.
4. Let $\mathbb{T} = \{T : T \subset O, T$ satisfies the conditions specified in q except distance related ones$\}$. $\mathcal{S} \in \mathbb{T}$ and $\forall T \in \mathbb{T}, d(q, \mathcal{S}) \leq d(q, T)$, where $d(q, \mathcal{S})$ and $d(q, T)$ depend on $d(\cdot, \cdot)$ in 3).

Here, the first condition requires the size of result set to be relatively small. Otherwise, it is inevitable to browse all objects in O and the advantage of index-view cannot be taken. The second condition requires the query to be decomposable. In other words, if we partition the dataset into two parts and perform the query on each part respectively, the final result can be derived from the two result sets without high overhead. It allows us to start from an index-view and expand it to its nearby subspaces to find the exact answers. The third condition defines a metric space. The last one requires q to be distance-sensitive, which means the final result is not far away from the position of query, and if there are two subsets satisfying the query constraint, the nearer one is better.

All conditions in Definition 1 are reasonable for query execution. And it is worth mentioning that all queries proposed over multi-dimensional datasets and spatial datasets listed above do satisfy the four conditions, and are index-view friendly queries. Due to the space limit, we will show how kNN is index-view friendly, as an example.

kNN is a query that finds the top-k nearest neighbors from a given point over a multi-dimensional space. Formally, given $q \in \mathbb{R}^d$ and $k \in \mathbb{N}^+$, return $\mathcal{S} \subset O$ such that $|\mathcal{S}| = k$, and $\forall o_i \in \mathcal{S}, \forall o_j \in O - \mathcal{S}, d(q, o_i) \leq d(q, o_j)$. $d(q, o_i), d(q, o_j)$ are distances between q and o_i, o_j, respectively. We show it is index-view friendly in Lemma 1.

Lemma 1. *kNN is index-view friendly query.*

Proof Sketch: It is obvious that it satisfies the first and the third condition. We show it is decomposable. Suppose O_1 and O_2 are two partitions as defined in Definition 1, and $\mathcal{S}_1 = \Pi(q, O_1), \mathcal{S}_2 = \Pi(q, O_2)$. Then, we show $\mathcal{S} = \Pi(q, \mathcal{S}_1 \cup \mathcal{S}_2)$ is the final result. From the definition, $\forall o_i \in \mathcal{S}_1, o_j \in O_1 - \mathcal{S}_1, d(q, o_i) \leq d(q, o_j)$. And, we have $\forall o \in \mathcal{S}, o_i \in \mathcal{S}_1 - \mathcal{S}, d(q, o) \leq d(q, o_i)$. Hence, $\forall o \in \mathcal{S}, o_j \in O_1 - \mathcal{S}, d(q, o) \leq d(q, o_j)$. In the same way, $\forall o \in \mathcal{S}, o_j \in O_2 - \mathcal{S}, d(q, o) \leq d(q, o_j)$.

Algorithm 1. index-view-centric-framework(v, q)

Input: v: an index-view defined on $\mathcal{I}$
      q: an index-view friendly query
Output: $\mathcal{S}$: result set
1 **begin**
2      $n = n_v$;
3      $\mathcal{S} = \Pi(q, n)$;
4   Initialize stop condition;
5   **while** *stop condition is not satisfied* **do**
6          $n' = $ sibling of n;
7          $\mathcal{S}' = \Pi(q, n')$;
8          $\mathcal{S} = $ merge $\mathcal{S}$ and $\mathcal{S}'$;
9   Update stop condition if necessary,
10         $n = n$.parent;
11   **end**
12 **end**

Since $O = O_1 \cup O_2$, $\mathcal{S}$ is the result set for q and can be derived from $\mathcal{S}_1$ and $\mathcal{S}_2$ directly. Next, we show the fourth condition holds. This is because for a kNN query, all subsets of O with size k are candidate results, and $\mathcal{S}$ is the nearest among all of them. $\qquad\qquad\qquad\qquad\qquad\qquad\qquad\qquad\qquad\qquad\qquad\qquad$ $\square$

3.2 The Index-View-Centric Framework

The index-view-centric framework we propose is illustrated in Algorithm 1. In the index-view-centric framework, given an index-view v and an index-view friendly query, it first performs the query in the data covered by v (line 3), which is the subtree of n_v. Then, it expands the search space in a bottom-up fashion. If the stop condition is not satisfied, it will conduct search at the parent of the current node n. For the reason $\mathcal{S}$ already contains the result from n, it is sufficient to query the sibling of n (line 7) and merge the two result sets (line 8). According to the second condition of index-view friendly query, it will be implemented efficiently. Here, in our algorithm, we take the same way to process a query at a node (e.g. n and n') using an existing filter-and-refinement based algorithm. We update the stop condition at every iteration (line 9).

Based on condition 4, the answer will not be far from the query position. Then, it is unlikely that it will backtrack from a node to the root given the fact that a query is near to an index-view. We confirm it in our experiment studies.

4 Index-View Selection

In this section, we will formulate the benefit of index-view and index-view selection problem in Section 4.1. We discuss how to select K index-views in Section 4.2.

4.1 Size Constraint Index-View Selection Problem

The height of index is of great importance for query processing. We define the *benefit* of a query q when accessing o with an index-view v as follows.

$$B_o(v) = \begin{cases} h_v & o \text{ is } \textit{covered} \text{ by } v \\ 0 & \text{otherwise} \end{cases} \tag{1}$$

where h_v is the height of the node, where the index-view v is, in the tree-structured index $\mathcal{I}$. Based on Eq. (1), given query distribution L, the benefit of an index-view v over an index $\mathcal{I}$ can be defined as follows.

$$B_L(v) = \sum_{o \in O} B_o(v) \cdot t_o = \sum_{o \in s_v} h_v t_o = h_v \sum_{o \in s_v} t_o = h_v t_v \tag{2}$$

where $t_v = \sum_{o \in s_v} t_o$. The benefit of v depends on h_v and the frequency of accessing objects covered by v.

It is obvious that an index-view v can cover more objects if the position of the index-view, n_v, is closer to the root, or in other words, the height of n_v is smaller. On the other hand, the more objects an index-view can cover, the less benefit it brings to every object in the subspace covered by the v.

As shown in Fig. 1(b), there are 3 index-views, $\mathcal{V} = \{v_0, v_1, v_2\}$. The index-view v_0 is represented by n_4 with height 2 in the KD-TREE, that covers $o_4 \sim o_7$. The benefit of v_0 is $B_L(v_0) = (21 + 21 + 10 + 14) \times 2 = 132$. Here, when it needs to access o_4 and/or o_5, using the index-view v_2 is more beneficial than using the index-view v_0, even though the query can be answered by both v_0 and v_2, because $B_o(v_2) > B_o(v_0)$ for o_4 and/or o_5. It implies we need to use the index-view with a larger height value when an object o can be covered by more than one index-view in $\mathcal{V}$. In general, if two index-views can access the same object, then the two index-views overlap. Recall that there is a nice property between two index-views in the tree-structured index if they overlap, which we show in Property 1.

Property 1. (Overlap of index-views) If two different index-views v_i and v_j defined on index $\mathcal{I}$ overlap, then either v_i is a subtree of v_j ($v_i \subset v_j$) or v_j is a subtree of v_i ($v_j \subset v_i$).

According to Property 1, if two index-views, v_i and v_j overlap, one must be a subtree of the other. Obviously, if $v_i \subset v_j$, then $B_o(v_i) > B_o(v_j)$ for all o in the overlapped region.

In general, every node in the tree-structured index $\mathcal{I}$ can be considered as an index-view. In this paper, we discuss selecting top-K index-views, that is a set of index-views of size K, denoted as $\mathcal{V}^K$ from a set of candidate index-views $\mathcal{V}$. In order to select top-K index-views, the benefit of accessing an object o using $\mathcal{V}$ is defined in Eq. (3). If it is impossible to find an index-view in $\mathcal{V}$ that covers o, $B_o(\mathcal{V})$ equals to 0. Otherwise, $B_o(\mathcal{V})$ is the height of the nearest index-view to o in $\mathcal{V}$.

$$B_o(\mathcal{V}) = \max_{v \in \mathcal{V}} B_o(v) \tag{3}$$

And the benefit of $\mathcal{V}$ to access objects in O for a given query distribution L is as follows.

$$B_L(\mathcal{V}) = \sum_{o \in O} B_o(\mathcal{V}) t_o \tag{4}$$

Here, the top-K index-view selection problem is a size-constraint problem. Formally, let $\mathcal{V}$ be a set of index-views, and let $\mathbb{V}_K$ denote all possible combinations of $\mathcal{V}^K$ (a subset of index-views of size K) over $\mathcal{V}$. The index-view selection is to find the max benefit $\mathcal{V}^K$, denoted as $\mathcal{V}_{\max}^K$ in $\mathbb{V}_K$, such that $B_L(\mathcal{V}_{\max}^K) \geq B_L(\mathcal{V}^K)$, for any $\mathcal{V}^K \in \mathbb{V}_K$.

It is worth noting that the size of $\mathbb{V}_K$ is large. Take KD-TREE as an example, assuming $|O| = 2^m$. $|\mathbb{V}_K| = \binom{N}{K} = \frac{N!}{K!(N-K)!}$, where $N = 2^{m+1} - 1$. In real application, N can be a large number and there are numerous $\mathcal{V}^K$ to be considered.

It is impossible to enumerate all $\mathcal{V}^K$ in $\mathbb{V}_K$, in order to compute $\mathcal{V}_{\max}^K$. In addition, a naive greedy algorithm does not work well if it tries to pick up the index-views with top-K largest benefit. The coverage of such $\mathcal{V}_{\max}^K$ can be not optimal because it is entirely possible that the index-views selected in $\mathcal{V}_{\max}^K$ overlap. When overlapping is considered in selecting index-views, the benefit of $\mathcal{V}_{\max}^K$ is not the sum of the benefits of all index-views in it. An index-view v_j already selected in $\mathcal{V}_{\max}^K$ becomes less beneficial if a new index-view v_i is selected into $\mathcal{V}_{\max}^K$ which is an ancestor of v_j. This is because v_i has already contributed a fraction of benefit to objects covered by v_j.

Table 1. Relationship

relationship	ΔB_1	ΔB_2
$h_{v_1} \geq h_{v_2} \geq h_{v_{\max}}$	0	$h_{v_2} - h_{v_{\max}}$
$h_{v_1} \geq h_{v_{\max}} \geq h_{v_2}$	0	0
$h_{v_{\max}} \geq h_{v_1} \geq h_{v_2}$	0	0
$h_{v_{\max}} \geq h_{v_2} \geq h_{v_1}$	0	0
$h_{v_2} \geq h_{v_{\max}} \geq h_{v_1}$	$h_{v_2} - h_{v_{\max}}$	$h_{v_2} - h_{v_{\max}}$
$h_{v_2} \geq h_{v_1} \geq h_{v_{\max}}$	$h_{v_2} - h_{v_1}$	$h_{v_2} - h_{v_{\max}}$

4.2 Index-View Selection

In this section, we prove our index-view selection problem to be submodular and give a greedy algorithm to select $\mathcal{V}_{\max}^K$. Due to Eq. (4), $B_L(\mathcal{V}) = \sum_{o \in O} B_o(\mathcal{V}) t_o$, we focus on $B_o(\mathcal{V})$ first that is to access a specific object o using $\mathcal{V}$. According to Property 1, the index-views that can be beneficial to access objects o are all along the path from the root r of the index $\mathcal{I}$ to the node where o is maintained.

Suppose that there are two index-views v_i, v_j in $\mathcal{V}$ to access the object o, where v_j is the descendant of v_i. Let h_{v_i}, h_{v_j} be the height of the nodes that represent v_i, v_j, then $h_{v_j} > h_{v_i}$. Let $h_{v_{\max}}$ denote the index-view with the largest height in $\mathcal{V}$. Here, $h_{v_{\max}}$ is h_{v_j}. Next, consider how to update $B_o(\mathcal{V})$ when there are new index-views to be considered. There are many possible index-views, for instance, v_0, v_1, v_2. Without loss of generality, suppose $h_{v_{v_0}} < h_{v_i} < h_{v_{v_1}} < h_{v_j} < h_{v_{v_2}}$. $B_o(\mathcal{V})$ changes if the height of the node, h_v, for the index-view v is larger than $h_{v_{\max}}$. It means v_2 does change the benefit of $B_o(\mathcal{V})$ but v_0 and v_1 do not. Here, $B_o(\mathcal{V} \cup \{v_1\}) = B_o(\mathcal{V} \cup \{v_1, v_0\}) = h_{v_{\max}}$, $B_o(\mathcal{V} \cup \{v_2\}) = B_o(\mathcal{V} \cup \{v_2, v_1\}) = B_o(\mathcal{V} \cup \{v_2, v_1, v_0\}) = h_{v_2}$. We have Lemma 2 based on such observations.

Lemma 2. $B_o(\mathcal{V})$ *is monotonic and submodular, i.e.*

1. $\forall v \notin \mathcal{V}, B_o(\mathcal{V}) \leq B_o(\mathcal{V} \cup \{v\})$;
2. $\forall v_1, v_2 \notin \mathcal{V}, B_o(\mathcal{V} \cup \{v_1\}) + B_o(\mathcal{V} \cup \{v_2\}) \geq B_o(\mathcal{V} \cup \{v_1, v_2\}) + B_o(\mathcal{V})$.

Proof Sketch: For a given object o, $\mathcal{V}_o = \{v \in \mathcal{V} : v \text{ covers } o\}$. The largest element in $\mathcal{V}_o$ is $v_{\max}$ such that $\forall v_i \in \mathcal{V}_o, h_{v_{\max}} \geq h_{v_i}$. Note that $h_{v_{\max}} = 0$ if $\mathcal{V}_o = \emptyset$, and $B_o(\{v\}) = 0$ if v does not cover o. To prove the first statement 1), according to Eq. (3), we have

$$B_o(\mathcal{V} \cup \{v\}) = \begin{cases} \max\{h_{v_{\max}}, h_v\} & o \text{ is } covered \text{ by } v \\ B_o(\mathcal{V}) & \text{otherwise} \end{cases}$$

Hence, $B_o(\mathcal{V}) \leq B_o(\mathcal{V} \cup \{v\})$. To prove the second statement 2), let $\Delta B_1 = B_o(\mathcal{V} \cup \{v_1, v_2\}) - B_o(\mathcal{V} \cup \{v_1\})$, $\Delta B_2 = B_o(\mathcal{V} \cup \{v_2\}) - B_o(\mathcal{V})$. The relationship between h_{v_1}, h_{v_2} and $h_{v_{\max}}$ can be categorized into 6 types. Table 1 illustrates how these relationships affect ΔB_1 and ΔB_2. Obviously, $\Delta B_2 \geq \Delta B_1$, then $B_o(\mathcal{V} \cup \{v_1\}) + B_o(\mathcal{V} \cup \{v_2\}) \geq B_o(\mathcal{V} \cup \{v_1, v_2\}) + B_o(\mathcal{V})$. □

According to the definition of $B_L(\mathcal{V})$, it is non-negative linear combination of $B_o(\mathcal{V})$. Based on Lemma 2, we have Corollary 1.

Corollary 1. $B_L(\mathcal{V})$ *is monotonic and submodular.*

For a submodular problem, [13] defines an efficient greedy strategy. The basic idea for this strategy is selecting index-view v that will increase the benefit of $\mathcal{V}$ most and add v into $\mathcal{V}$. We denote the marginal benefit ΔB of an index-view v to be $B_L(\mathcal{V} \cup \{v\}) - B_L(v)$. It is hard to use this greedy algorithm directly, for the reason marginal benefit of each index-view v is changing as $\mathcal{V}$ is changing. When an index-view v is added into $\mathcal{V}$, two categories of index-views' marginal benefit will be affected. The first part is those defined on ancestors of n_v. For every view v_i defined on the path, $h_{v_i} < h_v$, for all objects covered by v, v_i will not be considered any more. On the other hand, for objects covered by v_i but not v, v_i is still useful. These ancestor index-views will not be too many, therefore we can afford to recalculate the benefit for them. The second category of index-views will be affected by the addition of v is every index-view v_i defined on the subtree of n_v. One thing need to be noticed is the number of descendant nodes in the subtree could be very large, but many of them will never be used, because the skew of query distribution. As a result, we can adopt a lazy policy: each time we are considering an index-view with the largest marginal benefit, we validate its benefit to check whether this value is real. The validation is implemented by inspecting if any of its ancestors is in $\mathcal{V}$. If so, we re-calculate its marginal benefit.

Algorithm 2. index-view-selection(r, K)

Input: r: root of search index $\mathcal{I}$

K: the number of expected index-views

Output: $\mathcal{V}_{\max}^K$: maximum index-view set

1 **begin**
2 $\quad$ $\mathcal{V}_{\max}^K = \emptyset$;
3 $\quad$ Initialize an empty max-heap $\mathcal{H}$;
4 $\quad$ **for** *Every node n in $\mathcal{I}$* **do**
5 $\quad\quad$ $\mid$ $\mathcal{H}$.insert(v) with $B_L(\{v\})$;
6 $\quad$ **end**
7 $\quad$ **while** *($|\mathcal{V}_{\max}^K| < K$) and ($\mathcal{H}$ is not empty)* **do**
8 $\quad\quad$ $v = \mathcal{H}$.extract_max();
9 $\quad\quad$ **if** *v is false positive* **then**
10 $\quad\quad\quad$ $\mid$ $\Delta B = B_L(\mathcal{V}_{\max}^K \cup \{v\}) - B_L(\mathcal{V}_{\max}^K)$;
11 $\quad\quad\quad$ $\mid$ $\mathcal{H}$.insert(v) with ΔB;
12 $\quad\quad$ **end**
13 $\quad\quad$ **else**
14 $\quad\quad\quad$ $\mid$ $\mathcal{V}_{\max}^K = \mathcal{V}_{\max}^K \cup \{v\}$;
15 $\quad\quad\quad$ $\mid$ **for** *every node n as an ancestor of n_v* **do**
16 $\quad\quad\quad\quad$ $\mid$ modify $B_L(v)$ of v defined on n in $\mathcal{H}$;
17 $\quad\quad\quad\quad$ $\mid$ adjust $\mathcal{H}$;
18 $\quad\quad\quad$ **end**
19 $\quad\quad$ **end**
20 $\quad$ **end**
21 **end**

Based on Corollary 1, we design an algorithm (Algorithm 2) to select $\mathcal{V}_{\max}^K$ out of $\mathcal{V}$. We use a max-heap $\mathcal{H}$ to keep track of ΔB, the marginal benefit for every possible v. Since every node can be regarded as a candidate index-view, when $\mathcal{V} = \emptyset$, we calculate $B_L(v)$ for every v defined on $n \in \mathcal{I}$ and insert it into $\mathcal{H}$ (line 5). Every time, we pick an index-view v with maximum ΔB and add v into $\mathcal{V}$, until the heap is empty or there are K index-views in $\mathcal{V}$. For a max-heap, the one at the top is the largest (line 8). In line 16, when v is added to $\mathcal{V}_{\max}^K$, we modify the values of nodes in path from n_v to r in $\mathcal{H}$. Line 10 recalculates ΔB for index-views failed validation.

Lemma 3. *index-view selection can be solved with approximate factor $1 - \frac{1}{e}$.*

Proof Sketch: We design Algorithm 2 following the greedy strategy used in [13] which proves the approximate ratio. Even though there are might be false positive in algorithm 2, those index-views will be updated and re-inserted into $\mathcal{H}$. And the property of $\mathcal{H}$ guarantees the first v that is not false positive will be the right choice. Hence, the approximate ratio is the same to [13], $1 - \frac{1}{e}$. $\quad\square$

5 Experiment

We have conducted extensive experiments on a PC with two Intel Xeon X5550@ 2.67GHz CPU and 48GB main memory. The algorithms are implemented in

C++. All experiments are repeated 10 times and average values are reported. We implement our proposed techniques with an in-memory KD-TREE [3].

We use 4 large datasets to test our algorithms. Table 2 shows the information about the 4 real datasets. In the fifth column, "trajectory" means the data is GPS locations sampled from taxi trajectories, and "checkin" means the data is checkin positions that we crawled from location based social network. To test the effect of different datasets and query sets, we also generate several synthetic datasets. Since the base index $\mathcal{I}$ is of little importance, we randomly generate datasets and all of them follow uniform distribution. For query set, we adopt the famous Zipf distribution.

Table 2. Dataset

Dataset	#obj	#query	#qu/#obj	Type
Boston [4]	3,685,923	4,186,461	1.136	trajectory
T-drive [19]	11,317,142	17,762,489	1.570	trajectory
Gowalla	1,723,545	16,592,696	9.627	checkin
Foursquare	2,358,283	5,088,817	2.157	checkin

The Selection of $\mathcal{V}_{max}^K$: According to Eq. (3), we compute the benefits for each dataset (divided by the total cost) by varying the number of K. The $\mathcal{V}^K$ are selected by Algorithm 2. Fig. 2(a) shows how benefit (normalized by the total cost) varies according to K using 4 real datasets. It demonstrates the monotonic property, i.e. benefit increases with K. In Fig. 2(a), it shows $B_L(\mathcal{V}^K)$ increases very fast for the first 100 index-views, and afterwards the rate of growth becomes smaller. The first 100 index-views covers the most skewed parts. We compare Algorithm 2 with building the index from scratch. Fig. 2(b) shows the difference. Even for the largest dataset T-drive, Algorithm 2 can be completed within 1 milliseconds, which can be neglected.

Query Processing for a Given $\mathcal{V}_{max}^K$: We conduct experiment on kNN queries. Here, we use an index-view v if and only if query $q \in s_v$. The algorithms used in the filter-and-refinement framework for kNN query is best-first search (BFS) proposed in [10]. We test it using 6 different K values: 16, 32, 64, 128, 256 and 512. The query processing time without index-view is denoted as $K = 0$. Fig. 2(c) and Fig. 2(d) are the average execution time for kNN query. In general, the query processing time reduces with more index-views that cover larger number of queries. However, when K becomes too large, the cost for selecting proper index-view becomes larger and the query execution time can possibly increase.

The Effect of k: We further show the value of k of kNN that has great influence on search space. To prove the adaptability of index-view based framework, We vary k with value 2, 5, 10, 20 and test on Foursquare and Gowalla. Fig. 2(e) and 2(f) show the results. A larger k means a wider search space with higher overhead. It is the reason why both curves increase when k gets larger. It is important to note that index-view-centric will not bring extra overhead when queries become more complex.

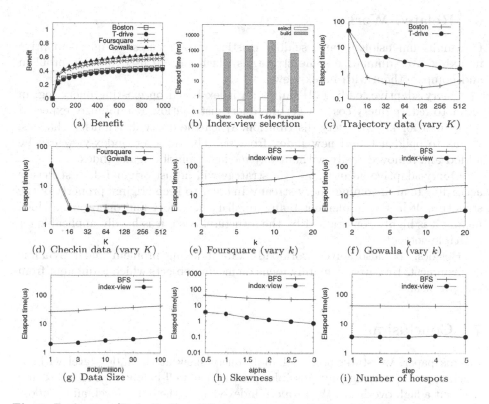

Fig. 2. Evaluation (Default: $K = 128$, $k = 10$, $\alpha = 1$, #obj = 10 million, #query = 1 million)

The Scalability: We randomly generate 5 datasets with uniform distribution. The sizes are, 1, 3, 10, 30, and 100 million.

Fig. 2(g) shows the results for normal best-first kNN search (BFS), and Algorithm 1 (index-view). The latter one outperforms the former one. BFS takes more than 10 times of execution time than index-view based algorithms. Another observation is the 3 lines are almost parallel, which demonstrates that the size of dataset does not has such influence as expected.

The Effect of Query Distribution: In Zipf distribution, the skewness is decided by the parameter α. The larger α is, the more skewed the distribution is. In addition to Zipf distribution, the number of selected hotspots also has impacts on query processing using $\mathcal{V}_{max}^{K}$. We test both using 1 million kNN queries. To generate different number of hotspots, we use random walk with different steps. A large step number leads to scattered hotspots. The results are shown in Fig. 2(h) and Fig. 2(i). When query distribution becomes more skewed while α becomes larger, our approaches can reduce the query processing time significantly. The number of hotspots does not have impacts on $\mathcal{V}_{max}^{K}$.

6   Related Works

For multi-dimensional data, seldom existing work focuses on query-adaptive indexing. We summarize some strategies which can be used for this problem and compare them with index-view.

Query-adaptive loading [1] builds an index based on existing query log, in order to reduce query cost for certain query distribution. It minimizes expected query cost for known query distribution. When the query distribution changes, it has to build a brand new index from scratch. However, index-view can be added and removed easily without any modification of original index.

Query-adaptive adjustment [15] strategy will adjust original data structure according to an incoming query stream. In order to keep original property of the structure, it has to implement many auxiliary accessory and operations, which means it is highly constrained by the structure. Also, it is hardly applicable for multiple hotspots.

Besides, [18] also solves workload-aware indexing problem under dynamic environment. However, it mainly focus on moving objects which is different from our problem.

7   Conclusion

In this paper, we study how to handle query skew at the index level without sacrifice of supporting any possible queries in a well-balanced tree index and without a high overhead. We propose index-view at the index level, and propose a new index-view-centric framework for query processing using index-views in a bottom-up manner. We study index-view selection problem, and we show the quality of the index-views selected is high and the overhead for selecting the index-views is very small which can be ignored. We confirm the effectiveness of our approach using large real and synthetic datasets.

Acknowledgments. This work was supported by grant of the Research Grants Council of the Hong Kong SAR, China 14209314.

References

1. Achakeev, D., Seeger, B., Widmayer, P.: Sort-based query-adaptive loading of r-trees. In: Proc. of CIKM 2012 (2012)
2. Arya, S., Mount, D.M., Netanyahu, N.S., Silverman, R., Wu, A.Y.: An optimal algorithm for approximate nearest neighbor searching. In: Proc. of SODA 1994 (1994)
3. Bentley, J.L.: Multidimensional binary search trees used for associative searching. Commun. ACM **18**(9) (1975)
4. Cudré-Mauroux, P., Wu, E., Madden, S.: Trajstore: an adaptive storage system for very large trajectory data sets. In: Proc. of ICDE 2010 (2010)
5. Felipe, I.D., Hristidis, V., Rishe, N.: Keyword search on spatial databases. In: Proc. of ICDE 2008 (2008)

6. Filho, Y.V.S.: Average case analysis of region search in balanced k-d trees. Inf. Process. Lett. **8**(5) (1979)
7. Finkel, R.A., Bentley, J.L.: Quad trees: A data structure for retrieval on composite keys. Acta Inf. **4** (1974)
8. Friedman, J.H., Bentley, J.L., Finkel, R.A.: An algorithm for finding best matches in logarithmic expected time. ACM Trans. Math. Softw. **3**(3) (1977)
9. Guttman, A.: R-trees: a dynamic index structure for spatial searching. In: Proc. of SIGMOD 1984 (1984)
10. Hjaltason, G.R., Samet, H.: Distance browsing in spatial databases. ACM Trans. Database Syst. **24**(2) (1999)
11. Levandoski, J.J., Sarwat, M., Eldawy, A., Mokbel, M.F.: Lars: a location-aware recommender system. In: Proc. of ICDE 2012 (2012)
12. Li, G., Feng, J., Xu, J.: Desks: direction-aware spatial keyword search. In: Proc. of ICDE 2012 (2012)
13. Nemhauser, G.L., Wolsey, L.A., Fisher, M.L.: An analysis of approximations for maximizing submodular set functionsi. Mathematical Programming **14**(1) (1978)
14. Papadias, D., Shen, Q., Tao, Y., Mouratidis, K.: Group nearest neighbor queries. In: Proc. of ICDE 2004 (2004)
15. Park, E., Mount, D.M.: A self-adjusting data structure for multidimensional point sets. In: Epstein, L., Ferragina, P. (eds.) ESA 2012. LNCS, vol. 7501, pp. 778–789. Springer, Heidelberg (2012)
16. Samet, H.: Foundations of multidimensional and metric data structures. Morgan Kaufmann (2006)
17. Sheng, C., Tao, Y.: Fifo indexes for decomposable problems. In: Proc. of PODS 2011 (2011)
18. Tzoumas, K., Yiu, M.L., Jensen, C.S.: Workload-aware indexing of continuously moving objects. PVLDB **2**(1) (2009)
19. Yuan, J., Zheng, Y., Zhang, C., Xie, W., Xie, X., Sun, G., Huang, Y.: T-drive: driving directions based on taxi trajectories. In: Proc. of GIS 2010 (2010)

Effective Spatial Keyword Query Processing on Road Networks

Hailin Fang[1], Pengpeng Zhao[1]($\boxtimes$), Victor S. Sheng[2], Jian Wu[1],
Jiajie Xu[1], An Liu[1], and Zhiming Cui[1]

[1] School of Computer Science and Technology, Soochow University,
Suzhou 215006, People's Republic of China
{hlfang,ppzhao,jianwu,xujj,anliu,szzmcui}@suda.edu.cn
[2] Computer Science Department, University of Central Arkansas, Conway, USA
ssheng@uca.edu

Abstract. Spatial keyword query plays an important role in many applications with rapid growth of spatio-textual objects collected. In this context, processing boolean spatial keyword query on road networks is one of the most interesting problems. When giving a query which contains a location and a group of keywords, our aim is to return k objects containing all the query keywords which are the nearest to the query location. Though the research on this problem has received extensive studies in Euclidean space, little is done to deal with it on road networks. We first propose novel indexing structures and algorithms that are able to process such query efficiently. Experimental results on multiple real-word datasets show that our methods achieves high performance.

Keywords: Road networks · Spatial keyword search · Spatial indexing

1 Introduction

With the increasing pervasiveness of the mobile devices and geo-location services, there are large amounts of spatio-textual data available in many applications. For instance, many applications (e.g, Twitter, Yellow page etc) provide target location information with a short text description. People can publish contents with geographical position information in these applications every day. For these huge spatial textual data, how to establish an effective real time query mechanism is a great challenge. The current approaches processing spatial keyword queries are mostly developed in Euclidean space [1–4,13,14,16]. In reality, our daily travels are usually constrained by the road networks, which leads to the network distance between two locations may be larger than their Euclidean distance. For example, the network distance between two hotels on the opposite banks of a river is completely different from their Euclidean distance. Then the result obtained under the Euclidean space may not be close to a query location on its road network.

For example, Fig.1 illustrates a part of road network and some spatio-textual objects residing on some roads. Fig.2 provides text descriptions of the all objects

© Springer International Publishing Switzerland 2015
M.A. Sharaf et al. (Eds.): ADC 2015, LNCS 9093, pp. 194–206, 2015.
DOI: 10.1007/978-3-319-19548-3_16

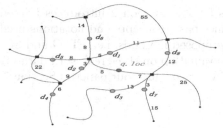

d_1	*steam-bath,laundry-service,lounge*
d_2	*steam-bath,restaurant*
d_3	*steam-bath,restaurant,safe*
d_4	*Wi-Fi,restaurant,laundry-service*
d_5	*steam-bath,safe,wellness*
d_6	*Wi-Fi,restaurant,lounge*
d_7	*steam-bath,restaurant,flowershop*
d_8	*casino,restaurant,fitness root*

Fig. 1. Road network

Fig. 2. The descriptions of objects on Road network

(represent hotels) shown in Fig.1. On this partial road network, there are 6 vertices(road intersections) and 8 objects, which are denoted as grids and circles, respectively. Each digit on edge presents the length of it. Given a query such as $q=\{q.loc, q.term, k\}$ in Fig.1, $q.loc$ indicates the location of the query, $q.term=\{steam-bath,restaurant\}$ presents a set of keywords, and k is the limited number of returns. Supposed that the user intends to find two nearest hotels to $q.loc$. According to the Euclidean distance, the result of this query is $\{d_2, d_3\}$. But due to the constrains of the road network, the road distance between $q.loc$ and the objects d_2 is 8, denoted as $\delta(q,d_2) = 8$. Similarly we have $\delta(q.loc,d_3) = 20$. However, we find d_7 also meets the text description of the query and its distance $\delta(q.loc,d_7) = 10$ is shorter than the distance to the object d_3. Therefore the best query result set is $\{d_2, d_7\}$ rather than $\{d_2, d_3\}$.

This example shows that the results of query on road networks are very different from Euclidean space. Compared with Euclidean space, it is much more challenging on road networks. The reason is that what we need is to compute the network distance between two objects rather than Euclidean distance. Thus existing methods in Euclidean space cannot be directly applied to road networks. Although spatial keyword query on road networks have attracted some research efforts in recent years [8,10], which are focused on different type of query like *range, top-k* query, but their solutions can not directly applied to boolean spatial keyword query(BSKQ) on road networks. To the best of our knowledge, there are no work studying the BSKQ on road networks.

In this paper, we propose two novel approaches to deal with boolean spatial keyword on road networks. The basic spatial keyword query method on road networks is to expanse the network from the query location using Dijkstra's algorithm, but during the process of traversing a network, the performance is very poor. So we propose a new method First Inverted file Then G-Tree, called FITG, by combining inverted index with spatial index G-Tree [15]. According to the principle of first text pruning then spatial pruning, it is avoid traversing the network in a point-to-point manner. Furthermore, we also propose a hybrid index named SG-Tree(Signature based G-Tree). The idea of SG-Tree is to create a signature for each node in G-Tree and partitions the whole road network into a group of interconnected sub-networks and organizes them in a hierarchy structure, which improves query efficiency tremendously.

The rest of the paper is organized as follows. Section 2 introduces related work in this field. Section 3 formally defines basic concepts and notations used in this paper. Section 4-5 elaborates the proposed methods, i.e., FITG and SG-Tree, respectively. Section 6 conducts the evaluation on three real-word datasets and shows our experimental results. Finally, we conclude our paper in Section 7.

2   Related Works

Currently there are three types of spatial keyword query(SKQ): boolean spatial keyword query (BSKQ), top-k spatial keyword query (Top-k SKQ) and range constrained spatial keyword query (RC-SKQ). In the past decades, researchers have done a lot of work on k-nearest neighbors(KNN) on road networks [5–7,9,11,15].

In recent years, SKQ has received extensive studies in Euclidean distance space [1–4,13,14,16], and achieved very significant results(e.g., [1,3] for a comprehensive survey). BSKQ is one of the most important query problems and many efficient query approaches have been proposed such as inverted R-tree [16], information retrieval R-tree [4], Inverted Linear Quad-tree [13], and some solve other types of query(e.g., top-k) method were proposed, like I^3 [14] and so on. However, all of these algorithms are based on Euclidean space, but real-life travel trajectories are constrained by road networks.

Recently, the problem of SKQ on road networks has been studied by [8–10, 12]. Shekhar et al. [12] proposed the CCAM method which reduces 2-dimensional data of a node to a single dimensional and effectively organizes the adjacent list of road nodes. We can take advantage of the access locality in the query processing that can reduce the I/O costs to improve query efficiency. Papadias et al. [9] proposed an efficient framework to store road networks and spatio-textual objects. Recently, Rocha-Junior et al. [10] raised efficient methods to address Top-k SKQ on road networks. They designed a framework of the index for using overlay networks to prune the regions of the network. Li et al. [8] proposed a new query that range-constrained SKQ on road networks and put forward several different indexing strategies to address this type of SKQ. Their proposed approaches are very similar, although addressing different types of problems. But all the above works have not been studied the BSKQ problem on road networks. Furthermore, we exploit a new elegant and efficient road networks index G-Tree[15] and the ubiquitous text index signature to propose a novel hybrid index called SG-Tree that can scale to large road networks.

3   Problem Statement

In this section, we introduce a graph model to represent a road network. Then, we define the related concepts which will be used in the following of this paper.

Road Networks: This paper uses a weighted graph to describe a road network, which is denoted as $G = (V, E, W)$, where V is a set of nodes that represent a road segment, the edge set is denoted as E and W is a set of weights denoting the cost on the corresponding edge, such as travel time or distance. $(v, \nu) \in E$ denotes an edge, and $w_{v,v}$ is a weight on this edge. The shortest path between two nodes v and ν is denoted as $|v, \nu|$, an $\|v, \nu\|$ is the minimum length between v and ν, i.e., $\|v, \nu\|=w_{v,v}$. Let $q.loc$ be a query, and o be a spatio-textual object on the road network, then the shortest distance between the query and o is denoted as follows:

$$\|q.loc, o\| = \begin{cases} \|q.loc, o\| & both\ q.loc\ and\ o\ are\ on\ the\ same\ edge \\ min\{\|q.loc, v\| + \|o, v\|, \|q.loc, \nu\| + \|o, \nu\|\} & otherwise \end{cases} \quad (1)$$

Spatio-Textual Object: A spatio-textual object is normally expressed by a point with coordinates and a set of keywords, which is described in a two dimensional space. For example, $\{loc, term\}$ presents a spatio-textual object, where $o.loc$ is the location of the object, including its latitude and longitude, and $o.term$ presents a set of keywords describing some text expressions, such as $term = \{t_1, t_2 .... t_f\}$. For simplicity, each object lies in its corresponding edge.

Spatial Keyword Query: According to the spatio-textual object definition, we use D to represent all objects in a spatial database as follows: $D = \{o | \forall o \in D, o = \{o.loc, o.term\}\}$. Given a road network as a weight graph G and a query point q, a spatial keyword search can retrieve k objects, each of which contains all keywords of the query and whose network distance is the shortest to the query location.

The Baseline Approach: Traditional spatial keyword query methods use the network expansion on road networks. On this basis, we combine the signature index on edge with network expansion and develop an enhance method called Signature based Network Expansion (SNE) as our baseline method. The structure of SNE we adopt a very popular data structure connectivity clustered access method (CCAM) [12] used to store a road network. We build a network R-tree to identify the road segment where the location of query is. In order to avoid loading a large number of irrelevant objects, we use the signature technology to organize all objects on each edge. We use $I(e, term)$ as the signature of each edge e. If there is at least one object contains the query keywords lying on the edge e, $I(e, term) = 1$; otherwise $I(e, term) = 0$, thus only partial edges containing query keywords can be loaded.

After providing the definitions of the related concepts, we will discuss the three proposed methods for spatial keyword query on road networks in details.

4   The First Inverted File Then G-Tree Approach

In the above baseline method, if the road network data is very complicated with a huge number of spatio-textual objects or locations of objects satisfying the condition are far away from the query, it needs to spend a lot of time on the edge expansion process. It means that, the time complexity is very high.

In order to solve this problem, we employ an excellent spatial road network index technique by combing G-Tree and a current traditional text inverted index. Using two separated indexes to present a new method called First Inverted file Then G-Tree (FITG in short). The index structure is as follows.

4.1 Index Structure

FITG method is executed according to the principle of first text pruning then spatial pruning. It avoids a point-to-point manner to traverse the network which can save much cost and improve query efficiency. In this algorithm, we first use a text index to find all objects which contain all the keywords of the query, and then set them as candidates to calculate the shortest network distance to query location. Since inverted index has been explained in many papers, we will not repeat it in this paper. G-Tree index is a novel and efficient spatial road network index structure, proposed by Zhong et al. [15]. G-Tree has two core features. The first feature is that it has a highly balanced tree structure which can recursively divide the road network into a plurality of sub networks and map each vertex to a corresponding sub network. The other is that it uses the best-first search algorithm, greatly improving the performance. For other details of G-Tree, please refer to the paper [15].

For example, given a query q whose location is show in Fig.1 and it contains the text description $q.term = \{steam\text{-}bath, restaurant\}$, whose aim is to find $k = 2$ spatio-textual objects. The steps of the FITG algorithm are as following:

step 1: *According to the inverted index, the returned candidates of the keyword "steam-bath" are:* d_1, d_2, d_3, d_5, d_7.

step 2: *According to the inverted index, the returned candidates of the keyword "restaurant" are:* $d_2, d_3, d_4, d_6, d_7, d_8$.

step 3: *After the intersection, three candidates are returned, which contain all the keywords:* d_2, d_3, d_7.

step 4: *Initialize the occurrence lists of* d_2, d_3, d_7 *by G-Tree.*

step 5: *Add* d_2 *to the result set* $R = \;< d_2, 8 >$.

step 6: *Add* d_7 *to the result set* $R = \;< d_2, 8 >, < d_7, 10 >$.

step 7: *Because the size of* $R.size = 2$ *is equal to* k, *so the process terminates and returns the result.*

4.2 Query Processing

In Algorithm 1, for lines 1-3 we use the inverted index technique to prune the whole text dimension first to find the object list L_i which contains any one of the keywords in a query. Then we make an intersection for these lists denoted as L. So the L is contain all objects which contains all query keywords. Then the priority queue, the result set and the occurrence list of candidates are initialized respectively in lines 4-6. Second, we use the spatial index G-Tree to calculate the network distance between any objects in L and the location of the query in lines 7-22, If a priority queue is null or the number of the result set is more than k, the process terminates.

Algorithm 1. $FITGTreeQueryProcessing(q.loc, q.term, k, I, T_g)$

Input: $q.loc$, $q.term$, k. I:is the Inverted index, T_g a G-Tree index.
Output: $\Re$ objects satisfying the query condition.
1: **for** each word t_i in $q.term$ **do**
2:　　$L_i \leftarrow$ I.getDocListByTerm(t_i);
3: $InvL \leftarrow$ the intersection of object pointers in L_i;
4: **Initialize** : $\Re := \emptyset; Q := \emptyset$;
5: $\Gamma \leftarrow InvL$; initialize the occurrence list from the candidate set
6: $Q.Enqueue(\langle T_{g_{root}}, 0\rangle)$
7: **while** $Q \neq \emptyset$ and $|\Re| \leq k$ **do**
8:　　$n = Q.\text{pop}()$;
9:　　**if** n is a leaf node **then**
10:　　　　**if** $q.loc \in n$ **then**
11:　　　　　　$mindist\_inside\_leaf(q.loc, n)$;
12:　　　　**else** $mindist\_outside\_leaf(q.loc, n)$;
13:　　　　**for** each $\tau \in \Gamma(n)$ **do**
14:　　　　　　$Q.Enqueue(\tau, SPDist(n, \tau))$;
15:　　**else if** n is a non-leaf node **then**
16:　　　　**for** each child node $c \in \Gamma(n)$ **do**
17:　　　　　　**if** $q.loc$ is in c **then**
18:　　　　　　　　$Q.Enqueue(c, SPDist(q.loc, c) = 0)$;
19:　　　　　　**else**
20:　　　　　　　　$mindist\_outside\_nonleaf(q.loc, c)$;
21:　　　　　　　　$Q.Enqueue(c, SPDist(q.loc, c))$;
22:　　**else if** n is is an object **then** insert then insert n into $\Re$
　　return $\Re$;

5　The Signature Based G-Tree Index Approach

From the above algorithm, the efficiency of the FITG approach is much higher than that of the baseline method SNE. FITG method can not only enhance the text pruning ability but also reduce the time on network distance computing. It dose not need to traverse the entire network. It only needs to calculate the network distance between the query and candidates, which reduces the time complexity and the computation cost. However, the FITG algorithm will cause a serious performance deterioration of algorithm if a great deal of object contain the query keywords or the query condition contains a large number of keywords. A large number of objects will be retrieved, which results in a high computation cost on finding the intersection from the candidates for each query keyword. So we propose an efficient and elegant method which can support an efficient spatial keyword query on large road networks. We integrate the popular textual index signature into spatial index G-Tree named Signature based G-Tree(SG-Tree). We will detail our index structure as follows.

5.1   Index Structure

In this approach, we create a signature for each node in G-Tree which node represents a sub-tree root node. The objects that each root node contains consist of those objects of its children nodes. We use the signature to determine whether the root node contains the query keywords. It will prune the entire sub-tree if it dose not match with query signature. The reason is that the signature of non-leaf node is composed of all children signatures. In order to enhance the spatial pruning ability, we integrate the traditional Incremental Nearest Neighbor(INE) method into the G-Tree. By integrating the traditional INE with the signature, the distance-first SG-Tree access nodes and spatio-textual objects have a minimum distance away from the location of the query, which contributes to improve the query efficiency of the distance-first spatial keyword.

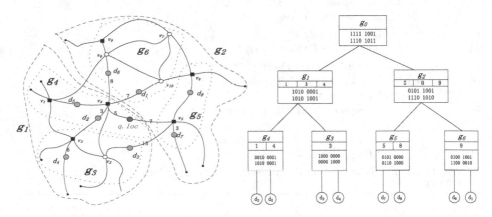

Fig. 3. Graph partition   **Fig. 4.** SG-Tree structure

For example, Fig.4 shows an example of SG-Tree which integrates the G-Tree of the partial road network in Fig.3 with the spatial-textual objects in Fig.2. Given a query point $q.loc$ as is shown in Fig.3 and its keywords description such as $q.term=\{steam\text{-}bath, restaurant\}$, k objects which are the nearest to the query location and containing all keywords of the query will be returned. The sequence of steps is as follows.

step 1: *Enqueue g_0: $Q = \{g_0, 0\}$;*
step 2: *Enqueue g_0: match the signature, Enqueue g_1, g_2;$Q = \{(g_1, 0), (g_2, 7)\}$;*
step 3: *Enqueue g_1: match the signature, Enqueue g_4, $Q = \{(g_4, 0), (g_2, 7)\}$;*
step 4: *Enqueue g_3: is a leaf node, d_2 contains keyword, Enqueue d_2, $Q = \{(d_2, 8), (g_2, 7)\}$;*
step 5: *Enqueue g_2: match the signature,Enqueue $g_5$$Q = \{(g_5, 7), (d_2, 8)\}$;*
step 6: *Enqueue g_5: is a leaf node then d_7 contains keywords, Enqueue d_7, $Q = \{(d_2, 8), (d_7, 10)\}$;*
step 7: *Enqueue d_2: d_2 is an object and added to $\Re$, $Q = \{(d_7, 10)\}$;*
step 8: *Enqueue d_7: d_7 is an object and added to $\Re$;*
step 9: *$\Re.size = 2$, then algorithm terminate and return $\Re$*

5.2   Query Processing

The SG-Tree query method has two advantages. One is using the text index signature to prune on text and the other is to conduct spatial pruning using the distance-first algorithm. It is SG-Tree that simultaneously prunes on the spatial and text two dimensions. In algorithm 2, all node signatures, occurrence lists and the priority queue are initialized in lines 1-4. For the objects in the queue, we iteratively dequeue and handle each element in the queue separately with three possibilities, such as a leaf node, a non leaf node and a spatial object respectively in lines 7, 15, 21 respectively. The priority queue is employed to keep objects accessed during the G-Tree traversing. It is used to determine the signature of a node whether it matches the query signature. If it does not match the whole sub-trees, it will be pruned in lines 11, 17. If the node matches the signature of the query, then it will be pushed into the priority queue according to their network distance. The algorithm will safely terminate when k answers are returned or the priority queue is empty.

Algorithm 2. *SignatureGtreeQueryProcessing(q.loc,q.term,k,T_g)*

Input: *q.loc, q.term, k, T_g.*
Output: $\Re$ objects satisfying the query condition.
1: **Initialize** : $S \leftarrow$ node; initialize all node signature
2: **Initialize** : $\Re := \emptyset; Q := \emptyset; Gamma - \emptyset;$
3: $Q.Enqueue(\langle T_{g_{root}}, 0 \rangle)$
4: $W \leftarrow$ signature$(q.term)$
5: **while** $Q \neq \emptyset$ **and** $|\Re| \leq k$ **do**
6:      $n = Q.pool();$
7:      **if then**n is a leaf node **then**
8:          **if** $q.loc \in n$ **then**
9:              $mindist\_inside\_leaf(q.loc, n);$
10:         **else** $mindist\_outside\_leaf(q.loc, n);$
11:     **if** S matches W **then**
12:         **for** each $o \in n$ **do**
13:             **if** o contain $q.term$ **then**
14:                 $\Gamma \leftarrow o;$ $Q.Enqueue(o, SPDist(q.loc, o));$
15:     **else if** n is a non-leaf node **then**
16:         **for** each child node $c \in \Gamma(n)$ **do**
17:             **if** S matches W **then** $\Gamma \leftarrow c;$
18:             **if** $q.loc$ is in c **then**
19:                 $Q.Enqueue(c, SPDist(q.loc, c)) = 0);$
20:             **else** $mindist\_outside\_nonleaf(q.loc, c);$
21:     **else if** n is is an object **then** insert then insert n into $\Re;$
        return $\Re;$

---

6 Experimental Evaluation

In this section, we will investigate our three approaches on three real-world road networks according to different evaluation criterions.

6.1 Setup

In this section, we evaluate the performance of our three approaches SNE, FITG and SG-Tree. The experiments are conducted on three real datasets which are road networks of CAL, NA, and SF(shown in Table 1) respectively. All datasets are obtained from the webset[1]. The spatio-textual objects are obtained from the US Board on Geographic Names[2] in which each objects is composed of a short text description with a geographic location. Table 1 summaries the detail of three datasets. Note that the scalability of G-Tree have been proved in paper [15], so do not repeat it. Our experiments were executed on Linux computer with 3.0 GHz CPU Inter processor and 4G RAM.

Table 1. Summary of the three Datasets

Data	Description	Vertices	Edge	Spatio textual object size(MB)
CAL	California	21,048	21,693	13.1
NA	North America	175,812	179,178	25.7
SF	San Francisco	174,956	223,001	62.9

6.2 Experimental Results

In this section, we will evaluate three approaches of spatial keyword query on road networks from different perspectives, such as index construction time, index size, a varied number of results, and a varied number of query keywords.

Evaluation on Index Construction: First, we evaluate the space and time consumption overhead of the index construction of three methods. Fig.5(a) illustrates the time consumption of the three index methods on the three different real datasets. From Fig.5(a), we can see that the index construction time of the baseline method is significantly more than that of the other two methods proposed in this paper. This is because SNE method utilizes the CCAM structure, it needs to take the longest time to partition the edges and to create the signature of each edge. As is shown in Fig.5(c), FITG method creating the spatial index for a road network dose not need to spend too much time so that the most of that time can be consumed for text indexing. The greater the spatio-textual object on a road network, the more time it requires. Note that its time spending on index creating index increases when the text data is relatively large. SG-Tree builds signatures based on a pseudo-document for each node on the G-Tree, so

---

[1] http://www.cs.utah.edu/~lifeifei/SpatialDataset.htm
[2] http://geonames.usgs.gov

it is more efficient than FITG. Both FITG and SG-Tree approaches spend less time creating index than of the baseline approach.

Evaluation on Index Size: Fig.5(b) shows that the index size of three approaches on three real different datasets. It can be seen that both baseline and FITG approaches need more external memory space to store index, which dues to that two methods need store more index information easy to search. The Baseline method, utilizing the CCAM structure, which needs more space to store adjacency information of each vertex and each edge. Besides, it also has to store signatures, so that its external memory consumption is very high. As we can see from Fig.5(d) that FITG needs more external memory because it keeps the inverted index. Fig.5(d) shows that the spatial index G-Tree only needs small extra space, while the Inverted File requires more space. In contrast, our proposed SG-Tree method only needs a very small space to store nodes and the signature on each node of G-Tree.

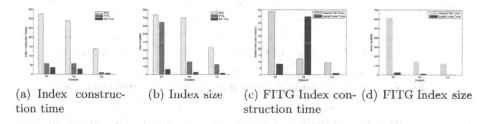

(a) Index construction time   (b) Index size   (c) FITG Index construction time   (d) FITG Index size

Fig. 5. Index Construction Size and Index Size

Evaluation on the varied Number of Results: Fig.6 depicts the response time within various numbers of results on the different datasets and the query has three keywords. We compare these three methods with varying number of results in our experiments. Fig.6(a), Fig.6(b) and Fig.6(c) show that both SG-Tree and FITG have better performance than that of SNE on three datasets. We can also find that the response time of SNE has little increment with the increment of the number of results. The reason is that SNE method in the query process needs to expand access edges. Thus, it needs to spend more time to check edges whether they contain the keywords when the result numbers are increased. However, the response time of both FITG and SG-Tree is almost unchanged. This is because both FITG and SG-Tree only need to calculate the shortest distance of few objects and this process does not need to spend much time.

Evaluation on the Number of Query Keywords: We evaluate the response time under the varying number of query keywords on different real datasets. The experimental results are shown in Fig.7. It shows that SG-Tree method performs much better than SNE and FITG. It can be ascribed to that it bypasses the hierarchy pruning on tree. With the keywords increasing, more and more nodes are pruned, so that the number of candidates becomes small. In Fig.7(a), we

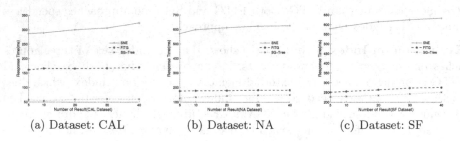

(a) Dataset: CAL   (b) Dataset: NA   (c) Dataset: SF

Fig. 6. Varying Numbers of Results on Different three Datasets

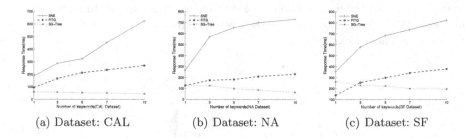

(a) Dataset: CAL   (b) Dataset: NA   (c) Dataset: SF

Fig. 7. Varying Numbers of Keywords on Different three Datasets

can see that with the increase in the number of keywords, the processing time in method FITG and SNE are growing gradually, while in method SG-Tree it is reducing. The same performance also can be seen in Fig.7(b) and Fig.7(c). The reason is that SG-Tree Method can prune lots of unrelated objects through eliminating many more G-Tree nodes. On the contrary, FITG and SNE need to check whether the object contains the keyword repeatedly. When the query has one keyword, SNE performance is reasonable. With the increment of the number of query keywords, its response time rises rapidly because it has to expand large edges. Compared to SNE, FITG performs better because it dose not expand edges and can save a lot of time.

In all, our experimental results show that our proposed two hybrid indexes and query algorithms outperform the baseline method. In particular, the SG-Tree index exploits the advantages of the hierarchical tree pruning on the text and spatial dimensions, and improves the query efficiency tremendously.

7   Conclusion

In this paper, we analyzed the advantages and disadvantages of the existing methods on road networks. After that, we proposed three novel approaches (SNE, FITG, and SG-Tree) to achieve rapid and efficient spatial keyword search. Our experimental results on real-world road networks show that the SG-Tree index structure is the most efficient method.

In the future, since real-time is quite important for urban planning, transportation planning, route planning with heavy traffic, we will integrate temporal information into spatial keyword query on road networks.

Acknowledgments. This work was partially supported by Chinese NSFC project (61170020, 61402311, 61440053), and the US National Science Foundation (IIS-1115417).

References

1. Cao, X., Chen, L., Cong, G., Jensen, C.S., Qu, Q., Skovsgaard, A., Wu, D., Yiu, M.L.: Spatial keyword querying. In: Atzeni, P., Cheung, D., Ram, S. (eds.) ER 2012 Main Conference 2012. LNCS, vol. 7532, pp. 16–29. Springer, Heidelberg (2012)
2. Cao, X., Cong, G., Jensen, C.S., Ooi, B.C.: Collective spatial keyword querying. In: Proceedings of the 2011 ACM SIGMOD International Conference on Management of Data, pp. 373–384. ACM (2011)
3. Chen, L., Cong, G., Jensen, C.S., Wu, D.: Spatial keyword query processing: an experimental evaluation. Proceedings of the VLDB Endowment **6**(3), 217–228 (2013)
4. De Felipe, I., Hristidis, V., Rishe, N.: Keyword search on spatial databases. In: IEEE 24th International Conference on Data Engineering, ICDE 2008, pp. 656–665. IEEE (2008)
5. Jensen, C.S., Kolářvr, J., Pedersen, T.B., Timko, I.: Nearest neighbor queries in road networks. In: Proceedings of the 11th ACM International Symposium on Advances in Geographic Information Systems, pp. 1–8. ACM (2003)
6. Lee, K.C., Lee, W.C., Zheng, B.: Fast object search on road networks. In: Proceedings of the 12th International Conference on Extending Database Technology: Advances in Database Technology, pp. 1018–1029. ACM (2009)
7. Lee, K.C., Lee, W.C., Zheng, B., Tian, Y.: Road: a new spatial object search framework for road networks. IEEE Transactions on Knowledge and Data Engineering **24**(3), 547–560 (2012)
8. Li, W., Guan, J., Zhou, S.: Efficiently evaluating range-constrained spatial keyword query on road networks. In: Han, W.-S., Lee, M.L., Muliantara, A., Sanjaya, N.A., Thalheim, B., Zhou, S. (eds.) DASFAA 2014. LNCS, vol. 8505, pp. 283–295. Springer, Heidelberg (2014)
9. Papadias, D., Zhang, J., Mamoulis, N., Tao, Y.: Query processing in spatial network databases. In: Proceedings of the 29th International Conference on Very Large Data Bases, vol. 29, pp. 802–813. VLDB Endowment (2003)
10. Rocha-Junior, J.B., Nørvåg, K.: Top-k spatial keyword queries on road networks. In: Proceedings of the 15th International Conference on Extending Database Technology, pp. 168–179. ACM (2012)
11. Samet, H., Sankaranarayanan, J., Alborzi, H.: Scalable network distance browsing in spatial databases. In: Proceedings of the 2008 ACM SIGMOD International Conference on Management of Data, pp. 43–54. ACM (2008)
12. Shekhar, S., Liu, D.R.: Ccam: A connectivity-clustered access method for networks and network computations. IEEE Transactions on Knowledge and Data Engineering **9**(1), 102–119 (1997)

13. Zhang, C., Zhang, Y., Zhang, W., Lin, X.: Inverted linear quadtree: efficient top k spatial keyword search. In: 2013 IEEE 29th International Conference on Data Engineering (ICDE), pp. 901–912. IEEE (2013)

14. Zhang, D., Tan, K.L., Tung, A.K.: Scalable top-k spatial keyword search. In: Proceedings of the 16th International Conference on Extending Database Technology, pp. 359–370. ACM (2013)

15. Zhong, R., Li, G., Tan, K.L., Zhou, L.: G-tree: an efficient index for knn search on road networks. In: Proceedings of the 22nd ACM International Conference on Conference on Information & Knowledge Management, pp. 39–48. ACM (2013)

16. Zhou, Y., Xie, X., Wang, C., Gong, Y., Ma, W.Y.: Hybrid index structures for location-based web search. In: Proceedings of the 14th ACM International Conference on Information and Knowledge Management, pp. 155–162. ACM (2005)

Cognition and Statistical-Based Crowd Evaluation Framework for ER-in-House Crowdsourcing System: Inbound Contact Center

Morteza Saberi[✉], Omar Khadeer Hussain,
Naeem Khalid Janjua, and Elizabeth Chang

School of Business, UNSW Canberra, Canberra, Australia
m.saberi.ie@gmail.com

Abstract. Entity identification and resolution has been a hot topic in computer science from last three decades. The ever increasing amount of data and data quality issues such as duplicate records pose great challenge to organizations to efficiently and effectively perform their business operations such as customer relationship management, marketing, contact centers management etc. Recently, crowdsourcing technique has been used to improve the accuracy of entity resolution that make use of human intelligence to label the data and make it ready for further processing by entity resolution (ER) algorithms. However, labelling of data by humans is an error prone process that affects the process of entity resolution and eventually overall performance of crowd. Thus controlling the quality of labeling task is an essential for crowdsourcing systems. However, this task becomes more challenging due to unavailability of ground data. In this paper, we address the above mentioned challenge and design and develop framework for evaluating performance of ER-In-house crowdsourcing system using cognition and statistical-based techniques. Our methodology is divided into two phases namely before-hand evaluation and in-process evaluation. In before-hand evaluation a cognitive approach is used to filter out workers with an inappropriate cognitive style for ER-labeling task. To this end, analytic hierarchy process (AHP) is used to classify the existing four primary cogitative styles discussed in the literature either as suitable or not-suitable for labelling task under consideration. To control the quality of work by crowd-workers, we extend and use the statistical approach proposed by Joglekar et al. during second phase i.e. in-process evaluation. To illustrate effectiveness of our approach; we have considered the domain of Inbound Contact Center and using Customer Service Representatives (CSRs) knowledge for ER-labeling task. In the proposed ER-In-house crowdsourcing system CSRs are considered as crowd-workers. Synthetic dataset is used to demonstrate the applicability of the proposed cognition and statistical-based CSRs evaluation approaches.

Keywords: Contact centers · Crowd evaluation · Entity resolution · Cognitive styles

© Springer International Publishing Switzerland 2015
M.A. Sharaf et al. (Eds.): ADC 2015, LNCS 9093, pp. 207–219, 2015.
DOI: 10.1007/978-3-319-19548-3_17

1   Introduction

Contact centers(CCs) are considered as an organization's touch point having a considerable effect on customer experience and retention [1]. Inbound contact centers (ICC) are one type of CC aiming to address the queries of customers effectively and efficiently. More than 70% of organizational business interactions are handled by inbound CCs and the companies with a focus on Total Customer Experience have seen an increase of 5% in customer retention which resulted in increased profit by 25% to 95% [2, 3]".

The emergence of ICT paradigm has led to the production of a huge amount of data (both structured and unstructured) in CC databases leading to number of challenges related to processing and using data to address customers' queries. Furthermore, customers use of diverse communication channels (email, phone call, voice recordings) to interact with ICC, make it challenging for contact center to efficiently run the business, which will eventually effect the performance of ICC [4]. According to Stringfellow et al., having access to rich channels of communication, such as the telephone, leads to data complexity (Stringfellow, Nie and Bowen 2004) which, in turn, can produce dirty data on the CC side. Missing data, wrong data and unusable data are the three main categories of dirty data explained in the literature [5]. Presence of dirty data in ICC's database leads to creation of multiple records of a single customer. This in turn, eventually results in delaying customer identification, leading to delay in addressing customer queries. This has a direct impact on customer satisfaction. To address this problem, on one hand we need efficient techniques for data cleansing, entity resolution etc.

To overcome the issues discussed above, human computing has played a significant role in providing solution to real world problems. The concept of human computing goes back to 1950 when a British scientist, Alan Turing, stated that "the idea behind digital computers may be explained by saying that these machines are intended to carry out any operations which could be done by a human" [6]. Today, human computing is a well-established discipline. Today, crowdsourcing [7], social computing [8] and collective intelligence [9] are some well-known fields of its application. Crowdsourcing is a problem-solving technique which combines the power of human and machine to solve the issues that are generally hard for the machine to address itself [10]. Jeff Howe and Mark Robinson proposed Crowdsourcing as a web-based business model [7]. It should be noted that human computing replaces human with machine, whereas crowdsourcing replaces machine with public human instead of local (insourced) human. Crowdsourcing has been utilized in different areas of research such as sentiment analysis, image processing [7, 11-14] labeling data [15], user studies [16], social and behavioral science [17, 18] etc. and has recently also been studied and applied in the database community [7, 10, 19-25] for entity resolution called as **ER-crowdsourcing**. Results of ER-crowdsourcing studies show that that crowdsourcing is capable of enhancing the entity resolution process [20] as generally human judgment is more accurate than machine, albeit expensive and slow [26]. One important element in each crowdsourcing system is its worker accuracy in performing human intelligence tasks (HITs). HIT in the ER-Crowdsourcing system is the process involved labeling of a given pair of records either it belongs to single customer or different customers.

In the literature there are **three mechanisms** have been deployed for controlling the quality of crowdsourcing system outputs : (a) recruiting good workers, (b) making incentivizing systems and (c) disregarding bad workers outputs[27]. Mason and Watts stated that incentivizing systems do not lead to better performance of crowdsourcing systems workers[28]. Thus it is concluded that the **Mechanism a** and **Mechanism c** are much important and we focus on them for controlling the quality of crowdsourcing system.

Examination of the literature demonstrates that most papers focused on the **Mechanism b and Mechanism c** of quality control of crowdsourcing systems [29]. However there are some works that focused on the first mechanism. As an example, Feldman and Bernstein suggest using cognitive-based task routing in crowdsourcing system. They focused on visual micro-tasks and use ETS cognitive tests to measure cognitive qualities of workers in visual tasks. They concluded that there is a positive correlation between performance on cognitive tests and correct answers of visual crowd-tasks.

In the current work, we propose ER-In-house crowdsourcing system using CSRs as crowd-worker. The proposed system will assist ICC in improving ER process efficiently to identify the customer and efficiently followed by the service delivery. To improve the effectiveness of ER-In-house crowdsourcing system we propose a cognition and statistical-based framework. The proposed framework is divided into two phases: before-hand evaluation and in-process evaluation. Figure 1 demonstrates conceptual representation of proposed framework. During before-hand evaluation, cognition based evaluation technique is used which assists ICC team to recruit workers who's cognitive style suits current labeling task for ER resolution. In other words, before-hand evaluation is applied as filtering stage which assists the CC team in recruiting workers with suitable cognitive style. To achieve this objective, AHP is used to classify the existing four primary cogitative styles discussed in the literature either as suitable or not-suitable for labelling task. During in-process evaluation phase, statistical-based evaluation technique is proposed that assists ICC team in estimating the crowd-worker *true*. The statistical based evaluation technique is used to monitor and control the quality of workers for labeling as it is not always a guarantee for high quality labeling as they are subject to making unintentional errors. To achieve this objective, we extend Joglekar et al. algorithm and use it as statistical based evaluation technique [30]. There are lot of works that focus on the third mechanism, disregarding bad workers outputs, [31-33]. The main feature of Joglekar et al. algorithm is that the true error is estimated along with confidence interval in their algorithm. We adopted this method by changing the criterion in selecting benchmark workers in estimation process and proposed new approach namely: heuristic estimation approach. The detail of this adoption is expressed in Section 4.2. In summary, the following contributions are made in the current study:

An in-house crowdsourcing system is defined in the ICC that use CSR experience in labeling records in entity resolution context. (Section 1).
We explain the relevance of cognitive styles and classify them to evaluate the performance of crowdsourcing systems. The most preferable style for

micro-tasks is found by suing AHP. To the best of our knowledge, this is the first study that considers cognitive style in determining appropriate crowds for labeling for ER (Section 3.1). Statistical-based crowd evaluation is presented that makes use of proposed Heuristic estimation algorithm. Result shows that it outperforms exhaustive selection strategy proposed by Joglekar et al. We explain proposed algorithm in Section 3.2.

We evaluated and analyzed the Heuristic estimation algorithm by using the synthetic dataset.

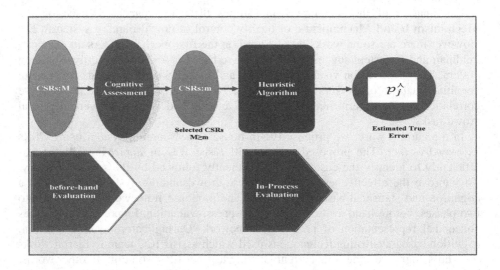

Fig. 1. Conceptual cognition and statistical-based crowd Evaluation framework

2   Cognitive Styles

Cognitive styles are defined as "acquired patterns or habits of information processing". Michael Driver and Kenneth Brousseau developed the model to categorize various cognitive styles. They considered two dimensions in this regard: amount of information that is used in decision making and type of focus. They categorized the people to two classes based on the first dimension: satisfiers and maximizers. *"People differ widely in the amount of information they use in decision-making. Some people reach conclusions on the basis of just a few facts, satisfier. Others reach conclusions only after gathering and studying large amounts of information, maximizers"*. Additionally they categorized type of focus to two classes: uni-focus and multi-focus. They call people either uni-focus or multi-focus decision maker. Uni-focus decision maker *"come up with one specific solution that they feel is the best or most feasible for the situation"*. Multi-focus decision makers, *"faced with the same situation, will quite predictably generate a variety of alternatives or options for dealing with the situation"*. It should be noted that there is no single style suitable for all situations. Each of them either uni-focus or multi-focus styles are suitable for a specific context and serve specific purpose. *"Uni-focus decision makers do well also in situations where following specific procedures, or decision rules, is required to arrive at conclusions, whereas multi-focus decision-makers are inclined to bend the rules needlessly or to invent new rules of their own"*.

Section 3.1.1 demonstrates which of these cognitive styles is suitable for labeling as an ER-labeling task in proposed in-house crowdsourcing system. It is further noted that Driver's Decision Style Exercise (DDSE) should be used in practice to measure concept of "cognitive style" and find the cognitive style of each CSR [34]. However as we used synthetic data to evaluate the proposed algorithm using DDSE is out of scope of the current study.

3   Cognition and Statistical-Based Crowd Evaluation Framework

A cognition and statistical-based crowd evaluation framework is proposed in this study to evaluate the quality of workers in in-house crowdsourcing systems. We use CSRs as crowds in the designed in-house crowdsourcing system. CSRs have a high level of experience and familiarity with the targeted database; thus distinguishing them from the ordinary crowd. In our previous work, we have proposed a probabilistic methodology for identifying a pair of record for crowd-worker to be labelled as duplicate record of a single customer (duplicate) or each of them represent different customer (non-duplicate)[35]. We proposed a probabilistic method that selects the most beneficial pair for labeling. In the current paper, we complete the proposed in-house crowdsourcing system by proposing evaluation framework that monitors the quality of in-house crowdsourcing system.

As we explained in the Introduction, The proposed framework is designed for two important timeframes: recruiting timeframe and in-process timeframe. We use cognitive styles as the criteria for recruiting (using) CSRs in in-house crowdsourcing system. We call this evaluation cognition based evaluation that act as a filter before posting of HITs. After filtering CSRs with inappropriate cognitive styles, we use statistical based evaluation techniques. The algorithm that is proposed by Jogeklar et al. is the point of departure for our proposed approach as statistical based evaluation approach [30]. We modified their approach and proposed a heuristic approach as a statistical based evaluation approach. Section 3.1.1 presents AHP technique that is used to find the most preferable cognitive style for labeling task. The result of AHP is used as cognitive base devaluation technique. Statistical based evaluation technique is also presented in Section 3.2.

3.1   Cognition Based Evaluation Technique: Before-Hand Evaluation

In this section we use AHP to determine the most suitable cognitive style for labeling tasks. The detail of AHP usage and associated results are explained in the next section.

3.1.1   Selecting Primary and Secondary Cognitive Style by Using AHP

The tasks that are posted by requester in crowdsourcing systems are called as human intelligence task (HIT). Although these tasks are generally tiny, their combination produces major accomplishments [29]. In this work, we focus on the following micro HIT task: *Labeling given pair of records as duplicate or non-duplicate pair.* We rank the performance of four cognitive styles in ten situations that are depicted in Table 1.

Table 1. Four cognitive styles performance on labeling ability features [1]

	Decisive	Flexible	Hierarchic	Integrative
Time pressure	7	5	3	1
Task overload	7	5	3	1
Task uncertainty	1	7	1	7
Involvement of others in a group	3	7	5	7
Amount of interaction with others	3	7	5	7
Preference for analytical reasoning	1	5	7	5
Preference for several methods	1	7	3	7
Using rules & scripts	7	3	7	1
Unfamiliarity	1	5	3	7

The following features from the above table have a considerable impact on the labeling quality in the following: time pressure, task overload, task uncertainty, preference for analytical reasoning, using rules & scripts.

Analytic Hierarchy Process (AHP)

AHP is a well-known technique in decision making that is introduced by Thomas L. Saaty[36]. In this technique, the model is presented in a hierarchy form that includes decision goal, solution alternative for defined goal and main criteria that describe appropriate and suitable goal. Our goal is to "select an appropriate cognitive style for labeling in crowdsourcing system". We consider nine features that are depicted in Table 1 as main criteria in AHP modeling. Table 2 demonstrates the associated score with each style as a result of AHP.

Table 2. AHP final results

Cognitive styles	Scores
Decisive Style	0.35
Flexible Style	0.25
Hierarchic Style	0.19
Integrative Style	0.21

According to AHP results, flexible and decisive styles are suitable primary and secondary styles for workers of our crowdsourcing system. We present this finding as hypothesis and study its validity by using gold standard tasks for workers with various cognitive styles. *Hypothesis: Flexible and decisive are two suitable cognitive styles for workers of our crowdsourcing system.*

---

[1] Poor performance:1, Excellent performance:7.

3.2   Statistical Based Evaluation Technique: In process Evaluation

We use statistical based evaluation as in-process evaluation phase. We explain 3-CSRs difference algorithm in Section 3.2.1 that is proposed by Joglekar et al. the generalized form of 3- CSRs difference algorithm is introduced in Section 3.2.2. We propose heuristic algorithm as generalized form of 3- CSRs difference algorithm in Section 3.2.2.2.

3.2.1   Three CSRs Difference Algorithm

In this section, we explain Joglekar et al. algorithm to estimate the true error of crowdsourcing system's worker [30]. The point of departure of their work is using Bernoulli statistical distribution. The Bernoulli random variable takes two values: zero and one. The zero value is taken when the event's result is failure and one value is taken when event's result is success. In the crowdsourcing context, the false labeling is considered as the success and they finally proposed the following formula for estimating the true error of first worker.

$$p_1^\wedge \leftarrow \frac{1}{2} - \sqrt{\frac{\left(a_{12} - \frac{1}{2}\right)\left(a_{13} - \frac{1}{2}\right)}{2\left(a_{23} - \frac{1}{2}\right)}} \tag{1}$$

When a_{ij} denotes the normalized number of times that worker$_i$ and worker$_j$ are agree the task. We explained Three CSRs difference algorithm by the numerical example in the following.

3.2.2   Generalized Three CSRs Difference Algorithm

The drawback of their approach is that it just gives the results of 3 workers. They proposed three strategies to extend the proposed approach: Exhaustive, Pruning and Greedy Search. These strategies assist the proposed HD algorithm in order to constructing two disjoint sets: S,T. these two disjoint sets are treated as the worker and HD algorithm use their result. To shape S and T as the worker, majority voting schema is used. In this paper, we use exhaustive strategy as generally numbers of CSR s are not too high. The strength of Exhaustive strategy is that it gives an optimum solution in view point of forming S and T. We take statistical confidence interval to find the most preferable S and T sets. The sets that make a minimum length (error) for confidence error are selected as the inputs of HD algorithm.

$$\text{Confidence Interval} = \left\{ p_j^\wedge - z_{\frac{\alpha}{2}} \sqrt{\frac{(1 - p_j^\wedge) p_j^\wedge}{m}}, p_j^\wedge + z_{\frac{\alpha}{2}} \sqrt{\frac{(1 - p_j^\wedge) p_j^\wedge}{m}} \right\} \tag{2}$$

Joglekar et al. proposed customized interval that is more time consuming in comparison with our traditional and well-known interval. Figure 2 demonstrates Exhaustive Selection strategy in order to find the most suitable disjoints sets for true error of worker estimation. As line 5 & 7 indicate, the most suitable disjoints sets are selected based on the minimum interval length of confidence interval.

3.2.2.1   Exhaustive Selection Strategy

There are *S(n,2)* different candidates for two disjoint sets : *S,T*. S(n,k) denotes Stirling numbers of the second kind:

S(n,k): the number of partitions of {1,2,...,n} into exactly k nonempty subsets.

S(n,k) value can be calculated by using the following formula that is derived based on combinatorial mathematics:

$$S(n,k) = \frac{1}{k!} \sum_{j=0}^{k} (-1)^{k-j} \binom{k}{j} j^n \qquad (3)$$

Exhaustive Selection Strategy is depicted in Figure 3. The best two disjoint sets; *S,T;* are determined among all *S(n,2)* candidates based on the *line* 5 criterion. In fact, two sets that produce minimum length of confidence interval are the best two disjoint sets.

	Exhaustive Selection (A, Worker$_j$)
	Input: Workers labels A,
	Output: Estimated of Worker$_j$'s True Error
1	$Sc, Tc \leftarrow partition(A - \{j\})$;
2	$k \leftarrow size(partition(A - \{j\}))$;
3	*For* $i = 1:k$
4	$P(i) \leftarrow$ **Hybrid Detection Algorithm** (Worker$_j$, Sc(i), Tc(i))
5	$E(i) \leftarrow z_{\frac{\alpha}{2}} \sqrt{\dfrac{(1 - P(i))P(i)}{m}}$
6	*End*
7	$Index \leftarrow Argmin(E)$;
8	**Return P**(Index);

Fig. 2. Exhaustive Selection Strategy

	Hybrid Detection Algorithm (Worker$_j$, S,T)
	Input: Two disjoints sets: **S,T**
	Agreement rate between j^{th} worker and S: a_{jS}
	Agreement rate between j^{th} worker and T: a_{jT}
1	$\hat{p_j} \leftarrow \dfrac{1}{2} - \sqrt{\dfrac{(a_{jS} - \frac{1}{2})(a_{jT} - \frac{1}{2})}{2(a_{ST} - \frac{1}{2})}}$
2	
	Return $\hat{p_j}$;

Fig. 3. Hybrid Detection Algorithm

3.2.2.2 Heuristic Estimation Algorithm

We slightly modified Exhaustive Selection Strategy and call it Heuristic Estimation algorithm. We demonstrate in examples that Heuristic Selection Strategy outperforms than Exhaustive Selection Strategy. Figure 4 presents Heuristic Selection Strategy.

	Heuristic Estimation algorithm (A, Worker$_j$)
	Input: Workers labels A,
	Output: Estimated of Worker$_j$'s True Error
1	$Sc, Tc \leftarrow partition(A - \{j\})$;
2	$k \leftarrow size(partition(A - \{j\}))$;
3	*For* $i = 1:k$
4	$P(i) \leftarrow$ **Hybrid Detection Algorithm** (Worker$_j$, Sc(i), Tc(i))
5	*End*
6	$P \leftarrow$ **Delete**(P)
7	$\hat{p_j} \leftarrow mean(P)$
8	**Return** $\hat{p_j}$;

Fig. 4. Heuristic Estimation algorithm

Note: *Delete* function deletes zero and complex numbers from its array input.

- Numerical Example

We simulate the current example for the crowdsourcing system with the following parameters:

Number of workers: **n: 7** *(that have Decisive cognitive style)*
Number of Tasks: **m: 10**

Table 4 depicts the labeling of seven workers for ten micro-tasks. We also assume the true error of workers area available to analyzing the performance of error estimation algorithms. We use the labeling of the other six workers in order to evaluate the quality of the given worker. There are $S(6,2)2$ partitions as the candidate for sets S and T. The heuristic approach is used to estimate workers' true error. Table 5 report the estimated value for the first worker by using proposed Heuristic approach and Exhaustive approach.

Table 3. Crowd labeling example

Workers	Tasks										True Error
	1	2	3	4	5	6	7	8	9	10	
1	1	1	2	2	2	1	2	2	1	1	0.3
2	2	1	1	2	1	2	2	2	1	1	0.3
3	1	1	2	2	2	1	2	2	1	1	0.4
4	1	2	2	2	1	1	2	2	1	1	0.1
5	1	2	2	2	1	1	2	2	1	1	0.1
6	1	1	2	2	1	2	2	2	2	1	0.2
7	2	2	2	2	1	2	1	2	1	1	0.2
True Labels	1	2	2	2	1	2	2	2	1	1	

Table 4. Comparison of Heuristic approach and Exhaustive approach estimations

	Heuristic approach	Exhaustive approach	True Error
$p_1^{\wedge}$	0.229	0.135	0.3

We compared these two approaches by using other four synthetic data that their true error rate is reported in Table 6. The dataset considers 7 CSRs with varying error ranges to represent their actual working. We compare the working of the heuristic and exhaustive approaches for worker 1 with an actual error range between 0.1 and 0.4. The determination of the error from both the approaches are show in Table 6.

From the table it can be seen that the heuristic approach outperforms the Exhaustive approach in all the considered cases. Furthermore, it can be seen that when the true error is very low, for example 0.1 there is a very slight difference in the error values given by both the approaches. However the accuracy of the heuristic approach increases as the true error value increases and when it reaches a value of 0.4, the error value given by heuristic approach is much more accurate than the exhaustive one.

Table 5. Five synthetic examples

		Five synthetic examples				
		1	2	3	4	5
Seven workers true error rate	1	0.1	0.2	0.3	0.4	0.1
	2	0.3	0.3	0.3	0.3	0.3
	3	0.4	0.4	0.4	0.4	0.1
	4	0.1	0.1	0.1	0.1	0.1
	5	0.1	0.1	0.1	0.1	0.1
	6	0.2	0.2	0.2	0.2	0.2
	7	0.2	0.2	0.2	0.2	0.2

Table 6. Comparison of Heuristic approach and Exhaustive approach estimations: Five synthetic examples

		Heuristic approach	Exhaustive approach	True Error
1	$p_1^\wedge$	0.007	0	0.1
2	$p_1^\wedge$	0.209	0	0.2
3	$p_1^\wedge$	0.229	0.135	0.3
4	$p_1^\wedge$	0.39	0.18	0.4
5	$p_1^\wedge$	0.001	0	0.1

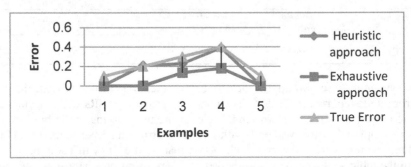

Fig. 5. Comparison of Heuristic approach and Exhaustive approach estimations

4   Conclusion

Customer Services provided by ICCs have a considerable effect on customer experience and retention. Due to diverse channels of communications available to customers, huge amount of data is being preceded and stored in ICCs databases. This leads to data quality issues leading to deterioration of ICCs performance. For example presence of dirty data in inbound ICC's database leads to creation of multiple records of same customer eventually leading to delay in customer recognition followed by addressing customer queries. To address this problem, ICCs should have access to the efficient technique for entity resolution. ER-In-house crowdsourcing system is proposed in this study to improve the process of entity resolution in ICCs. We design and develop framework for evaluating performance of ER-In-house crowdsourcing system using cognition and statistical-based techniques. The proposed framework has two phases: before-hand evaluation and in-process evaluation. AHP is used to classify and identified decisive style as the most suitable cognitive style for ER-labeling task. The Heuristic estimation algorithm is proposed for in-process evaluation and its performance results on five synthetic examples demonstrate its applicability and superiority over exhaustive algorithm.

References

1. Hart, M., Mwendia, K., Singh, I.: Managing knowledge about customers in inbound contact centres. In: Proceedings of the European Conference on Knowledge Management. ECKM 2009
2. Reichheld, F.F.: Loyalty rules!: How today's leaders build lasting relationships. Harvard Business Press (2001)
3. Millard, N.: Learning from the 'wow' factor—how to engage customers through the design of effective affective customer experiences. BT Technology Journal 24(1), 11–16 (2006)
4. LaValle, S., Lesser, E., Shockley, R., Hopkins, M., Kruschwitz, N.: Big Data, Analytics and the Path From Insights to Value. MIT Sloan Management Review 52(2), 21–32 (2011)
5. Kim, W., Choi, B.-J., Hong, E.-K., Kim, S.-K., Lee, D.: A taxonomy of dirty data. Data mining and knowledge discovery 7(1), 81–99 (2003)
6. Turing, A.M.: Computing machinery and intelligence. Mind, 433–460 (1950)
7. Davidson, S.B., Khanna, S., Milo, T., Roy, S.: Using the crowd for top-k and group-by queries. In: Book Using the Crowd for Top-k and Group-by Queries, pp. 225–236. ACM (2013)
8. Wang, F.-Y., Carley, K.M., Zeng, D., Mao, W.: Social computing: From social informatics to social intelligence. Intelligent Systems, IEEE 22(2), 79–83 (2007)
9. Szuba, T.M.: Computational collective intelligence. John Wiley & Sons, Inc. (2001)
10. Sarma, A.D., Parameswaran, A., Garcia-Molina, H., Halevy, A.: Finding with the crowd. In: Book Finding with the Crowd (2012)
11. Brabham, D.C.: Crowdsourcing as a model for problem solving an introduction and cases. Convergence: the international journal of research into new media technologies 14(1), 75–90 (2008)
12. Yi, J., Jin, R., Jain, A.K., Jain, S.: Crowdclustering with sparse pairwise labels: a matrix completion approach. In: Book Crowdclustering with Sparse Pairwise Labels: A Matrix Completion Approach, pp. 1–7 (2012)
13. Bigham, J.P., Jayant, C., Ji, H., Little, G., Miller, A., Miller, R.C., Miller, R., Tatarowicz, A., White, B., White, S.: Vizwiz: nearly real-time answers to visual questions. In: Book Vizwiz: Nearly Real-Time Answers to Visual Questions, pp. 333–342. ACM (2010)
14. Pak, A., Paroubek, P.: Twitter as a corpus for sentiment analysis and opinion mining. In: Book Twitter as a Corpus for Sentiment Analysis and Opinion Mining, pp. 1320–1326 (2010)

15. Snow, R., O'Connor, B., Jurafsky, D., Ng, A.Y.: Cheap and fast—but is it good?: evaluating non-expert annotations for natural language tasks. In: Book Cheap and Fast—but is it Good?: Evaluating Non-Expert Annotations for Natural Language Tasks, pp. 254–263. Association for Computational Linguistics (2008)
16. Kittur, A., Chi, E.H., Suh, B.: Crowdsourcing user studies with mechanical turk. In: Book Crowdsourcing User Studies with Mechanical Turk, pp. 453–456. ACM (2008)
17. Mason, W., Suri, S.: Conducting behavioral research on Amazon's Mechanical Turk. Behavior research methods **44**(1), 1–23 (2012)
18. Schmidt, L.: Crowdsourcing for human subjects research. In: Proceedings of CrowdConf (2010)
19. Whang, S.E., Lofgren, P., Garcia-Molina, H.: Question selection for crowd entity resolution. Proceedings of the VLDB Endowment **6**(6), 349–360 (2013)
20. Doan, A., Franklin, M.J., Kossmann, D., Kraska, T.: Crowdsourcing applications and platforms: A data management perspective. Proceedings of the VLDB Endowment **4**(12), 1508–1509 (2011)
21. Feng, A., Franklin, M., Kossmann, D., Kraska, T., Madden, S., Ramesh, S., Wang, A., Xin, R.: Crowddb: Query processing with the vldb crowd. Proceedings of the VLDB Endowment **4**(12) (2011)
22. Gokhale, C., Das, S., Doan, A., Naughton, J.F., Rampalli, R., Shavlik, J., Zhu, X.: Corleone: hands-off crowdsourcing for entity matching. In: Book Corleone: Hands-Off Crowdsourcing for Entity Matching
23. Jiang, L., Wang, Y., Hoffart, J., Weikum, G.: Crowdsourced entity markup. In: Book Crowdsourced Entity Markup, pp. 1–10 (2013)
24. Demartini, G., Difallah, D.E., Cudré-Mauroux, P.: ZenCrowd: leveraging probabilistic reasoning and crowdsourcing techniques for large-scale entity linking. In: Book ZenCrowd: Leveraging Probabilistic Reasoning and Crowdsourcing Techniques for Large-Scale Entity Linking, pp. 469–478. ACM (2012)
25. Yang, Y., Singh, P., Yao, J., Au Yeung, C.-m., Zareian, A., Wang, X., Cai, Z., Salvadores, M., Gibbins, N., Hall, W., Shadbolt, N.: Distributed human computation framework for linked data co-reference resolution. In: Antoniou, G., Grobelnik, M., Simperl, E., Parsia, B., Plexousakis, D., De Leenheer, P., Pan, J. (eds.) ESWC 2011, Part I. LNCS, vol. 6643, pp. 32–46. Springer, Heidelberg (2011)
26. Mozafari, B., Sarkar, P., Franklin, M.J., Jordan, M.I., Madden, S.: Active learning for crowd-sourced databases, CoRR, abs/1209.3686 (2012)
27. Venetis, P., Garcia-Molina, H.: Quality control for comparison microtasks. In: Book Quality Control for Comparison Microtasks, pp. 15–21. ACM (2012)
28. Mason, W., Watts, D.J.: Financial incentives and the performance of crowds. ACM SigKDD Explorations Newsletter **11**(2), 100–108 (2010)
29. Feldman, M., Bernstein, A.: Cognition-based Task Routing: Towards Highly-Effective Task-Assignments in Crowdsourcing Settings (2014)
30. Joglekar, M., Garcia-Molina, H., Parameswaran, A.: Evaluating the crowd with confidence. In: Proceedings of the 19th ACM SIGKDD International Conference on Knowledge Discovery and Data Mining (2013)
31. Khattak, F.K., Salleb-Aouissi, A.: Improving crowd labeling through expert evaluation. In: Book Improving Crowd Labeling Through Expert Evaluation (2012)
32. Su, H., Zheng, K., Huang, J., Liu, T., Wang, H., Zhou, X.: A crowd-based route recommendation system-CrowdPlanner. In: Book A Crowd-Based Route Recommendation System-CrowdPlanner, pp. 1178–1181. IEEE (2014)

33. Lease, M.: On Quality Control and Machine Learning in Crowdsourcing. In: Book On Quality Control and Machine Learning in Crowdsourcing (2011)
34. Driver, M.J.: Decision style: Past, present, and future research, International perspectives on individual differences, pp. 41–64 (2000)
35. Saberi, M., Hussain, O.K., Janjua, N.K., Chang, E.: In-house crowdsourcing-based entity resolution: dealing with common names. In: Book In-House Crowdsourcing-Based Entity Resolution: Dealing with Common Names, pp. 83–88. IEEE (2014)
36. Saaty, T.L.: The analytic hierarchy process: planning, priority setting, resources allocation. McGraw, New York (1980)

Community Based Information Dissemination

Zhengwei Yang[1], Ada Wai-Chee Fu[1(✉)], Yanyan Xu[1],
Silu Huang[2], and Ho Fung Leung[1]

[1] Chinese University of Hong Kong, Hong Kong, China
{zwyang,adafu,yyxu,hfl}@cse.cuhk.edu.hk
[2] University of Illinois at Urbana-Champaign, Champaign, USA
shuang86@illinois.edu

Abstract. Given a social network, we study the problem of finding k seeds that maximize the dissemination of information. Based on the principle of homophily, communities play an important role since information can be disseminated to communities via the seeds. We introduce a new mechanism for detecting communities satisfying the pertinent criteria for communities and information dissemination. We demonstrate the effectiveness of our approach by an application of the results for influence maximization.

Keywords: Information dissemination · Community detection

1 Introduction

With the growth in social networks and other massive networks, network analysis has emerged as an important research topic. The detection of communities or the listing of *cohesive subgraphs* for a given graph has been of great interest. From studies in sociology, communities are a powerful channel for the dissemination of information [14,16]. Our problem can be described as follows. In a social network, each vertex corresponds to an individual. We are given a limited amount of resources to inform a *seed* set of vertices (individuals) of size k and the problem is how to choose the seed set to maximize the spread from this seed set to other individuals in the network. We call this the IDM problem (IDM stands for Information Dissemination Maximization).

Inspired by the study in sociology about the role of homophily in information dispersal, we propose a model on the IDM problem based on communities. As in previous works on community discovery or cohesive subgraphs, we assume that we are given a simple undirected unweighted graph [9]. Our community definition is related to two basic criteria for cohesive subgraphs, namely, the concept of a clique (i.e. a set of vertices that induces a complete subgraph), and the density of the subgraph. In addition, we consider the distances of vertices from the seed. A key idea in the community search is that we look for the seeds in the process. Each community search begins with a potential seed vertex.

We show that the related optimization problems are NP-hard. We propose efficient algorithms for finding good core-based communities. We then apply the

© Springer International Publishing Switzerland 2015
M.A. Sharaf et al. (Eds.): ADC 2015, LNCS 9093, pp. 220–231, 2015.
DOI: 10.1007/978-3-319-19548-3_18

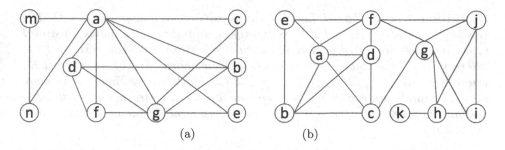

Fig. 1. Two Example Graphs

solution to the problem of influence maximization [11], which has important applications in viral marketing, and the results on a real dataset show that our solution outperforms other state-of-the-art methods.

2 Problem Definition

We study the problem of information dissemination maximization (IDM) in an undirected simple graph, $G = (V, E)$, where V is the set of vertices and E is the set of edges of G. An edge in E between vertices u, v in V is denoted by (u, v) or (v, u), u is a neighbor of v, and vice versa. $adj(u)$ is the set of neighbors of u. Degree $d(u) = |adj(u)|$. A subgraph of G induced by vertex set V' is denoted by $G(V')$. We state the general problem definition as follows.

Problem Definition (IDM): Given a graph $G = (V, E)$, and a positive integer k, information dissemination maximization (IDM) aims to find a set S of k seeds, where $S \subseteq V$, that maximizes the number of vertices that are informed by the seed vertices according to some information dissemination model.

A complete undirected simple graph $G = (V, E)$ is a graph such that every pair of vertices u, v in V is linked by an edge (u, v) in E. A subset of vertices, $C \in V$, is called a **clique** if the subgraph of G induced by C is a complete subgraph. The size of C is given by the number of vertices in C. The edges $(v_0, v_1), (v_1, v_2)...(v_{\ell-1}, v_\ell)$ in G form a path $v_0, v_1, ..., v_\ell$, the length of this path is ℓ.

Definition 1. Density: *Given a set of vertices $S \subseteq V$ and the induced subgraph $G_S = (V_S, E_S)$, the density of S is denoted as $den(S)$ and $den(S) = \frac{2|E_S|}{|V_S|*(|V_S|-1)}$.*

Definition 2. Radius: *Given a core set C and a set S, where $C \subseteq S \subseteq V$, the radius of S regarding C is defined as the maximum shortest path from $u \in S$ to core C, denoted as $R_C(S)$ and $R_C(S) = max_{u \in S}\{min_{v \in C}|SP(u,v)|\}$, where $|SP(u,v)|$ is the length of a shortest path from u to v in $G(S)$.*

We say that $S \subseteq V$ satisfies a density constraint of γ if $den(S) \geq \gamma$. Let $C \subseteq S \subseteq V$, we say that S and C satisfy a radius constraint of r if $R_C(S) \leq r$.

Definition 3. *Core-based Local Community: Given a graph $G = (V, E)$. Given a clique size threshold of K, a density constraint γ, and a radius constraint r, a candidate core-based local community of a vertex $u \in V$ is a vertex set V', where $V' \subseteq V$, such that (1) there exists a clique (**core**) $c(u)$ of size at least K, where $u \in c(u) \subseteq V'$, (2) $R_{c(u)}(V') \leq r$ and (3) $den(V') \geq \gamma$. There can be more than one candidate core-based local community for a vertex u, one of them is assigned to u and we refer to it as the core-based local community of u, denoted by $LC(u)$.*

For simplicity we may refer to a core-based local community as local community or simply LC. The computation of a core-based local community for a vertex u can be broken down into 2 steps, the core $c(u)$ is first located, followed by an extension to a neighborhood of $c(u)$ within the radius constraint and the density constraint.

LC based Information Dissemination Model: Given a seed set S where each vertex w in S is assigned a local community LC_w under the constraints of K, γ, and r, LC_w forms a base for information dissemination by w. The spread base by seed set S is denoted as $I_S = \bigcup_{w \in S} LC_w$. The size of the spread base is given by $g(S) = |I_S|$.

Example 1. Figure 1 (a) shows a graph with 9 vertices. Let $K = 3$, the density constraint, γ, be 1, and the radius constraint, r, be 1. Suppose the seed set $S = \{g, m\}$ and $LC_g = LC(g) = \{a, b, d, g\}$, $LC_m = LC(m) = \{a, m, n\}$, then the spread base size is $g(S) = |\{a, b, d, g, m, n\}| = 6$. Note that there are other possible candidates for $LC(g)$, such as $\{a, b, g, c\}$ and $\{a, d, f, g\}$. One such candidate is set as $LC(g)$.

Example 2. Consider the graph in Figure 1 (a) again, let $K = 3$, and the radius constraint, r, be 1. If the density constraint $\gamma = 0.6$, then we may set $LC(g) = \{g, a, b, d, f, c, e\}$ since there are 15 edges in the induced graph. If $\gamma = 0.8$, then we may set $LC(g) = \{g, a, b, d, f, c\}$, since there are 12 edges in the induced subgraph.

IDM-LC Maximization Problem: The IDM problem under the LC based information dissemination model is to select k local communities so as to give the maximum value of $g(S)$, where S is the set of k seeds to which the k local communities are assigned.

It is easy to show that this problem is NP-hard since the classical maximum clique problem can be reduced to this problem by setting $k = 1$, $\gamma = 1$. In the next sections we shall examine the sub-problems involved.

3   Core-Based Local Community

From previous discussions, computing a core-based local community for u consists of two steps: finding the core and extending the core. In the following, we show that these sub-problems are hard, we propose greedy algorithms for getting feasible solutions and analyze the corresponding complexity for each of the two sub-problems.

3.1 Finding the Cores

We are given a threshold of the core size K, a radius constraint r, and a density constraint of γ. First we consider the problem of finding the core for vertex u with a size of K or above. We show that this is NP-hard by showing that the decision problem of whether there exists a clique of size K is NP-hard. The proof is by a reduction from the classical maximum clique problem. A clique of size k exists in a graph G if and only if there exists in G a vertex v such that the maximum clique containing v has size k.

The maximum clique containing a vertex v is desirable for the core of v because a clique is the most densely connected subgraph with the smallest diameter. This is related to the NP-hard problem of computing the maximum clique of a graph. However, known algorithms for maximum clique cannot be adopted for two reasons. Firstly, existing heuristic algorithms are not scalable to very large graphs [3,18]. Typically they deal with denser graphs, and their targeted graphs are small, e.g. $|V| \leq 1000$, whereas social networks have very low average degrees but $|V|$ is very large. Secondly, our problem is to find the maximum clique containing a vertex for each vertex in the graph, which is different from finding a single maximum clique for the entire graph. Here, we deal with this problem with an efficient greedy algorithm as shown in Algorithm 1.

In Algorithm 1, we maintain a clique $c(u)$ which contains u for each vertex u. Initially $c(u)$ contains only u, and more vertices are added to $c(u)$ iteratively. A vertex v is a candidate to be added to $c(u)$ if and only if v is a neighbor to every vertex in the current $c(u)$. Thus, for each vertex u, initially $c(u) = \{u\}$ and the initial candidate set $cand$ consists of the neighbors of u (*line 4*). After the initialization, we iteratively select the vertex u' such that the candidate set in the next iteration is maximized (*line 6*) and update the candidate set by intersecting with the neighbors of u' (*line 7*). This maintains the invariant that each candidate is a neighbor to every vertex in $c(u)$. We repeat until no more candidate remains. Algorithm 1 selects a clique in a way to maximize the potential clique size at each vertex selection in *line 7*.

Though there can be more than one eligible LC for a vertex, only one of them will be selected. In Figure 2 (a), $\{a, b, c, g\}$ and $\{a, b, d, g\}$ are eligible and one of them will be returned as the core $c(a)$. We show that this algorithm has a scalable time complexity.

Lemma 1. *Given $G = (V, E)$, the time complexity of Algorithm 1 is given by $O(|c_{max}||V|d_{max}^2)$, where $|c_{max}|$ is the maximum core size, and d_{max} is the maximum degree of a vertex.*

Proof. In each while loop, *line 6* is the most costly operation compared to *lines 7,8*. Hence we calculate the complexity of *line 6*, which involves the intersection of two sorted sets, for each while loop of $\forall u \in V$. Note that $d(u) = |adj(u)|$.

$$\Sigma_{v \in cand}(d(u) + d(v)) < \Sigma_{v \in adj(u)}(d(u) + d(v)) = d^2(u) + \Sigma_{v \in adj(u)}d(v)$$

The time complexity is analyzed as follows: $\Sigma_{u \in V}|c_{max}| * (d^2(u) + \Sigma_{v \in adj(u)}d(v)) = |c_{max}|(\Sigma_{u \in V}d^2(u) + \Sigma_{u \in V}\Sigma_{v \in adj(u)}d(v)) = |c_{max}|(\Sigma_{u \in V}d^2(u) + \Sigma_{u \in V}d^2(u)) = 2|c_{max}|(\Sigma_{u \in V}d^2(u))$. Thus, the complexity is $O(|c_{max}||V|d_{max}^2)$. $\square$

Algorithm 1. SelectMC(G,K)

  Input : A graph $G = (V, E)$, parameter K
  Output: $C = \{c(u): c(u)$ is an approximate maximum clique containing $u \in V\}$
 1 **begin**
 2    $C \leftarrow \emptyset$;
 3   **foreach** *vertex* $u \in V$ **do**
 4       $cand \leftarrow adj(u)$; $c(u) \leftarrow \{u\}$;
 5   **while** $cand \neq \emptyset$ **do**
 6          $u' \leftarrow argmax_{v \in cand}\{|cand \cap adj(v)|\}$;
 7          $cand \leftarrow cand \cap adj(u')$;
 8          $c(u) \leftarrow c(u) \cup \{u'\}$;
 9   **if** $|c(u)| < K$ **then**
10          $c(u) \leftarrow \emptyset$;
11       $C \leftarrow C \cup \{c(u)\}$;
12   **return** C;

Most existing social networks have been found to be scale-free [1,8] and they have a highly scalable time complexity as shown below.

Lemma 2. *For a scale free network G with a parameter of γ, $2 < \gamma < 3$, the time complexity of Algorithm 1 is $O(|c_{max}||V|d_{max})$.*

Proof. From the proof of Lemma 1, the time complexity of SelectMC(G,K) is $O(2|c_{max}|(\Sigma_{u \in V}d^2(u))$. For a scale-free network, the degree distribution follows a power law. The fraction $p(k)$ of vertices in the network having degree k is given by $p(k) \approx k^{-\gamma}$, where typically $2 < \gamma < 3$.

$\Sigma_{u \in V}d^2(u) \approx \sum_{k=1}^{d_{max}} p(k)|V|k^2 \approx |V|\sum_{k=1}^{d_{max}} k^{2-\gamma} = |V|\sum_{k=1}^{d_{max}} k^{-\alpha}$, where $\alpha = \gamma-2$ and $0 < \alpha < 1$. Since the summation in the above expression is a monotonically increasing function of k, we can bound it by $\int_{k=0}^{d_{max}} k^{-\alpha} \leq \sum_{k=1}^{d_{max}} k^{-\alpha} \leq \int_{k=1}^{d_{max}+1} k^{-\alpha} \leq \frac{1}{1-\alpha}(d_{max} + 1)^{1-\alpha} = O(d_{max})$. Thus, $\Sigma_{u \in V}d^2(u)=O(|V|d_{max})$, and the complexity of $O(|c_{max}||V|d_{max})$ follows.   □

Many social networks have a d_{max} much smaller than $|V|$, typically less than $\sqrt{V}$. The core size c_{max} is also very small. As reported in [19], the maximum clique sizes in their experiments are below 100. The core sizes in our experiments with real datasets are similarly small.

3.2 Extending the Cores

After getting the core for each vertex u, we next extend the core to get a core-based local community that is within the density and radius constraints. Since the goal is to maximize the spread, in this step we consider to maximize the size of the local communities. However, we show that the maximum local community problem is NP-hard by a reduction from the maximum quasi-clique problem

---

Algorithm 2. SelectLC(G, C, γ, r)

Input : A graph $G = (V, E)$, $C = \{c(u) : u \in V\}$, γ and r
Output: Set LC of core-base local communities $LC(u)$ $\forall u \in V$

1 **begin**
2    $LC \leftarrow \emptyset$;
3   **foreach** *vertex* $u \in V$ **do**
4      $cand \leftarrow \{v | R_{c(u)}(\{v\}) \leq r, \forall v \in V - c(u)\}$; $LC(u) \leftarrow c(u)$;
5   **while** $cand \neq \emptyset$ **do**
6        $u' \leftarrow argmax_{v \in cand}\{|LC(u) \cap adj(v)|\}$;
7   **if** $den(LC(u) \cup u') < \gamma$ **then**
8   break;
9        $cand \leftarrow cand - u'$; $LC(u) \leftarrow LC(u) \cup u'$;
10     $LC \leftarrow LC \cup \{LC(u)\}$;
11   return LC;

---

which is NP-complete [17]. Given a simple undirected graph $G = (V, E)$ and a constant $0 \leq \gamma' \leq 1$, a subset of V is called a γ'-quasi-clique if it induces a subgraph with a density of at least γ'. We skip the proof for interest of space, the proof can be found in [23].

Theorem 1. *Given a graph $G = (V, E)$, computing the maximum local community $LC(u)$ for a vertex $u \in V$, given a core of C and a density constraint of γ, a radius constraint of r, with the clique threshold $K = 1$, is NP-hard.*

The above shows that the problem of maximizing the local community under the special condition of $K = 1$ is NP-hard, hence, the general problem where $K \geq 1$ is also NP-hard.

We propose a heuristic algorithm (Algorithm 2) to extend the core for each vertex. First, find the candidate set from the radius constraint (*line 4*). Next, iteratively select the vertex u' such that u' has the largest neighborhood size with respect to the current $LC(u)$ (*line 6*). If the density is still above the threshold after adding u' (*line 7*), we include u' into the local community of u $LC(u)$, and update the candidate set (*lines 9, 10*). Note that if two vertices have the same core, we will not do the extension redundantly.

Example 3. Consider Figure 1 (b). We extend the core of $c(a) = \{a, b, c, d\}$ to get $LC(a)$. Let $\gamma = 0.7$ and $r = 1$. Initially $LC(a) = c(a)$ and $cand = \{e, f, g\}$. Pick vertex e to extend the core, since $|LC(a) \cap adj(e)| = 2$ and $den(\{a, b, c, d\} + \{e\}) = 2 * (6 + 2)/(5 * 4) = 0.8 > \gamma$. Then, $cand = \{f, g\}$. Next, pick vertex f to extend the core since $|LC(a) \cap adj(f)| = 3$ and $den(\{a, b, c, d, e\} + \{f\}) \approx 0.73 > \gamma$. $cand$ is now $\{g\}$. Finally, $den(\{a, b, c, d, e, f\} + \{g\}) = < 0.7$ if g is added. Hence, $LC(a) = \{a, b, c, d, e, f\}$. Similarly, we can get $LC(g) = \{f, g, h, i, j\}$. Note that b, c, d have the same LCs as a, and h, i, j have the same LCs as g.

Algorithm 3. SelectSeedSet(LC, k)

Input : $LC = \{LC(u) = (V_{LC(u)}, E_{LC(u)}) : u \in V\}$ and k
Output: Top-k seed set S
1 **begin**
2      $cand = V$, $S = \emptyset$, $I_S = \emptyset$, $g(S) = |I_S|$, $i = 0$;
3   **while** $i < k$ **do**
4          $u \leftarrow argmax_{v \in cand}\{|V_{LC(v)} \cup I_S| - g(S)\}$; $cand = cand - u$;
5          $w \leftarrow$ highest degree vertex in $LC(u)$;
6   **if** $w \notin S$ **then**
7             $S = S \cup w$; $I_S = V_{LC(u)} \cup I_S$; $g(S) = |I_S|$; $i + +$;

8   **return** S;

Lemma 3. *The time complexity of Algorithm 2 is given by $O(|V||cand_{max}|(|lc_{max}| + d_{max}))$, where $|cand_{max}|$ is the maximum candidate set size, c_{max} is the maximum core size, and d_{max} is the maximum degree of a vertex, lc_{max} is the maximum size of a LC.*

PROOF: Let $cand(u)$ be the initial $cand$ set. For each vertex u, the initialization at *line 4* costs $O(\sum_{v \in cand(u)}(|c(u)| + d(v)))$ time. The set $cand$ can be computed by a breadth first search from u. After the initialization, The intersection size at *line 6* is computed by adding u' (in $O(1)$ time) to the intersecting set $LC(u) \cap adj(v)$ of the neighbors v of the selected u' in the previous iteration. The maximum size is obtained by a scan of the candidate set. Let $M = \min(|LC(u)|, |cand(u)|)$, processing u takes $O(\sum_{v \in cand(u)} d(v) + \sum_{j=1}^{M}(|cand(u)| - j + 1))$ time. Note that $M \leq lc_{max}$ and $c_{max} \leq lc_{max}$. Hence, summing up the above processing time for all $u \in V$ gives $O(\sum_{u \in V} \sum_{v \in cand(u)}(|c(u)| + d(v))) = O(\sum_{u \in V} |c(u)||cand(u)| + \sum_{u \in V} \sum_{v \in cand(u)} d(v)) = O(|V||cand_{max}|(lc_{max} + d_{max}))$. The time complexity is $O(|V||cand_{max}|(lc_{max} + d_{max}))$. □

4   Seed Selection

After calculating $LC(u)$ for each vertex u, we aim to select a seed set S to maximize the information spread base, i.e., $g(S) = |I_S| = \bigcup_{u \in S} LC(u)$. This problem can be shown to be NP-hard by a transformation from the *Maximum coverage problem*. Here we use a greedy algorithm to select the top k seeds. At each iteration, we choose a vertex u where $LC(u)$ contains the largest number of uncovered vertices. We examine $LC(u)$ and choose the highest degree vertex, w, as the next seed, provided that w has not been chosen before. The corresponding pseudocode is shown in Algorithm 3. Algorithm 3, adding a vertex w to S that maximizes $g(S + w) - g(S)$ in each iteration, can be shown to attain $(1 - 1/e)$ approximation. This is because g is monotone ($g(S + v) \geq g(S)$) and submodular(diminishing return: $g(S + v) - g(S) \geq g(T + v)g(T)$, $\forall S \subseteq T$) [15].

Finally, we show that our seed selection algorithm is also efficient and can be scalable to large graphs.

Lemma 4. *Assume a new seed is picked in each while iteration, the time complexity of Algorithm 3 is given by $O(k|V|(|LC(avg)| + |I_S|))$, where $|LC(avg)|$ is the average LC size for the seeds in S.*

PROOF: Each seed is selected by scanning the candidate set of LC's and the current spread base. $I_S(i)$ and $cand(i)$ below refer to the spread base and candidate set before the i-th iteration, respectively. Since

$$\Sigma_{i=1}^{i=k} \Sigma_{v \in cand(i)} (|LC(v)| + |I_S(i)|) < k|V|(|LC(avg)| + |I_S|)$$

thus, the time complexity is $O(k|V|(|LC(avg)| + |I_S|))$.   □

5   Application in Influence Maximization

We consider the application of information dissemination for the problem of influence maximization. One important use of influence maximization is viral marketing [7,21]. The problem of influence maximization (IM) can be defined as follows: Given a graph $G = (V, E)$, with vertex set V and edge set E. Given a model for quantifying the influence of a vertex set. The problem is to choose a set $S \subseteq V$ of k' vertices, or seeds, to target so that the influence of S is maximized.

In Section 7, we describe the issues found with the prevalent models based on probabilistic propagations. We propose to model influence from the perspective of communities instead. Our assumption is that influence is related to information dissemination. We apply the seed set solution of the IDM problem to the IM problem. The rationale is that influence increases with information spread. Unlike previous models, we do not predict the influence since it is application dependent, e.g. the kind of products, or how contagious a disease is, etc. Instead, we aim to maximize the base for the influence.

6   Experimental Results

Our experiments are conducted on a computer with an Intel i7 CPU, 16GB RAM with Ubuntu 12.04 and implemented in C++. For the algorithms that involve randomization we repeat each experiment 1000 times and report the average result. Let us call our method Corebased. For Corebased, the minimum core size K is set to 4, and the radius r is set to the default value of 1. The value of r can be set to 1 because setting r to 2 or more has little effect on the results. This is because the diameter of the LC becomes at most 5 for $r = 2$, and the diameter of a social network is typically small. Hence, when $r = 2$, the reaches become too far and the sparsity of the graph will lead to violation of the density constraint. K can be set to 4 without affecting the results. This is because 2-cliques and 3-cliques (triangles) are numerous in our datasets but for size 4 or above the clique number decreases and these cliques become significant as cohesive components.

6.1   Results on Running Time

In our first set of experiments, we verify the efficiency of our method as predicted by the runtime complexity analysis. We have tested on real graphs from Koblenz Large Network Collection (KONECT), Stanford Large Network Dataset Collection (SNAP), and Max Planck Institute collection. For our method, we convert directed graphs into undirected graphs by making each edge undirected.

	Amazon	DBLP	Epinions	Facebook	Arxiv
$|V|$	403394	317080	75879	63731	34546
$|E|$	3387388	1049866	405740	1545686	421578
$|d_{max}|$	2752	343	3044	1098	864

The runtime results are shown in Figure 2. The time to compute cores and extend cores are dominating. With higher degree vertices in Epinion, more vertices in the candidate set of the core can be used to extend the core, hence, more set intersection operations in Algorithm 2, and bigger LC sizes. For seed selection, we scan the communities to determine the next seed, the time complexity grows linearly with the total community size. Overall, the results show that our algorithm can efficiently handle large graphs.

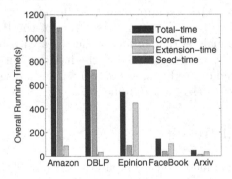

Fig. 2. Runtime of CoreBased for 5 datasets with $\gamma = 0.4$ and 50 seeds

6.2   Results on Application: Influence Maximization

For the experiment on influence maximization, we follow the methodology in [10]. It is a known issue in the study of IM that it is difficult to obtain ground truth. To the best of our knowledge, [10] provides the only method for this issue. It takes a dataset with a social graph recording the friendship among a set of users and also an action log, which consists of triples (u, a, t) recording an action a by user u at time t. The action log concerning an action a is called the propagation trace associated with a. E.g., an action is the rating of a movie. If user v rates a movie, and at a later time a friend u of v also rates the movie,

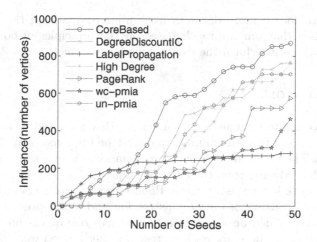

Fig. 3. Comparison of influence for different algorithms

there is a propagation from v to u. Hence, we would find (v, a, t) and (u, a, t') in the action log, where $t' > t$. In [10], such propagation from a user to the friends of the user is considered the ground truth.

For the Facebook dataset of New Orleans from socialnetworks.mpi-sws.org, there are two components, one contains a list of user-to-user links and the second is from a list of wall postings, both lists contain a UNIX timestamp for each link. The second component is thus the action log.

We consider each wall posting an action. The friendships among users are obtained from the user-to-user links. Suppose u is in the seed set, then on the wall of u, we take the posting by u to be the initiating action. The influence of u is the number of friends of u that have posted on u's wall after u's own posting.

Since our solution is related to community search, we compare with existing community search methods. We consider the highest degree vertex as a potential seed in each computed community. In the recent survey in [6], 40 community discovery methods are listed, however, very few of these methods have scalable complexity. We have selected the only two scalable algorithms reported to handle large graphs under the category of "Diffusion", which is related to information dissemination. The two methods are Label Propagation [20] and DegreeDiscountIC [5]. Our method is tailored for maximizing the spread base, it is not fair to compare the spread base. Instead, we compare by the measure of influence.

We also compare with the following methods: High Degree, PageRank, un-pmia [4], and wc-pmia [4]. High Degree is a baseline approach by select k vertices with the highest degrees as the seeds. pmia is a scalable algorithm that is shown to be effective for the IC model for the IM problem [4], wc-pmia adopts the WC model while un-pmia is pmia with uniform probability of 0.01 at each edge. We adopt the settings for PageRank and DegreeDiscountIC from [4].

The results are shown in Figure 3. Our method consistently produces the highest influence among all the other methods in the experiment. Although it is difficult to locate more datasets for similar comparison, the results here provide evidence that our approach based on the principles of homophily has great potential to outperform the existing models and methods.

7   Related Work

For LCs, we compute a clique for each vertex, this is related to the maximum clique problem, since the maximum clique must be the maximum clique for one of the vertices. The maximum density-based quasi clique problem is proved to be NP-complete by reducing from the classical clique problem in [17]. [2] utilizes an existing framework known as a Greedy Randomized Adaptive Search Procedure (GRASP), which consists of initial construction and local optimization in each iteration. However, such an algorithm only returns one quasi-clique, while our problem requires one clique for each vertex. Another related work [22] is to find a subgraph of G that contains a given set of query nodes and which is densely connected. The problem is to maximize the minimum degree with size restriction. The authors propose a greedy algorithm to solve their problem in $O(|V| + |E|)$ time.

IM has been studied under the IC and LT models [11]. [4] proposes a scalable algorithm for the IC model with a parameter on the influence probability for early stopping. Graph sparsification is introduced in [13] with edge pruning during influence propagation. However, existing IM models are not validated with ground truth. In the viral marketing study in [12], it is shown that the real world does not match these models. Similar findings are reported in [10], a scatter plot between influence predicted by the models and the actual influence on a real dataset shows a big discrepancy where the predicted influences are many times higher than the actual values.

8   Conclusion

For future work, we may consider the issue of overlapping communities. The LC model can be extended to include multiple communities for each seed. Another extension is to consider directed graphs in community search, which is an interesting problem in general. To our knowledge the IDM problem is a new problem and has important applications in IM. Most existing IM studies rely on propagation models shown in some recent works such as [10,12] to be problematic. Most works assumed that such a model is the ground truth and results were compared only based on measurements defined by such models. Our study deviates from this trend and shows that the homophily based IDM model for IM can produce better results in a case study. More study will be needed for this approach, but the methodology is sound and we believe that this can be a promising new approach.

References

1. http://knoect.uni-koblenz.de/networks
2. Abello, J., Resende, M.G.C., Sudarsky, S.: Massive quasi-clique detection. In: Rajsbaum, S. (ed.) LATIN 2002. LNCS, vol. 2286, pp. 598–612. Springer, Heidelberg (2002)
3. Bomze, I., Budinich, M., Pardalos, P., Pelilo, M.: The maximum clique problem. Handbook of Combinatorial Optimization A, pp. 1–74 (1999)
4. Chen, W., Wang, C., Wang, Y.: Scalable influence maximization for prevalent viral marketing in large-scale social networks. In: 16th SIGKDD, pp. 1029–1038 (2010)
5. Chen, W., Wang, Y., Yang, S.: Efficient influence maximization in social networks. In: 15th SIGKDD, pp. 199–208. ACM (2009)
6. Coscia, M., Giannoti, F., Pedreschi, D.: A classification for community discovery methods in complex networks. Journal of Statistical Analysis and Data Mining 47(11), 41–45 (1997)
7. Domingos, P., Richardson, M.: Mining the network value of customers. In: 7th SIGKDD (2001)
8. Faloutsos, M., Faloutsos, P., Faloutsos, C.: On power-law relationships of the internet topology. In: SIGCOMM (1999)
9. Fortunato, S.: Community detection in graphs. Physical Reports 486, 75–174 (2010)
10. Goyal, A., f. Bonchi, Lakshmanan, L.: A data-based approach to social influence maximization. In: VLDB (2011)
11. Kempe, D., Kleinberg, J., Tardos, É.: Maximizing the spread of influence through a social network. In: Ninth SIGKDD, pp. 137–146. ACM (2003)
12. Leskovec, J., Adamic, L.A., Huberman, B.A.: The dynamics of viral marketing. ACM Transactions on the Web 1(1), 1–39 (2007)
13. Mathioudakis, M., Bonchi, F., Castillo, C., Gionis, A., Ukkonen, A.: Sparsification of influence networks. In: 17th SIGKDD, pp. 529–537. ACM (2011)
14. Michael, J.: Labor dispute reconciliation in a forest products manufacturing facility. Forest Products Journal 47(11), 41–45 (1997)
15. Nemhauser, G., Wolsey, L., Fisher, M.: An analysis of the approximations for maximizing submodular set functions. Mathematical Programming 14, 265–294 (1978)
16. de Nooy, W., Mrvar, A., Batagelj, V.: Exploratory Social Network Analysis with Pajek. Cambridge University Press, Cambridge, UK (2005)
17. Pattillo, J., Veremyev, A., Butenko, S., Boginski, V.: On the maximum quasi-clique problem. Discrete Applied Mathematics 161(1), 244–257 (2013)
18. Pullan, W., Franco, M., Mauro, B.: Cooperating local search for the maximum clique problem. J. Heuristics 17, 181–199 (2011)
19. Pullan, W., Hoos, H.H.: Dynamic local search for the maximum clique problem. Journal of Artificial Intelligence Research 25, 159–185 (2006)
20. Raghavan, U., Albert, R., Kumara, S.: Near linear time algorithm to detect community structures in large-scale networks. Physical Review E 76 (2007)
21. Richardson, M., Domingos, P.: Mining knowledge-sharing sites for viral marketing. In: KDD (2002)
22. Sozio, M., Gionis, A.: The community-search problem and how to plan a successful cocktail party. In: 16th SIGKDD, pp. 939–948. ACM (2010)
23. Yang, Z., Fu, A., Xu, Y., Huang, S., Leung, H.: Community based information dissemination. Technical Report, CSE, CUHK (2015). http://www.cse.cuhk.edu.hk/~adafu/paper/cbid.pdf

A Fast and Effective Image Geometric Verification Method for Efficient CBIR

Ling-Bo Kong[1]([✉]), Ling-Hai Kong[2], Tao Yang[3], and Wei Lu[1]

[1] School of Software Engineering, BeiJing JiaoTong University,
S.S.E of BJTU, BeiJing 100044, China
mlinking@126.com, luwei@bjtu.edu.cn
[2] IAPCM, BeiJing 100088, China
kong_linghai@iapcm.ac.cn
[3] TshingHua University, BeiJing 100084, China
ytao@mail.tsinghua.edu.cn

Abstract. Along with the widespread use of IT techniques, the requirements for CBIR (Content-Based Image Retrieval) is attractive for researchers from diverse areas. CBIR's challenge is still how to ensure the meaningfulness of the retrieved images, for which the geometric consistency should be considered. And RANSAC and its variants are popular in the post-verification stage for that. This paper presents a Delaunay triangulation (DT) based method for that, some properties of which ensure its stability to capture the local structures. By converting the geometric verification into DT mapping, our method could not only catch invariant local structure points, but also is much more efficient ($O(Nlog(N))$). We evaluate our approach on common image benchmark and demonstrate its effectiveness for image geometric verification problem.

Keywords: CBIR · Geometric verification · SIFT · RANSAC · Delaunay triangulation

1 Introduction

Along with the widespread use of IT techniques, the development of the Internet, and the availability of image capturing devices, mass of multimedia data (including images) are produced. Efficient image searching, browsing and retrieval tools are required by users from various domains, including remote sensing, fashion, crime prevention, publishing, medicine, architecture, etc. For this reason, many general purpose image retrieval systems have been developed since the early 1980s, which are generalized as CBIR. According to [11,16], CBIR usually follows similar framework as Search Engine now, shown in Figure 1. There are three major processing stages, which are (1) "Feature Collection", (2) "Feature based Image Filtering" and (3) "Image Verification".

L.-B. Kong — This work was supported by the China Fundamental Research Funds for the Central Universities under Grant No.2011JBM320, and the National Natural Science Foundation of China (NSFC) under Grant No.61272353.

© Springer International Publishing Switzerland 2015
M.A. Sharaf et al. (Eds.): ADC 2015, LNCS 9093, pp. 232–243, 2015.
DOI: 10.1007/978-3-319-19548-3_19

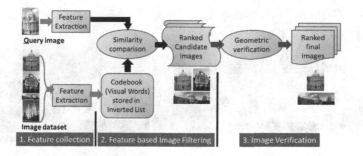

Fig. 1. The general framework of CBIR or large scale image retrieval. Three stages are pipelined, namely "1. Feature Collection", "2. Feature based Image Filtering" and "3. Image Verification".

For stage 1, most of the recent image retrieval systems rely on local features, in particular the SIFT (Scale Invariant Feature Transform) descriptor [12] and its variants. Those descriptors are typically used jointly with a bag-of-words (BOW) or bag-of-visual-word (BOV or BOVW) approach, which is the core of stage 2. After the retrieval of candidate images by feature filtering, a geometric verification is needed because the feature filtering only returns the candidate images which contain many similar features of the given query image. The spatial information is usually considered in the post-processing step, through a spatial verification like RANSAC (RANdom SAmple Consensus) and its variants[6,7,9,15].

Fig. 2 demonstrates the processing sketch of CBIR, in which (a) "bodleian106" and (b) "bodleian36" are the query and sample images from Oxford image dataset[1,13]. Fig. 2 (c) demonstrates the filtered SIFT points in query and sample images - called putative points. Those putative points are connected with lines for clarification, in which we can easily see the trend of the corresponding points. Fig. 2 (d) shows the SIFT points verified by RANSAC, which are well corresponded with lines.

However, there are mainly two limitations of using RANSAC (also its many variants) to act as the geometric verification method. One is that the performance of RANSAC (and its variants) is computationally expensive to find the spatial verification according to [13,21]. Therefore, they are applied only to the top candidate images filtered by stage 2. The other limitation is that RANSAC can only be effective to find the linearly correspondent points. This could be clearly understood by the two circles and rectangles in Fig. 2 (c) and (d). From Fig. 2 (c), some points enclosed by the circle and rectangle should be believed as well matched points. But, we can see from Fig. 2 (d) that no SIFT points in the circle and rectangle are marked as the verified points.

In this paper we present a new geometric verification method. It is based on an intuitive logic. With the putative points derived from two images, we could construct the Delaunay triangles. And the stable properties of Delaunay triangulation ensures that if the two images are similar, the two triangles should share many triangles with same vertices. Based on this intuition, we further propose

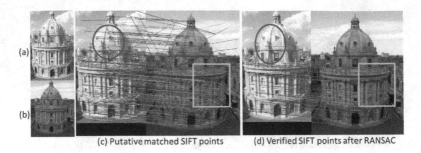

(c) Putative matched SIFT points   (d) Verified SIFT points after RANSAC

Fig. 2. Demonstration of CBIR processing sketch. (a) and (b) are the query and sample images from Oxford dataset. (c) is the matched key points based on SIFT, which are used to get candidate images. (d) is the verified key points after RANSAC with 150 loops. The two circles and rectangles in (c) and (d) are used to illustrate that many interesting SIFT points could not be correctly verified by RANSAC.

a method which converts the image geometric verification into the intersection computation. Compared with the related state-of-art methods, our approach could not only be more efficient, but also be powerful to find valid well matched points. Fig. 3 demonstrates the approach, and more details could be found in later Section 2.

The rest of this paper is organized as follows. Section 2 describes our method in more details. The experiments for evaluating our approach are described in Section 3. Section 4 gives an overview of related research. Final section is our conclusion.

2 Our Approach

In this section, our approach is presented in details. Since our proposal takes usage of Delaunay triangulation technique, the short review about it is first presented in Section 2.1. Then the logic of our method is shown in Section 2.2. And at Section 2.3 the algorithm following our approach is given and some attributes of our approach are discussed.

2.1 Delaunay Triangulation

In mathematics, Voronoi diagram refers to the diagram, which partitions a plane with N points into convex polygons such that each polygon contains exactly one generating point and every point in a given polygon is closer to its generating point than to any other. The Delaunay triangulation is a triangulation, which takes a region of any dimension space and divides it into subregions and which is equivalent to the nerve of the cells in a Voronoi diagram. The complexity of the algorithm is $O(Nlog(N))$[3].

Delaunay triangulation has certain properties, which ensure that noise affects the Delaunay triangulation only locally. And therefore, the Delaunay triangulation was found to have the best structural stability under random positional perturbations. In this paper, we use the stability property of Delaunay triangulation, shown in the next section.

2.2 The Idea of Our Proposal

Our proposal is based on the Delaunay triangulation and an intuitive logic. By Delaunay triangulation, it means we can construct two concrete Delaunay triangulations based on the putative points in the query image and the sample image, marked as DT_Q and DT_S respectively. Fig. 3 (a) and (c) demonstrate this.

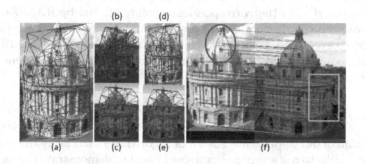

Fig. 3. The demonstration of our proposal. (a) and (c) show the Delaunay triangulations of the putative SIFT points in query and sample images - marked as DT_Q and DT_S respectively. (b) illustrates the counterpart triangulations by connecting the corresponding points of the query's DT_Q in the sample image - marked as T_{QS}. (d) and (e) demonstrate the correspondence triangulations between DT_Q and DT_S which are derived with the help of T_{QS} - green color for query image and red color for sample image. Finally (f) shows the effectiveness of the correspondence points derived by our approach with the help of connection lines. The circle and the rectangle cover the correspondence points missed by RANSAC compared with the Fig. 2 (d).

Besides, we can also map the DT_Q into the sample image to build a counterpart triangulation, marked as T_{QS}. The procedure is as follows. For each triangulation Q_i in DT_Q, its three vertexes are labelled as vq_{i1}, vq_{i2} and vq_{i3}. Clearly there are corresponding matched SIFT points in sample image, and we use vs_1, vs_2 and vs_3 without loss of generality. The triangle based on vs_1, vs_2 and vs_3 is then the counterpart triangle in the sample image for Q_i in the query image. We use T_{QS} to label the connections derived from the mapping of the DT_Q. Generally, this counterpart triangle will not be a Delaunay one only if the two images are same, however, the mapping from DT_Q into T_{QS} is still useful. Fig. 3 (b) demonstrates the counterpart triangles in the sample image, from which we can see they are chaotically connected in the sample image.

However, the gold should be somewhere, and we need a simple logic here. Based on the above description, T_{QS} could be understood as the mapping of DT_Q in the sample image. While the DT_S is the Delaunay triangles controlled by the real structures of the corresponding points. If there are some triangles in both T_{QS} and DT_Q with same points as vertexes, we could believe this implies those triangles have same local structures. This also means the corresponding points of those triangulations in query and sample images respectively should be the correspondence points with high confidence.

Here we use Fig. 3 to demonstrate the effect first, and present the algorithm and the related analysis in later section. Fig. 3 (b) and (c) show the T_{QS} and DT_S in the sample image. While Fig. 3 (d) and (e) illustrate DT_Q in query and DT_S in the sample respectively. The triangles of the correspondence matched points in query and sample images are shown in different colors - green for the query image and red for the sample image. Fig. 3 (f) connects the correspondence points with lines, which is popular to show the effectiveness of the matched points in related papers. Compared with the correspondence points filtered by RANSAC in Fig. 2 (d), we can see clearly that our approach could locate other correspondence points, which are enclosed with a circle and a rectangle. Especially, those correspondence points should be intuitively believed as real correspondence points from the connection lines.

2.3   Algorithm and Analysis

The algorithm of our approach is shown in Algorithm 1. To clarify the processing steps, here I'd like to use a simple example of Fig. 4 to demonstrate the algorithm.

Algorithm 1. GEOMATCH ALGORITHM

Input: The putative points of query and sample images - P_Q and P_S. Their cardinality is same, marked as N
Output: All the verified point pairs between P_Q and P_S
Begin
1. DT_Q and DT_S ← Construct the Delaunay triangulations from P_Q and P_S for the query and sample images respectively;
2. T_{QS} ← Construct the counterpart triangulations in sample image;
3. MT_{QS} ← Find all ternary points in DT_S which are also contained in T_{QS};
4. MT_{SQ} ←Derive the corresponding triangulations of the MT_{QS} in the query image;
5. Res ← $MT_{QS} \cap MT_{SQ}$;
    /* Notice: duplicate pair should be discarded. */
6. **Return** Res.
End

The input of the algorithm contains two point sets in which their correspondences are marked with the positions, as shown in Fig. 4 (a). P_Q is $\{1, 2, 3, 4\}$, while P_S is $\{a, b, c, d\}$. The coordinates of those points are enclosed with squared brackets, and their correspondence is shown in the Fig. 4 (a). We should remind here that the "1-4" numbers and "a-d" characters are not explicitly recorded. They are just used here to make the demonstration clearly. In fact the paired items in correspondence matrix are the indices of the matched points in P_Q and P_S.

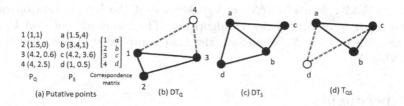

1 (1,1)	a (1.5,4)	
2 (1.5,0)	b (3.4,1)	
3 (4.2, 0.6)	c (4.2, 3.6)	
4 (4, 2.5)	d (1, 0.5)	
P_Q	P_S	Correspondence matrix

(a) Putative points   (b) DT_Q (c) DT_S (d) T_{QS}

Fig. 4. The demonstration of the algorithm. (a) shows the putative points - P_Q and P_S - together with their positions in corresponding images. (b) and (c) are the Delaunay triangulations of those putative points in query and sample images - DT_Q and DT_S. (d) illustrates the counterpart triangulations of the DT_Q in sample image - T_{QS}.

The positions of those putative points are shown in Fig. 4 (b) and (c) respectively, in which the corresponding Delaunay triangles are also illustrated. According to the properties of Delaunay triangulation, the local structures of the points are reserved in the connected triangles. From Fig. 4 (b) and (c) we can easily distil the vertexes of the triangles. We still use DT_Q and DT_S for the distilled ternary points, namely DT_Q is $\{(1,2,3),(1,3,4)\}$, while DT_S is $\{(a,d,b),(a,b,c)\}$ (Step 1 in Algorithm 1).

Since the correspondence of the P_Q and P_S is already recorded as shown in Fig. 4 (a), it's easy to distil the counterpart triangles in sample image from DT_Q, that is $T_{QS} = \{(a,b,c),(a,c,d)\}$, shown in Fig. 4 (d) (Step 2 of Algorithm 1). This also implies that the structures of P_Q are mapped into the sample image. With the comparison of DT_S and T_{QS} now, it's easy to find the same ternary point items. We get $\{(a,b,c)\}$ here - MT_{QS} (Step 3). And with the help of the correspondence between the putative points in Fig. 4 (a), it's also simple to distil the counterpart in the query image of MT_{QS}. We get $\{(1,2,3)\}$ as MT_{SQ} (step 4). As for step 5, it's just to do intersection between MT_{QS} and MT_{SQ}, and pair the corresponding vertexes. Finally, we get $\{(1,a),(2,b),(3,c)\}$ as Rcs.

The simple analysis of our methods covers two issues as follows.

1. Our proposal could find meaningful points compared with RANSAC series methods.

    From the description of our approach we can see that the verification is not based on the distance comparison, but the local structure correspondence. While RANSAC series generally use distance comparison to find inliers, which is limited to find non-linear matched points (See more details in Section 4). This could also be vividly found by comparing the Fig. 2 (d) and Fig. 3 (f).

2. The complexity of our algorithm is $O(Nlog(N))$ where N is the number of putative points.

    For step 1, the construction of the Delaunay triangles is $O((Nlog(N))$ according to [3]. While for step 2, the distillation of the counterpart is $O(N)$. This is because any set P of N points has $2N - h - 2$ triangles, where h is the number of vertices of $ch(P)$ (Convex Hull of the P). As for step 3, its complexity is also $O((Nlog(N))$ because of the intersection operation with the help of Binary Search algorithm. The complexity of step 4 and 5 is also linear with N. Therefore, the final complexity of our algorithm is $O(Nlog(N))$.

3   Experiments

To verify our proposal, extensive experiment is carried out. Section 3.1 describes the configuration of our experiment. The experiment result and some analysis are shown in Section 3.2.

3.1   Experiment Configuration and Datasets

Two methods are compared in our experiment - the RANSAC from VLFEAT[20], and the LO+-RANSAC[2,7]. The reason we choose them is that RANSAC is the standard baseline to understand its variants, while LO+-RANSAC is reported as one of the most efficient RANSAC variants[6]. The limitation of RANSAC and its variants could be clearly demonstrated with them, which is that they are generally not sensitive to the more complicated structure verification, especially for the deformation situation like that shown in Fig. 2 (d) and Fig.3 (f). You can also read later Section 4 for this, or other related papers directly[6,13] etc.

Two major issues are considered to evaluate our algorithm - the performance and the effectiveness. With the selected query images, those three methods are executed 6 times for each filtered sample images. The running-time for the first loop is discarded and the average time of the rest 5 loops is recorded as the performance of those methods. As for the effectiveness, we first sort the retrieved images according the their verified points, and then check them manually with our colleague. The default loop number for RANSAC and LO+-RANSAC to try samplings is 100 so as to demonstrate the performance advantage of our proposal, which is far less than the loop number.

The experiment is carried out in a Dell Vostro 1400 laptop, in which CPU is 2-DUO with 2.0 GHz, and main memory is 4 GB. The operating system is Windows 7 Professional version. We use Matlab R2012a (7.14) to implement our algorithm, and other codes for comparison. The dataset used here is the Oxford Buildings images[1], which is popularly taken as the standard benchmark to test CBIR techniques. Eleven of those query images are shown in Fig. 5, in which the words under the images are the image directory for the query image.

3.2   Discussion of the Results

To discuss the experiment result, we first use one instance to demonstrate our style in Section 3.2. Then the running-time results of those methods are shown and discussed in Section 3.2.

Explanation with One Instance. Different from the related papers, we use Fig. 6 style to demonstrate the result, because this style could show more details about the effectiveness. And since the other results of different query images are similar, this paper shows only one, and we will publish all the results in our website. There are three semantic regions in Fig. 6:

1. 3 images at the left side are the images used as query and top samples by different methods. Here, the query image is "bodleian 291", while "116" and "244" are the bodleian images returned by RANSAC, LO+-RANSAC and our approach as the top verified images.
2. The table at the center contains the major information of those methods, which will be discussed in details later;
3. The 4 images at the right side are the images in "bodleian" which are not successfully filtered by all those three methods. In fact, many images are missed in other query processing. This manifests the limitation of current CBIR methods, which is usually not clearly mentioned in related papers.

Fig. 5. 11 of those used query images. Two rectangles mean that those corresponding images are from "bodleian" and "oxford".

Now we'll present more details about the result in Fig. 6. First is about the effectiveness of those methods to discriminate the correct target images with the given query. In the central table, the experiment result is organized into three major parts - the ordered "Image ID" at the left, the number of verified points at the center, and the average running-time at the right. In fact, the descendant order of the top 20 images is based on the number of verified points. From the table we can see all those three methods could correctly recognize the query image itself. However, RANSAC and LO+-RANSAC return "116" as the 2nd top image.

We also can see that LO+-RANSAC later could return "244" as the 3rd top one. However, this still demonstrates the reliability limitation of LO+-RANSAC, which means its result is a little chaotic. As for our approach, we find that if the number of verified points is more than 6, the sample image has high probability to be well similar with the query. This is also found in other queries. By contrast,

Image ID			# of Verified points			Time (mille, sec)		
RANSAC	LORANSAC	Ours	RANSAC	LORANSAC	Ours	RANSAC	LORANSAC	Ours
291	291	291	690.2	3769	3335	90.91	31.32	66.19
116	116	244	11.8	9	23	29.39	47.77	3.04
166	244	226	10.6	8	6	29.35	40.17	1.82
325	367	62	10	7.8	3	29.91	83.35	2.79
473	473	77	9.4	7.6	3	29.1	31.66	1.6
244	417	83	7.6	7.4	3	61.75	71.23	1.73
300	52	134	7.4	7.2	3	34.15	70.46	1.96
105	70	163	7.2	7	3	30.08	31.62	1.88
108	96	361	6.8	7	3	31.79	70.46	2.02
52	102	380	6.6	7	3	29.49	75.28	1.91
192	105	386	6.4	7	3	33.9	47.67	1.79
319	303	395	6	7	3	29.23	69.04	2.19
357	363	410	5.8	7	3	29.35	13.88	1.97
27	408	0	5.6	7	0	29.32	39.57	2.14
60	409	1	5.6	7	0	29.57	62.87	2.06
167	13	2	5.6	6.8	0	28.99	30.8	1.6
236	60	7	5.6	6.8	0	29.61	74.22	1.86
324	109	9	5.6	6.8	0	32.5	83.38	1.68
396	253	10	5.6	6.8	0	68	67.71	1.89
409	357	12	5.6	6.8	0	34.28	61.56	2.05
					Avg time:	34.2	56.46105	2.019

Fig. 6. Comparison of three methods - RANSAC, LORANSAC and our proposal. 3 images at the left are the query (Q=bodleian291) and two top images derived by those methods respectively - 116 and 244. The center table shows the experiment data - "Image ID", "number of verified points" and "Time" in millisecond, and only top 20 images are listed here. The 4 images at the right are the ones which could not be , while our human being may correctly figure them out.

there is no clear threshold for RANSAC and LO+-RANSAC to assign the sample images as the very target images. You can understand this from Fig. 6, in which except for the query image itself, the number of the verified points for RANSAC and LO+-RANSAC is almost continuous - from 5.6 to 11.8, and from 6.8 to 9. The observation of taking 6 as the experiential threshold for our approach is also applicable in other queries. This is obviously helpful for CBIR to return more meaningful result.

As for the performance, the first row is special because it's the running-time to compare the query image itself. In our experiment we find LO+-RANSAC has the best performance for this. This is also true in other queries. We further compute the average running-time of those three methods except for the query image itself, and the result is shown beneath the table.

Performance in Total. We record the running-time of the three methods for those 11 query images. The data is organized into two figures - the running-time on the query image itself and the other images, shown in Figure 7(a) and Figure 7(b) respectively.

From Figure 7(a), the LO+-RANSAC usually has best performance for the query image itself. There is also an interesting observation, that is, if the number of putative points is larger than 1000, the best choice is LO+-RANSAC. However, this situation is only true for the query image itself. All the rest images generally have less number of putative points. While according to Figure 7(b), the max value of the average running-time (namely except for the query image itself) of our approach is no more than 5 milliseconds.

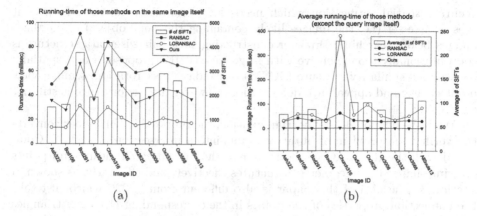

Fig. 7. Running-time comparison of three methods - RANSAC, LO+-RANSAC and our proposal - 7(a) on the query image itself, and - 7(b) on the images except for the query image itself. The lines with scattered dots use the left Y-axis, whose unit is millisecond. The vertical bars use the right Y-axis, which demonstrates the number of the SIFT points.

4   Related Work

In this section, we briefly introduce the methods to cope with the related problem mentioned in previous sections, i.e., how to improve the effectiveness and efficiency for better result in CBIR.

After the introduction of CBIR in 1980s, it always attracts the researchers' attention from different areas, like Computer Vision, DBMS (Database Management System), Machine Learning etc. By combining local invariant features (like SIFT, Harris Corner Detector, etc.) into BoW framework (or BOV, or BOVW. as shown in Fig. 1), recent endeavours mainly focus on how to adopt geometric properties into CBIR to improve the effectiveness and efficiency.

The straightforward idea for this is to use spatial information in a post-verification step. [18] considers the nearest neighbour layout support way to refine the putative points. While [13] adopts RANSAC variant - LO+-RANSAC in the post-verification step. To some extent, all the RANSAC variants could be used in this step. However they usually come with high computational cost, and are consequently only used to verify a limited number of top-ranked images.

Later on, many various approaches are proposed to encode relatively weak spatial constraints in the initial search step without sacrificing much retrieval efficiency. Feature locations are probably the most frequently used spatial information as they can be easily integrated into the inverted file representation[5, 10, 13, 19, 21]. Unfortunately, these spatial constraints are either too restrictive so that only translation can be handled, or too loose to capture enough information.

Another way to compensate the deficiency in feature matching is to automatically expand the query[8, 14, 17]. However, the performance of query expansion tends to be degraded by false positive search results. Therefore it requires accurate spatial verification which needs high computational cost. In [14], it uses a close set (i.e. the images likely containing the same object) of database images before searching. However constructing such pair-wise data structure is computationally too expensive with large dataset. [17] proposes a new spatially constrained similarity measure (SCSM) to handle object rotation, scaling, view point change and appearance deformation. However, its performance is still not competitive.

With the help of Delaunay triangulation, the proposal in this paper converts the verification problem of putative points into a intersection procedure. Compared with the RANSAC methods, our method could catch meaningful points obeying similar local geometric structures effectively and efficiently, as shown in Section 3. The idea of this paper is also different from [4,22], which just take the connection properties of the points in the corresponding Delaunay triangles to improve the matching veracity.

5   Conclusion

This paper presents a fast and effective approach to verify the candidate images for CBIR. Based on the intuitive observation of the correspondence between the Delaunay triangles of the putative points, we could convert the verification problem as the mapping problem between two Delaunay triangle sets, and could further use an intersection procedure of the ternary vertexes for that. With the support of the DT's stability of capturing local structure, our approach could not only catch invariant local structure points, but also is much more efficient. The experiments on standard benchmark data demonstrates its effectiveness for the geometric verification, and also has competitive performance.

One of our future work lies to discuss the rationale of our approach, even we have experimental confirmation here. The considerations may include the relationship between the verified points of our approach and RANSAC variants, the usage of the connectivity among those verified points for better transformation matrix, etc. Besides, the evaluation of our proposal on more datasets will also be carried out.

Acknowledgments. Thanks to the anonymous reviewers to check the idea of this paper. We also appreciate the help from Karel Lebeda about the LO+-RANSAC code.

References

1. http://www.robots.ox.ac.uk/~vgg/data/oxbuildings/index.html
2. http://lebeda.sk/DP/index.xhtml
3. Attali, D., Boissonnat, J.-D., Lieutier, A.: Complexity of the delaunay triangulation of points on surfaces: the smooth case. In: 19th Annual Symposium on Computational Geometry, pp. 201–210 (2003)
4. Bhattacharya, P., Gavrilova, M.: DT-RANSAC: A Delaunay Triangulation Based Scheme for Improved RANSAC Feature Matching. In: Gavrilova, M.L., Tan, C.J.K., Kalantari, B. (eds.) Transactions on Computational Science XX. LNCS, vol. 8110, pp. 5–21. Springer, Heidelberg (2013)

5. Cao, Y., Wang, C., Li, Z., Zhang, L., Zhang, L.: Spatial bag-of-features. In: Proc. CVPR, pp. 3352–3359 (2012)
6. Choi, S., Kim, T., Yu, W.: Performance evaluation of RANSAC family. In: Proc. BMVC, pp. 110–119 (2009)
7. Chum, O., Matas, J., Kittler, J.: Locally optimized RANSAC. In: Michaelis, B., Krell, G. (eds.) DAGM 2003. LNCS, vol. 2781, pp. 236–243. Springer, Heidelberg (2003)
8. Chum, O., Mikulík, A., Perdoch, M., Matas, J.: Total rrecall II: Query expansion revisited. In: Proc. CVPR, pp. 889–896 (2011)
9. Fischler, M.A., Bolles, R.C.: Random sample consensus: A paradigm for model fitting with applications to image analysis and automated cartography. Comm. of the ACM **24**, 381–395 (1981)
10. Jégou, H., Zisserman, A.: Triangulation embedding and democratic aggregation for image search. In: Proc. CVPR, pp. 3310–3317 (2014)
11. Liu, Y., Zhang, D.S., Lu, G.J., Ma, W.Y.: A survey of content-based image retrieval with high-level semantics. Pattern Recognition **40**, 262–282 (2007)
12. Lowe, D.: Distinctive image features from scale-invariant keypoints. IJCV **60**(2), 91–110 (2004)
13. Philbin, J., Chum, O., Isard, M., Sivic, J., Zisserman, A.: Object retrieval with large vocabularies and fast spatial matching. In: Proc. CVPR, pp. 1–8 (2007)
14. Qin, D.F., Gammeter, S., Bossard, L., Quack, T., Van Gool, L.: Hello neighbor: accurate object retrieval with k-reciprocal nearest neighbors. In: Proc. CVPR, pp. 777–784 (2011)
15. Raguram, R., Frahm, J.-M., Pollefeys, M.: A comparative analysis of RANSAC techniques leading to adaptive real-time random sample consensus. In: Forsyth, D., Torr, P., Zisserman, A. (eds.) ECCV 2008, Part II. LNCS, vol. 5303, pp. 500–513. Springer, Heidelberg (2008)
16. Rui, Y., Huang, T.S., Chang, S.F.: Image retrieval: current techniques, promising directions, and open issues. J. Visual Commun. Image Representation **10**(4), 39–62 (1999)
17. Shen, X.H., Lin, Z., Brandt, J., Avidan, S., Wu, Y.: Object retrieval and localization with spatially-constrained similarity measure and k-nn re-ranking. In: Proc. CVPR, pp. 3013–3020 (2012)
18. Sivic, J., Zisserman, A.: Video google: a text retrieval approach to object matching in videos. In: Proc. ICCV, pp. 1470–1477 (2003)
19. Tolias, G., Furon, T., Jégou, H.: Orientation covariant aggregation of local descriptors with embeddings. In: Fleet, D., Pajdla, T., Schiele, B., Tuytelaars, T. (eds.) ECCV 2014, Part VI. LNCS, vol. 8694, pp. 382–397. Springer, Heidelberg (2014)
20. Vedaldi, A., Fulkerson, B.: VLFeat: An open and portable library of computer vision algorithms (2008)
21. Zhang, Y.M., Jia, Z.Y., Chen, T.: Image retrieval with geometry-preserving visual phrases. In: Proc. CVPR, pp. 809–816 (2011)
22. Zhao, X.Y., He, Z.X., Zhang, S.Y.: Improved keypoint descriptors based on delaunay triangulation for image matching. Optik **125**, 3121–3123 (2014)

Efficient Mining of Non-derivable Emerging Patterns

Vincent Mwintieru Nofong[(✉)], Jixue Liu, and Jiuyong Li

School of Information Technology and Mathematical Science,
University of South Australia, Adelaide, Australia
vincent.nofong@mymail.unisa.edu.au

Abstract. Emerging pattern mining is an important data mining task for various decision making. However, it often presents a large number of emerging patterns, most of which are not useful as their emergence are derivable from other emerging patterns. Such derivable emerging patterns most often are trivial in decision making. To enable mine the set of non-derivable emerging patterns for decision making, we employ deduction rules in identifying the set of non-derivable emerging patterns. We subsequently make use of a significance test to identify the set of significant non-derivable emerging patterns. Finally, we develop NEPs, an efficient framework for mining the set of non-derivable and significant non-derivable emerging patterns. Experimentally, NEPs is efficient, and the non-derivable emerging pattern set which is smaller than the set of all emerging patterns, shows potentials in trend prediction on a Twitter dataset.

Keywords: Frequent patterns · Emerging patterns · Non-derivable emerging patterns · Significance test

1 Introduction

Emerging Patterns (EPs), the set of patterns whose frequencies increase from one dataset to another, are vital in various decision making. In static datasets such as those with classes (male vs. female, cured vs. not cured), EPs can reveal useful and hidden contrast patterns between datasets to support decision making such as classifier construction [4,15], disease likelihood prediction [13], discovering patterns in gene expression data [16], and so on. In sequential datasets, EPs are useful in decision making such as, studying and understanding customers' behaviour [19], predicting future purchases [17] and so on.

Though EP mining is an important data mining task, it is a difficult task as the downward closure property in frequent pattern mining is not applicable in EP mining [3,18]. Various studies have thus been proposed for efficient mining of, emerging patterns [5,13,16] and interesting emerging patterns [6–8,15,20,21]. However, these works often present a too large or a too small number of EPs for decision making. Reporting a large number of EPs makes it difficult to identify the set of useful ones as some EPs are redundant. On the other hand, reporting a

© Springer International Publishing Switzerland 2015
M.A. Sharaf et al. (Eds.): ADC 2015, LNCS 9093, pp. 244–256, 2015.
DOI: 10.1007/978-3-319-19548-3_20

small number of EPs may miss some useful EPs needed in decision making. For example, in our experiments on a Twitter dataset, though [6] (closely related to our work) reports a smaller number of EPs for decision making, it misses some interesting and useful emerging hashtags such as, "#tcot[1]#romneyryan2012", "#tcot#Obama", and "#news#Syria". The emergence of these useful emerging hashtags missed in [6] are however not implied by any emerging hashtag in their set of reported "interesting" emerging hashtags.

Motivated by the above mentioned challenges in existing works on EP mining, we study a method to mine the set of non-derivable EPs for decision making. We achieve this as follows. We initially employ deduction rules in mining our proposed non-derivable emerging patterns and show how all EPs can be recovered from the set of non-derivable EPs. We subsequently make use of a significance test to enable identify the set of significant non-derivable EPs and finally develop NEPs, an efficient non-derivable EP mining framework.

We make the following contributions to the discovery of emerging patterns.

1. We introduce the non-derivable EP set which achieves a major size reduction in the number of reported EPs.
2. We propose and develop NEPs, an efficient framework for mining the set of non-derivable EPs.

2   Related Works

The concept to EP mining was introduced by Dong and Li in [4] where they proposed an EP detection technique for static datasets with classes. They referred to an emerging pattern as an itemset whose support increases significantly from one dataset to another. More specifically, they defined an emerging pattern as an itemset whose growth rate is greater than a given threshold. Emerging pattern mining has since been researched on in works such as [6–9,13,15,20,21].

Over the past years however, some researchers argued that the EP definition proposed in [4] often generates too many EPs making it difficult identifying the set of interesting and useful EPs for decision making. Various constraints and techniques were thus proposed to enable mine interesting categories of EPs. Such works include, but not limited to: jumping EPs [7,15,20], essential EPs [8], and interesting EPs [6,21].

Though the above mentioned works may result in a reduction in size of reported emerging patterns, they are not lossless as all EPs cannot always be derived from their reported EP sets. Additionally, they often either miss some non-derivable emerging patterns, or report some trivial emerging patterns. Missing these non-derivable EPs or reporting some trivial EPs would most often be detrimental if not trivial in decisions making with EPs.

To the best of our knowledge, since no work exists on mining the set of non-derivable EPs for decision making, we focus on how these non-derivable EPs can be efficiently mined for decision making.

---

[1] The hashtag #tcot, "Top Conservatives On Twitter" provides a way for conservatives and Republicans to locate and follow the tweets of their like-minded brethren.

3 Preliminaries

The problem of frequent pattern mining and its associated notation can be given as follows. Let $I = \langle i_1, i_2, ..., i_n \rangle$ be a set of literals, called items. Then, a transaction is a nonempty set of items. A pattern S is a set of transactions satisfying some conditions of measures like frequency. A pattern is of length-k if it has k items, for example, $S = \{a, b, c\}$ is a length-3 pattern.

Given a database of n transactions, $\mathbf{D} = <d_1, d_2, d_3, \ldots, d_n>$, where each d_i in $\mathbf{D}$ is identified by i called TID, the *cover* of a pattern S in $\mathbf{D}$, $cov_\mathbf{D}(S)$, is the set of *TIDs* of events that contain S. That is, $cov_\mathbf{D}(S) = \{i : d_i \in \mathbf{D} \wedge S \subseteq d_i\}$. The exclusive domain of a pattern S in $\mathbf{D}$, $ed_\mathbf{D}(S)$, is the set of *TIDs* for records in $\mathbf{D}$ which do not contain S. That is, $ed_\mathbf{D}(S) = \{1, \ldots |\mathbf{D}|\} \setminus \{cov_\mathbf{D}(S)\}$. The *support* of a pattern S in $\mathbf{D}$, $sup_\mathbf{D}(S)$, is defined as, $sup_\mathbf{D}(S) = \frac{|cov_\mathbf{D}(S)|}{|\mathbf{D}|}$ where $|cov_\mathbf{D}(S)|$ is called the *support count* of S in $\mathbf{D}$.

Frequent pattern mining is the process of discovering all patterns in a database, $\mathbf{D}$, whose frequencies are larger than a user specified minimum support threshold. That is, given ε as the minimum support threshold, a pattern S in $\mathbf{D}$ is a frequent pattern if, $sup_\mathbf{D}(S) \geq \varepsilon$.

Based on the inclusion/exclusion (deduction) rules of non-derivable itemsets [1], a pattern $S_2 \in \mathbf{D}$ is said to be derivable from another $S_1 \in \mathbf{D}$ if, $S_1 \subset S_2$ and $sup_\mathbf{D}(S_1) = sup_\mathbf{D}(S_1)$.

3.1 Emerging Patterns

Given two datasets, $\mathbf{D_i}$ and $\mathbf{D_{i+1}}$, the growth rate of a pattern S, $GR(S)$, from $\mathbf{D_i}$ to $\mathbf{D_{i+1}}$ is defined as [4]:

$$GR(S) = \frac{sup_{D_{i+1}}(S)}{sup_{D_i}(S)} \tag{1}$$

Based on the growth rate, Dong and Li in [4] introduced the concept of emerging pattern mining. Formally, they defined an emerging pattern as follows.

Definition 1. *[4] Given $\rho > 1$ as the growth-rate threshold, a pattern S is said to be a ρ-emerging (ρ-EP or simply EP) from $\mathbf{D_i}$ to $\mathbf{D_{i+1}}$ if $GR(S) \geq \rho$.*

Though Definition 1 has been accepted and used in mining EPs in works such as [9,13,16,20], it has the following challenges. Firstly, a large number of EPs are often reported and this makes it difficult to comprehend and identify the set of useful EPs. Secondly, the emergence threshold ρ largely determines the number of discovered EPs. If ρ is set low, a large set of EPs will be discovered, most of which might be trivial. However, if ρ is set high, some useful EPs needed in decision making will be missed.

Over the past years, some researchers argued that finding all EPs above a minimum growth rate constraint as proposed in [4] generates too many EPs to be analysed. Soulet et al. [21] thus proposed a condensed representation approach for mining EPs based on closed patterns. Fan and Ramamohanarao [6] however proposed a way of selecting the set of *interesting* EPs. They define an interesting EP as follows.

Definition 2. *[6] Given $\rho > 1$ as the growth rate threshold, a pattern S is an interesting emerging pattern from $\mathbf{D_i}$ to $\mathbf{D_{i+1}}$ if:*

1. *S is frequent in both $\mathbf{D_i}$ and $\mathbf{D_{i+1}}$,*
2. *$GR(S) \geq \rho$,*
3. *$\forall Y \subset S, GR(Y) < GR(S)$, and,*
4. *$|S| = 1$, or $|S| > 1 \;\wedge\; (\forall Y \subset S || Y| = |S| - 1, \chi^2[sup_{\mathbf{D_i}}(S), sup_{\mathbf{D_{i+1}}}(S),$ $sup_{\mathbf{D_i}}(Y), sup_{\mathbf{D_{i+1}}}(Y)] \geq \eta).$*

In Definition 2, the authors aimed at identifying the set of EPs that:

- Cover both datasets-Condition 1.
- Have sharp discriminating powers-Condition 2.
- Are not subsumed by their emerging subsets-Condition 3, and
- Have significantly different supports from their immediate subsets to ensure the items of an EP are highly correlated-Condition 4.

Though Definition 2 will report a set of EPs as interesting, due to Conditions 3 and 4, some EPs which capture and reflect vital contrasts or emerging trends will be missed. For example, in a Twitter dataset[2] for the month of November 2012, given a minimum support of 0.04%, $\rho = 1.0$ and $\eta = 3.841$, the following observations were made. Some non-derivable emerging hashtags, such as; #tcot#romneyryan2012, #tcot#Obama, and #news#Syria, which reflect important emerging trends from 1st to 2nd, and from 2nd to 3rd of November, were missed by Definition 2 as their growth rates are less than those of their emerging subsets (#tcot, #romneyryan2012, #Obama and #news respectively). However, the emergence of these subsets do not indicate the emergence of their supersets. That is, though the emerging hashtag #tcot could be easily associated with #romneyryan2012 in #tcot#romneyryan2012, so cannot be said of the emergence of #Obama in #tcot#Obama. Similarly, the emergence of #news does not in anyway imply that news about Syria, #news#Syria, is also emerging.

4   Problem Statement and Definitions

Though with Definitions 1 and 2 the set of EPs and interesting EPs can be identified, some trivial EPs may be reported or some vital EPs pruned. To avoid this situation, we begin by defining an EP as follows.

Definition 3. *Given ε as the minimum support, a pattern S is an emerging pattern from $\mathbf{D_i}$ to $\mathbf{D_{i+1}}$ if it is frequent in both $\mathbf{D_i}$ and $\mathbf{D_{i+1}}$, and $GR(S) > 1$.*

For simplicity, we represent the set of emerging patterns from $\mathbf{D_i}$ to $\mathbf{D_{i+1}}$ as, $E_i^{i+1} = \{(S_1, GR(S_1)), (S_2, GR(S_2)), \dots, (S_k, GR(S_k))\}$.

Property 1. Given a two EPs, $(S_1, S_2) \in E_i^{i+1}$ such that, $S_2 \supset S_1$, S_2 is a derivable EP of S_1 if, $sup_{D_i}(S_1) = sup_{D_i}(S_2)$ and $sup_{D_{i+1}}(S_1) = sup_{D_{i+1}}(S_2)$.

---

[2] Obtained from CNetS (http://carl.cs.indiana.edu/data/).

Property 1 is based on the inclusion/exclusion rules of non-derivable itemsets proposed in [1]. From Property 1, we define a non-derivable EP as follows.

Definition 4. *Given an EP set, $E_i^{i+1} = \{(S_1, GR(S_1)), \ldots, (S_k, GR(S_k))\}$ from $\mathbf{D_i}$ to $\mathbf{D_{i+1}}$, an EP S_m is a non-derivable EP if $\nexists S_t \in E_i^{i+1} | S_t \subset S_m$, $sup_{D_i}(S_m) = sup_{D_i}(S_t)$ and $sup_{D_{i+1}}(S_m) = sup_{D_{i+1}}(S_t)$.*

For EP regeneration, if S_2 is a derivable EP from S_1, S_1 is represented as $(S_1, GR(S_1), \{X_1\})$, where $X_1 = S_2 \setminus S_1$.

We represent the non-derivable EP set in this work as: $nE_i^{i+1} = \{(S_1, GR(S_1), X_1), (S_2, GR(S_2), X_2), \ldots, (S_k, GR(S_k), X_k)\}$, where $X_i = \{\{X_{i,a}\}, \{X_{i,b}\} \ldots \{X_{i,z}\}\}$ is the set of items in the EPs that are derivable from S_i. For example, an EP $S_1 = \{a, d\}$ and $X_1 = \{\{e\}, \{f\}\}$, imply the emergence of $S_2 = \{a, d, e\}$ and $S_3 = \{a, d, f\}$ are derivable from S_1. If any EP is not derivable from a non-derivable EP, S_3, then $X_3 = \{\emptyset\}$.

To ensure information on all EPs are not lost, we propose a method

---

Algorithm 1. Regen_All_EPs(nE_i^{i+1})

Input: Non-derivable EP set, nE_i^{i+1}
Output: All EP set, E_i^{i+1}

1: Create set $E_i^{i+1} = \emptyset$
2: **for each**$(S_n, GR(S_n), X_n) \in nE_i^{i+1}$ **do**
3:     Add S_n to E_i^{i+1}
4:     **if** $X_n \neq \emptyset$ **then**
5:         **for each** $X_{n,m} \in X_n$ **do**
6:             Create $C_k = S_n \cup X_{n,m}$
7:             $GR(C_k) = GR(S_n), X_k = \emptyset$
8:             Add C_k to E_i^{i+1}
9: Return E_i^{i+1}

---

called Regen_All_EPs (see Algorithm 1) which recovers all EPs from nE_i^{i+1}. For $\mathbf{D_i}$ and $\mathbf{D_{i+1}}$, and a minimum support ε, all the EPs from $\mathbf{D_i}$ to $\mathbf{D_{i+1}}$ can always be regenerated from nE_i^{i+1} using Regen_All_EPs(nE_i^{i+1}) as follows. Firstly, the set of all EPs, E_i^{i+1} is created Line 1. In Line 3, each S_n in nE_i^{i+1} is added to E_i^{i+1}. From Lines 4 to 8, if S_n contains any derivable EP, that is, $X_n \neq \emptyset$, for each set $X_{n,m} \in X_n$, a candidate C_k is created as $C_k = S_n \cup X_{n,m}$ where, $sup_{D_i}(C_k) = sup_{D_i}(S_n)$, $sup_{D_{i+1}}(C_k) = sup_{D_{i+1}}(S_n)$, $GR(C_k) = GR(S_n)$ and $X_k = \emptyset$. C_k (with its properties) is then added to E_i^{i+1}. This process continues until all EPs in nE_i^{i+1} are assessed and E_i^{i+1} returned in Line 9.

Though with Definition 4, a set of non-derivable EPs will be reported, some might be insignificantly emerging. To enable the detect the set of significant non-derivable EPs, we propose a significance test on the detected non-derivable EPs. We hypothesize that the supports are independent from time stamped datasets, that is, they do not change. We use chi-square test [12] to test the hypothesis. If the hypothesis is rejected, there is a significant change in the supports. We formally define a significant non-derivable EP as follows.

Definition 5. *An EP, S, from $\mathbf{D_i}$ to $\mathbf{D_{i+1}}$ is a significant non-derivable EP if S is a non-derivable EP and $\chi^2(S) \geq \eta$.*

where, $\chi^2(S)$ is the chi-square test value of the support distributions of S in $\mathbf{D_i}$ and $\mathbf{D_{i+1}}$, evaluated based on contingency Table 1 as:

$$\chi^2(S) = \frac{n(ad - bc - \frac{n}{2})^2}{(a+c)(b+d)(a+b)(c+d)} \tag{2}$$

Equation 2 (refer to [12] for a proof) evaluates the statistical significance level of a patterns' support change in the two datasets with the Yattes correction continuity. We choose 3.841 as the critical value for our chi-square test since it gives us 95% confidence of support difference.

EPs with chi-square values greater than or equal to η (η = 3.841) are identified as significantly emerging for the two datasets.

Table 1. Contingency Table

	D_i	D_{i+1}	
S	$a=|cov_{D_i}(S)|$	$c=|cov_{D_{i+1}}(S)|$	$a+c$
$\neg S$	$b=|ed_{D_i}(S)|$	$d=|ed_{D_{i+1}}(S)|$	$b+d$
	$a+b$	$c+d$	$n=a+b+c+d$

With our definitions, our EP mining problem can be defined as the process of mining all non-derivable and significant non-derivable EPs from dataset, $\mathbf{D_i}$ to $\mathbf{D_{i+1}}$, given a minimum support (ε) and a critical significance value (η).

5 Mining the Set of Non-derivable Emerging Patterns

We initially present a definition and a vital property used in designing our proposed non-derivable EP mining algorithm (NEPs). We later introduce and discuss the vital steps in NEPs.

Definition 6. *For a given database* $\mathbf{D}$ *and a minimum support, a pattern* S, *is a frequent generator in* $\mathbf{D}$ *if for all frequent patterns,* $S_n \in \mathbf{D}$, *there does not exist any frequent pattern* S_m *such that* $S_m \subset S$ *and* $sup_{\mathbf{D}}(S) = sup_{\mathbf{D}}(S_m)$.

Property 2. If A *is not a generator, then* $\forall A' \supset A$, A' *is not a generator.*

From Property 2, if an itemset A is a frequent generator, then all of its subsets are generators. The reason being that if one of the subsets of A is not a generator, then A cannot be a generator. From Definition 4, we observe that for S to be a non-derivable EP, it must be a frequent generator in both datasets. We employ this anti-monotone property of non-generators as our pruning strategy in NEPs. This reduces the sizes of the datasets and that of the pattern tree built by NEPs.

5.1 The NEPs Algorithm

Our non-derivable EP detection framework, NEPs, is as shown in Algorithm 2. NEPs is an extension of the frequent pattern tree structure [11] in discovering our proposed set of non-derivable EPs. To enable detect our proposed non-derivable EPs, we extend the frequent pattern tree [11] to find frequent generators in the two datasets. We construct our pattern tree PT from two datasets using the detected frequent generators. The set of non-derivable EPs are then mined from our pattern tree. The structure of our pattern tree is as discussed below.

Pattern Tree Structure. Our pattern tree, PT, is a tree structure consisting of a tree of patterns, made of one main root (R), a frequent item header table and a set of item prefix subtrees as children of the root. Each node in the item prefix subtree consists of the following fields:

- *Item*: This registers the item that this node represents.
- *Count_D_i*: Stores number of records in $\mathbf{D_i}$ containing items in the path reaching this node.
- *Count_D_{i+1}*: Records such number as *Count_D_i* in $\mathbf{D_{i+1}}$.
- *Node link*: Links to the next node of the *PT* carrying the same item or null.

In the frequent item header table, *HT*, each entry consists of the following fields:

- *Item*: This registers the item that the header node link represents.
- *Total D_i*: Represents the *count* in the item's corresponding node-link in $\mathbf{D_i}$.
- *Total D_{i+1}*: Represents same number as *Total D_i* but for $\mathbf{D_{i+1}}$.
- *Header Node link*: This points to the first node in the *PT* carrying the item.

Algorithm 2. NEPs($D_i, D_{i+1}, \varepsilon, \eta$)

Input:D_i, D_{i+1}, minimum support ε, and critical value η.
Output: Non-derivable EP set, nE_i^{i+1} and significant noderivable EP set, snE_i^{i+1}.

1: Create $L_1 = \emptyset$ ▷ The set of frequent length-1 generators in both datasets
2: Scan datasets D_i and D_{i+1} with ScanData(D_n, ε) to return F_i and F_{i+1}
3: Create Set $F_1 = (F_i \cap F_{i+1})$
4: **for each** item $a_y \in F_1$ **do**
5:      **if** $sup_{D_i}(a_y) \neq sup(\emptyset) \wedge sup_{D_{i+1}}(a_y) \neq sup(\emptyset)$ **then** Add a_y to L_1
6: Sort L_1 in support-descending order as in F_i
7: Create header table HT with L_1
8: Create Root, R, of pattern tree
9: Const_PT(D_i, D_{i+1}, L_1)
10: Mine_EPs(PT, ε, η)
11: Recheck nE_i^{i+1} and update snE_i^{i+1} with nE_i^{i+1}
12: Return nE_i^{i+1} and snE_i^{i+1}

Our proposed non-derivable EP mining framework, *NEPs* (Algorithm 2), employs four major steps: finding the set of frequent length-1 items in both datasets; identifying frequent generators; pattern tree construction; and mining EPs from the pattern tree. We discuss each step in the following sections.

5.2 Finding Frequent Length-1 Items

This step finds the set of frequent length-1 items in both $\mathbf{D_i}$ and $\mathbf{D_{i+1}}$ with regards to the minimum support using Algorithm 3. In Line 2 of Algorithm 2, ScanData(D_n, ε) (Algorithm 3) is called to scan both $\mathbf{D_i}$ and $\mathbf{D_{i+1}}$. As shown in Line 5 of Algorithm 3, only length-1 items whose supports in $\mathbf{D_n}$ are greater than the minimum support

---

Algorithm 3. ScanData(D_n, ε)

Input: Dataset, D_n, minimum support, ε
Output: Frequent length-1 item set, F_n

1: Create HashMap h_n; Create set F_n
2: **for each** transaction $T \in D_n$ **do**
3:      **for each** length-1 item $a \in tr$ **do**
4:         $h_n.add(a, h_n(a) + 1)$
5: $F_n = \{a, count = h_n(a) | sup_{D_n}(a) \geq \varepsilon\}$
6: Return F_n

---

are added to F_n. For datasets, $\mathbf{D_i}$ and $\mathbf{D_{i+1}}$, Algorithm 3 returns the sets F_i and F_{i+1} which contain the frequent length-1 items in $\mathbf{D_i}$ and $\mathbf{D_{i+1}}$ respectively.

5.3   Identifying Frequent Length-1 Generators

This step (from Lines 3 to 6 of Algorithm 2) finds the set of frequent length-1 generators in both $\mathbf{D_i}$ and $\mathbf{D_{i+1}}$. As shown in Line 3 of Algorithm 2, the set F_1 is created from an intersection of F_i and F_{i+1}. For each item, a_y in F_1, a_y is added to L_1 (in Line 5) if it is a generator in both $\mathbf{D_i}$ and $\mathbf{D_{i+1}}$. That is, if $sup_{D_i}(a_y) \neq sup(\emptyset)$ and $sup_{D_{i+1}}(a_y) \neq sup(\emptyset)$. This is the first step of pruning based on Property 2. In Line 6 of Algorithm 2, the set L_1 is then sorted in support descending order of F_i, that is, based on their frequency in $\mathbf{D_i}$. L_1, which contains all frequent length-1 generators and their supports in $\mathbf{D_i}$ and $\mathbf{D_{i+1}}$ is then used in creating the header table, HT, of the pattern tree in Line 7 of Algorithm 2. This is followed by creating the root of PT in Line 8.

5.4   Pattern Tree Construction

This step constructs the PT from $\mathbf{D_i}$ and $\mathbf{D_{i+1}}$ by calling Const_PT(D_i, D_{i+1}, L_1) in Line 9 of Algorithm 2. PT is constructed from $\mathbf{D_i}$ and $\mathbf{D_{i+1}}$ in the following steps. Starting with $\mathbf{D_i}$, for each transaction (T) in $\mathbf{D_i}$, Line 3 in Algorithm 4 selects the set of common frequent items in T and L_1 as CF_n. In Line 4, CF_n is sorted

---
Algorithm 4. Const_PT(D_i, D_{i+1}, L_1)

Input: D_i, D_{i+1}, Frequent length-1 generators, L_1
Output: Pattern tree, PT

1: **for each** $D_n \in (D_i, D_{i+1})$ **do** ▷ Starts with D_i
2:     **for each** transaction $T \in D_n$ **do**
3:         Create set $CF_n = T \cap L_1$
4:         Sort CF_n as $[a|F]$ based on order of L_1
5:         Ins_PT($[a|F]$,R)
6:         **while** $F \neq \{\emptyset\}$ **do** Ins_PT(F,C)
7: Return PT

---

in support descending order as $[a|F]$ (sorting based on the order in L_1, that is, their frequency in $\mathbf{D_i}$), where a is the first element and F the remaining list.

The function $Ins\_PT([a|F],R)$ in Line 5 inserts $[a|F]$ into PT following the similar insertion technique in [11] as follows: If the the root, R, has a child C such that $C.item = f.item$, then increment $C's$ $count\_D_i$ by 1. Else a new node C is created and its $count\_D_i$ set as 1, its parent link be linked to root R and its node link be linked to the node with the same item-name via the node link structure. If F is non-empty, $Ins\_PT(F, C)$ is called recursively (where C is the last created or visited child in the pattern tree).

Dataset $\mathbf{D_{i+1}}$ is also scanned following the same process as the second scanning of $\mathbf{D_i}$. The only difference here is in $Ins\_PT([a|F], R)$ and $Ins\_PT(F, C)$ in Lines 5 and 6 where $count\_D_{i+1}$ are incremented. The PT construction is complete after scanning the last transaction in $\mathbf{D_{i+1}}$. In Fig. 1, we illustrate the complete PT constructed from toy datasets $\mathbf{D_1}$ and $\mathbf{D_2}$.

5.5   Mining Non-derivable EPs

This step mines all non-derivable EPs from the PT by calling Mine_EPs(PT, ε, η) (Algorithm 5) in Line 10 of Algorithm 2. Algorithm 5 mines the set of non-derivable EPs from PT as follows.

For each item a_k in HT starting from the last item to the first item in the HT, its conditional databases, $cd_{D_i}(a_k)$ and $cd_{D_{i+1}}(a_k)$ are created in Lines 3 and 4. In Lines 6 and 7, the set of items, Ps, (from node a_k to R) in the paths containing a_k are added to $cd_{D_i}(a_k)$ and $cd_{D_{i+1}}(a_k)$, l and t number of times respectively, where l and t are the counts of a_k in D_i and D_{i+1} at node a_k. For instance, considering item $\{e\}$ in Fig. 1, $\{e\}$ will be added to its two conditional database 6 times each since it has a count of 6 in both $\mathbf{D_1}$ and $\mathbf{D_2}$ at the node of concern.

Algorithm 5. Mine_EPs(PT, ε, η)

Input: Pattern Tree, PT, minimum support, ε, and critical value η
Output: nE_i^{i+1} and snE_i^{i+1}

1: Create $nE_i^{i+1} = \emptyset$; Create $snE_i^{i+1} = \emptyset$
2: **for each** item $a_k \in HT$ **do**   ▷ From $k=|HT|$ to 1
3:      Create conditional database, $cd_{D_i}(a_k)$ of a_k in D_i
4:      Create conditional database, $cd_{D_{i+1}}(a_k)$ of a_k in D_{i+1}
5:      **for each** set of items, Ps, in paths containing a_k **do**▷ items from node a_k to R
6:          Add Ps to $cd_{D_i}(a_k)$, l number of times   ▷ $l = \text{count}_{D_i}(a_k)$ at node a_k
7:          Add Ps to $cd_{D_{i+1}}(a_k)$, t number of times ▷ $t = \text{count}_{D_{i+1}}(a_k)$ at node a_k
8:      $Mine(cd_{D_i}(a_k), cd_{D_{i+1}}(a_k), \eta, \varepsilon)$
9:      Let FPs= frequent pattern set returned by $Mine(cd_{D_i}(a_k), cd_{D_{i+1}}(a_k), \eta, \varepsilon)$
10:     **for each** pattern $S_n \in FPs$ **do**
11:         **if** S_n is EP **then**
12:             Update nE_i^{i+1} with S_n in the form $(S_n, GR(S_n), X_n)$
13: Return nE_i^{i+1}

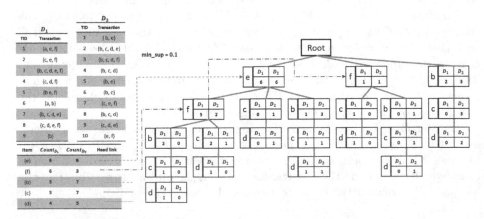

Fig. 1. Pattern Tree Constructed from $\mathbf{D_1}$ and $\mathbf{D_2}$ at $\varepsilon = 0.1$

For any two of conditional databases, $cd_{D_i}(a_k)$ and $cd_{D_{i+1}}(a_k)$ of a_k, the function $Mine(cd_{D_i}(a_k), cd_{D_{i+1}}(a_k), \eta, \varepsilon)$ in Line 8 builds the subtrees for the conditional databases and mines the set of frequent patterns (FPs) from the subtrees similar to the subtree mining in [11]. For any frequent pattern S_n in FPs, it is tested for emergence and if emerging, it is used in updating nE_i^{i+1} in Line 12 based on Definition 4. This process continues until all items in the HT are visited. The set of non-derivable EPs is returned in Line 12 of Algorithm 5.

The set of non-derivable EPs are re-examined in Line 11 of Algorithm 2 to ensure no derivable EP is reported. The significant non-derivable EPs are then obtained from nE_i^{i+1} based on Definition 5. NEPs terminates after nE_i^{i+1} and snE_i^{i+1} are returned in Line 12 of Algorithm 2. For lack of space, we do not illustrate the EP mining process on Fig. 1. However, $(\{b\}, 1.26, \{\phi\})$, $(c, 1.26, \{\phi\})$, $(d, 1.13, \{c\})$, $(\{b, c\}, 2.25, \{\phi\})$ and $(\{b, d\}, 1.80, \{c\})$ will be reported as the non-derivable EPs.

6   Empirical Analysis

We compared the performance of NEPs and our implementation of the method proposed [6] (iEP-miner) on: (i.) the number of reported EPs, (ii.) runtime and (iii.) effectiveness of decision made with the detected EPs.

For this comparison, let $aEPs, iEPs, nEPs$, and $snEPs$ refer to: all EPs, interesting EPs (detected by iEP-miner), non-derivable EPs and significant non-derivable EPs respetively. All methods are implemented in Java and experiments carried on a 64-bit Win. 7 PC, Intel Core i5, CPU 2.50GHz, 4GB memory.

Reported EPs: Table 2 shows the number of reported EPs (from edible class to poisonous class for mushroom, and class 1 to 2 for breast) for two datasets[3]. As can be seen, NEPs drastically reduces the number of reported EPs for decision making. Though iEP-miner always reports a smaller size of EPs, as explained in previous sections, it misses some non-derivable and vital EPs needed in decision making.

Runtime: Fig. 2 shows the runtime of NEPs and iEP-miner. Though NEPs reports higher number of EPs, for high minimum supports, NEPs slightly outperforms iEP-miner in dense datasets as seen in Fig. 2. However, at low minimum supports in dense datasets, iEP-miner slightly outperform NEPs since more EPs which do not satisfy Conditions 3 and 4 of Definition 2 are pruned. Most of these EPs pruned are however non-derivable making iEP-miner handle smaller datasets and smaller trees compared to NEPs, hence the slight out performance of iEP-miner at low minimum supports in dense datasets. In Fig. 3, we show the performance of NEPs and the naive way of detecting non-derivable EPs (without the non-frequent generator pruning).

Table 2. Reported Emerging Patterns

Dataset	#cls	$\varepsilon = 6\%, \eta = 3.841, \rho = 1$			
		aEPs	iEPs	nEPs	snEP
Mushroom	2	21352	24	1912	1694
Breast	2	664	6	69	54

---

[3] From http://cgi.csc.liv.ac.uk/~frans/KDD/Software/LUCS-KDD-DN/DataSets

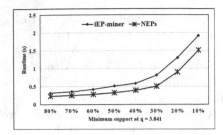
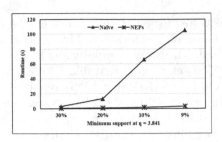

Fig. 2. Mushroom ($\rho=1.0, \eta=3.841$)   **Fig. 3.** Mushroom ($\rho=1.0, \eta=3.841$)

Decision Making: Table 3 shows a preliminary application of EPs in trend prediction. For lack of space we show the tweet trend prediction for only the first four days in November 2012. For any three datasets from consecutive days, for example, D_i, D_{i+1} and D_{i+2}, we use D_i and D_{i+1} as the training set, with D_{i+1} and D_{i+2} as our tests set. Based on a detected emerging hashtags from the training set, we predict their continuous emergence in the test set. We employed the $F1$-measure as the overall goodness measure. The precision and recall used in evaluating the $F1$-measure are obtained respectively as:

Table 3. EPs in Trend Prediction

		$\varepsilon = 0.05\%$ at $\eta = 3.841$		
Days	EP	Prec	Recall	F1-mesure
D1	iEPs	64.44	25.66	36.71
D2	nEPs	70.16	59.29	**64.27**
D3	snEPs	69.64	34.51	46.15
D2	iEPs	61.96	29.08	39.58
D3	nEPs	73.02	70.41	**71.69**
D4	snEPs	72.92	35.71	47.95

$$Precision = \frac{\#EPs\ Correctly\ Predicted}{\#All\ Predicted\ EPs}, and\ Recall = \frac{\#EPs\ Correctly\ Predicted}{\#All\ EPs}$$

As seen in Table 3, of our preliminary results, $nEPs$ turns out as the best set for trend prediction.

7   Conclusions and Future Works

Non-derivable EPs are EPs whose emergence from one dataset to another cannot be derived from other EPs. We make use of deduction rules to propose a non-derivable interestingness measure for EPs. We further employ a significance test to identify the set of significant non-derivable EPs. Finally, we develop NEPs, an efficient non-derivable EP mining framework. Our experimental results shows NEPs is efficient and the set of non-derivable EPs which achieves a size reduction in the number of reported EPs shows potential in trend prediction. Our future works include, classifier formation and trend prediction modelling based on non-derivable EPs. We anticipate the significant non-derivable EPs will be more reliable in classifier formation since they reflect reliable contrasting class features.

References

1. Calders, T., Goethals, B.: Mining all non-derivable frequent itemsets. In: Elomaa, T., Mannila, H., Toivonen, H. (eds.) PKDD 2002. LNCS (LNAI), vol. 2431, pp. 74–86. Springer, Heidelberg (2002)
2. Calders, T., Goethals, B.: Non-derivable Itemset Mining. Data Mining and Knowledge Discovery 14(1), 171–206 (2007)
3. Cheng, M.W.K., Choi, B.K.K., Cheung, W.K.W.: Hiding emerging patterns with local recoding generalization. In: Zaki, M.J., Yu, J.X., Ravindran, B., Pudi, V. (eds.) PAKDD 2010, Part I. LNCS, vol. 6118, pp. 158–170. Springer, Heidelberg (2010)
4. Dong, G., Li, J.: Efficient mining of emerging patterns: discovering trends and differences. In: 5th ACM SIGKDD International Conference on Knowledge Discovery, pp. 43–52. ACM (1999)
5. Dong, G., Li, J.: Mining Border Descriptions of Emerging Patterns from Dataset Pairs. Knowl. Inf. Syst. 8(2), 178–202 (2005)
6. Fan, H.: Efficiently mining interesting emerging patterns. In: Dong, G., Tang, C., Wang, W. (eds.) WAIM 2003. LNCS, vol. 2762, pp. 189–201. Springer, Heidelberg (2003)
7. Fan, H., Ramamohanarao, K.: Fast Discovery and the Generalization of Strong Jumping Emerging Patterns for Building Compact and Accurate Classifiers. IEEE Transactions on Knowledge and Data Engineering 18(6), 721–737 (2006)
8. Fan, H., Kotagiri, R.: An efficient single-scan algorithm for mining essential jumping emerging patterns for classification. In: Chen, M.-S., Yu, P.S., Liu, B. (eds.) PAKDD 2002. LNCS (LNAI), vol. 2336, pp. 456–462. Springer, Heidelberg (2002)
9. Garcia-Borroto, M., Martnez-Trinidad, J.F., Carrasco-Ochoa, J. A.: A Survey of Emerging Patterns for Supervised Classification. Arti. Int. Review, 1–17 (2012)
10. Goethals, B., Muhonen, J., Toivonen, H.: Mining Non-Derivable Association Rules. In: 5th SIAM International Conference on Data Mining, pp. 239–249. SIAM (2005)
11. Han, J., Pei, J., Yin, Y.: Mining Frequent Patterns Without Candidate Generation. In: ACM SIGMOD Record, vol. 29, No. 2, pp. 1–12. ACM (2000)
12. Kurtz, A.K., Mayo, S.T.: Chi Square. In: Kurtz, A.K., Mayo, S.T. (eds.) Statistical Methods in Education and Psychology, pp. 362–391. Springer, New York (1979)
13. Li, J., Liu, H., Downing, J.R., Yeoh, A.E.J., Wong, L.: Simple Rules Underlying Gene Expression Profiles of More than Six Subtypes of Acute Lymphoblastic Leukemia (ALL) Patients. Bioinformatics 19(1), 71–78 (2003)
14. Li, J., Dong, G., Ramamohanarao, K., Wong, L.: Deeps: A New Instance-Based Lazy Discovery and Classification System. Machine Learning 54(2), 99–124 (2004)
15. Li, J., Dong, G., Ramamohanarao, K.: Making Use of the Most Expressive Jumping Emerging Patterns for Classification. Knowl. Inf. Syst 3(2), 131–145 (2001)
16. Li, J., Wong, L.: Emerging Patterns and Gene Expression Data. Genome Informatics, 3–13 (2001)
17. Nofong, V.M., Liu, J., Li, J.: A study on the applications of emerging sequential patterns. In: Wang, H., Sharaf, M.A. (eds.) ADC 2014. LNCS, vol. 8506, pp. 62–73. Springer, Heidelberg (2014)

18. Poezevara, G., Cuissart, B., Crèmilleux, B.: Extracting and Summarizing the Frequent Emerging Graph Patterns from a Dataset of Graphs. Journal of Intelligent Information Systems **37**(3), 333–353 (2011)
19. Tsai, C.Y., Shieh, Y.C.: A Change Detection Method for Sequential Patterns. Decis. Support Syst. **46**(2), 501–511 (2009)
20. Terlecki, P., Walczak, K.: Jumping Emerging Patterns with Negation in Transaction Databases - Classification and Discovery. Information Sciences **177**(24), 5675–5690 (2007)
21. Soulet, A., Crémilleux, B., Rioult, F.: Condensed representation of emerging patterns. In: Dai, H., Srikant, R., Zhang, C. (eds.) PAKDD 2004. LNCS (LNAI), vol. 3056, pp. 127–132. Springer, Heidelberg (2004)

Using Word Embeddings to Enhance Keyword Identification for Scientific Publications

Rui Wang, Wei Liu$^{(\boxtimes)}$, and Chris McDonald

Department of Computer Science and Software Engineering,
The University of Western Australia,
35 Stirling Highway, Crawley, WA 6009, Australia
rui.wang@research.uwa.edu.au,
{wei.liu,chris.mcdonald}@uwa.edu.au

Abstract. Automatic keyword identification is a desirable but difficult task. It requires considerations of not only the extraction of important words or phrases from a text, but also the generation of abstractive ones that do not appear in the text. In this paper, we propose an approach that uses *word embedding vectors* as an external knowledge base for both keyword extraction and generation. Our evaluation shows that our approach outperforms many baseline algorithms, and is comparable to the state-of-the-art algorithm on our chosen dataset. In addition, we also introduce a new approach for evaluating the task of keyword extraction, that overcomes a common problem of overly strict matching criteria. We show that using *word embedding vectors* is a simpler, yet effective, method for both keyword extraction and generation.

Keywords: Keyword extraction · Keyword generation · Word embeddings

1 Introduction

Keywords provide a high level description of a document, which is important for many areas of text processing, including document classification and clustering, information retrieval, and question-answering. In this paper, we use the term *keyword* uniformly represents both single keyword and multi-word keyphrase, for example, *neural network* is considered as one keyword. However, manual annotation does not scale for large document sets. Thus, automatically identifying keywords from large document sets is highly desirable.

Most previous studies [1–4] focus on extracting important words or phrases that can describe the core theme of a document. A shortcoming is that they cannot identify any words that do not appear in the actual content of a document. In reality, however, keywords serve different proposes. In many cases, such as assigning keywords for scientific articles, it requires choosing highly abstractive words that may only describe the field of the study to which an article belongs. Identifying keywords, therefore, is a task that requires not only extracting representative words from a document but also suggesting abstractive words that

© Springer International Publishing Switzerland 2015
M.A. Sharaf et al. (Eds.): ADC 2015, LNCS 9093, pp. 257–268, 2015.
DOI: 10.1007/978-3-319-19548-3_21

may not appear in the document. So in the context of this paper, by keyword identification, we mean both keyword extraction and generation.

Contemporary state-of-the-art keyword extraction algorithms are only able to achieve a precision of around 35%, even on their chosen datasets [5]. However, the evaluation may be biased due to a common issue of using overly strict matching criteria. A standard evaluation technique used by most of the studies is to compare the stems and the order of a sequence of tokens extracted by computers with the tokens selected by human, namely *exact matching*. A match is counted only if the computer extracted tokens are in the exact same order and stems as human assigned ones. For example, *computer science* matches *comput scienc*, but not *comput* or *scienc comput*. The problem of this evaluation technique is the inability of measuring a correct match when one phrase is a sub-phrase of another, but have the same semantic meaning, such as *iPad* and *iPad tablet*.

In this paper, we focus on these issues. We first propose an unsupervised graph ranking approach using *word embedding vectors* as an external knowledge base for extracting keywords that appear in a document, and evaluate our approach on both the commonly used *exact matching* and our proposed *corpus-aware matching* criteria. We then introduce a keyword generation approach built on top of our extraction model using the power of *word embedding vectors* to suggest abstractive keywords not appearing in a document.

For keyword extraction, we adopt a graph-based rank algorithm to identify the importance of words. A document is represented as a weighted undirected graph, where vertices represent words, and edges are the co-occurrence relations between pairs of words constrained by a window size. The weight of an edge is computed by our proposed weighting scheme that analyses the strength of the semantic relatedness between two words. We evaluated our approach against three baseline algorithms and the state-of-the-art algorithm on our chosen dataset. Our experiments show that it outperforms all baseline algorithms and is comparable to the state-of-the-art algorithm without any customised word filters.

For keyword generation, we compute the mean vector of the most important words identified by the extraction model, then apply the *approximate nearest neighbour* algorithm [6] to identify the nearest words to the mean vector. Our experiments show very promising results.

Training word embedding vectors, however, is a time-consuming task. In this study, we simply made use of three publicly available pre-trained word embedding vectors: SENNA [7], HLBL [8], and Turian [9], which are trained using unsupervised machine learning algorithms, enabling the use of large-scale training sets, such as a Wikipedia snapshot.

This paper is organised as follows: Section 2 briefly summarises the related work, and Section 3 is an overview of our proposed keyword extraction approach. Section 4 introduces our weighting scheme and Section 5 describes the graph-based ranking algorithm. Section 6 introduces the keyword generation approach. Section 7 introduces our corpus-aware matching evaluation. Experiments are described in Section 8. We draw our conclusion with an outlook to future work in Section 9.

2   Related Work

Since our approach is unsupervised, in this section we focus only on unsupervised techniques. Most unsupervised keyword extraction is a two-step task: representing words and documents, and ranking. Early approaches [1,2] treat words as solely *statistical elements* or *atomic symbols*. When words are represented as *statistical elements*, documents are typically represented as matrices, such as *term-document* or *co-occurrence* matrices. Ranking is conducted by applying statistical techniques combined with linguistic theories to score the information carried by each word. Term Frequency – Inverse Document Frequency (TF-IDF) [1] is the most well-known approach. In *atomic symbol* word representation, each word represents itself or the stem of the word. A document is represented as a graph, in which nodes represent words and two words are connected by an edge if they co-occur in a given window size. Graph ranking algorithms, such as PageRank [10], are applied to score the words.

However, words represented as *statistical elements* or *atomic symbols* do not carry any semantic knowledge about themselves. This means no prior knowledge is given before ranking starts. Since all the knowledge can only be obtained from the document, it restricts the opportunity to identify important words or phrases capturing a document's core theme. Further improvements have been made by recently proposed knowledge based approaches [11–13] that use external knowledge bases, such as WordNet or Wikipedia, to provide background knowledge. Many of these approaches use *distributional word representation* to encode the semantic relatedness of words, which enables words containing some semantic knowledge of themselves on top of a given document. Words that carry important information then can be extracted by analysing the semantic relatedness between words, and how they actually present in the given document. In recent years, *word embeddings* have shown the power of encoding both semantic and syntactic features of words [14]. Therefore, we propose using word embeddings as an alternative resource to measure the semantic relatedness of words. To the best of our knowledge, this is the first attempt of using word embeddings in keyword identification task.

Many studies on *keyword generation* fall into the field of search engine advertising, specifically focused on query and advertiser log mining, proximity searches, and meta-tag crawlers. This is a different task from our study of *keyword generation*, where we are interested in suggesting abstractive keywords that even do not appear in the actual content of a document. To our best knowledge, the only study which is similar to ours is presented by Liu et al. [15], which takes keyword extraction and generation as a translation problem based on the intuition that since a document and its keywords describe the same topic, they can be seen as written in two different languages. They use IBM Model 1 [16] to train the translation model by computing the product of each word's TF-IDF score and translation probability.

3   Keyword Extraction

Our approach uses a pipeline architecture. For each document in the corpus, we firstly preprocess the document, and then construct a weighted undirected graph. Weights are assigned to edges using the weighting scheme described in Section 4. We rank the graph using the algorithm described in Section 5. In post-processing, words collapse as phrases if they are adjacent in the text, and the phrases with highest scores are extracted as the keywords for the document.

Preprocessing. In preprocessing, we first split a text into sentences and tokenise them, and then run part-of-speech (POS) tagging using the Stanford parser [17]. We next remove plurals for each tagged noun and, finally, only adjectives and nouns are considered as valid words for ranking because the majority of keywords only contain adjectives and nouns.

Word Embedding Dataset. Training word embeddings is computationally expensive. In this study, we use three pre-trained publicly available embeddings: SENNA, HLBL, and Turian's embeddings. The three embeddings are available with different dimensions from 25 to 100. We choose 50 dimensions for all the embeddings. SENNAs were trained over an entire Wikipedia snapshot covering 130,000 words. HLBL's embeddings were trained over 10 million words covering about 246,000 words, and Turian's embeddings cover 268,000 words. It is worth noting that there are a small number of words (about 2.5%) in our evaluation dataset are not covered by the pre-trained embeddings; we assigned them the *unknown* values provided by each embeddings.

Graph Construction and Ranking. After preprocessing, all valid words are represented as vertices in the graph. Two words that co-occur in a given window size (we use 2, 5, and 10 in this study), will be connected and a weight is assigned using the weighting scheme described in Section 4. Once the graph is constructed, we use the ranking algorithm described in Section 5 to score each word.

Post-processing. After ranking, all the words and their associated scores are saved into a dictionary D. We adopt the phrase formation approach of [3]: words that occur in a sequence in the document are collapsed into candidate phrases, and only phrases ended with nouns are extracted. Each extracted phrase is scored using:

$$PhraseScore(p) = \sum_{w_i \in p} S(w_i) \tag{1}$$

where p is a phrase, w_i is a word contained in p, and $S(w_i)$ is the word score from the ranking. Finally, the top n scored phrases are stemmed and extracted as keywords (keyphrases) for a document.

4   Proposed Weighting Scheme

In this section, we introduce our weighting scheme. It computes the strength of relation of a pair of words in a given document using the information supplied both by word embeddings and local statistical information.

The task of keyword extraction is to identify a few important words that represent the core theme of an article such that they tie and hold the entire text together. Intuitively, the words having stronger semantic relatedness tend to tie each other in a given document, constituting the frame of the document. However, how words actually co-occur in the document also assists in identifying their importance. Two words can be very important if they frequently co-occur in a document despite having weaker semantic relatedness. Therefore, we consider that the importance of words can be identified by two factors: their semantic relatedness in general, and co-occurrence statistics in a document.

We thus propose to compute the strength of relation of a pair of words as the product of the semantic relatedness and local co-occurrence co-efficient, as:

$$sr(w_i, w_j) = semantic(w_i, w_j) \times cooccur(w_i, w_j) \tag{2}$$

The detailed calculations of semantic relatedness and local co-occurrence co-efficient are as follows. We use the values contained in word embedding vectors to measure the semantic distance between words. Word embeddings are dense, low-dimensional (typically 50 -1000 dimensions), and real-valued vectors, which encode both semantic and syntactic information. Commonly used approaches to compute the semantic relatedness or similarity between two points in a vector space are *Euclidean Distance* and *Cosine Similarity*. We wish to assign a higher semantic relatedness score to a pair of semantically related words, conversely two semantically unrelated words should be penalised by assigning a lower score.

Thus, the semantic relatedness between w_i and w_j ($w_i \neq w_j$) using is Euclidean Distance computed as:

$$semantic(w_i, w_j) = \frac{1}{Euclidean(w_i, w_j)} = \frac{1}{\sqrt{\sum_{k=1}^{n}(v_{ik} - v_{jk})^2}} \tag{3}$$

where n is the vector's length, with embedding vector $v_i = \{v_1, v_2, ..., v_n\}$ corresponding to word w_i. If using *Cosine Similarity*, semantic relatedness is computed as:

$$semantic(w_i, w_j) = \frac{1}{1 - cosine(w_i, w_j)} = \frac{||v_i|| \cdot ||v_j||}{||v_i|| \cdot ||v_j|| - v_i \cdot v_j} \tag{4}$$

To compute the co-occurrence co-efficient, we choose *Point-wise Mutual Information*, and *Dice Coefficient*.

We evaluate our work using four combinations: $\frac{Dice(w_i, w_j)}{Euclidean(w_i, w_j)}$, $\frac{Dice(w_i, w_j)}{1 - Cosine(w_i, w_j)}$, $\frac{PMI(w_i, w_j)}{Euclidean(w_i, w_j)}$, and $\frac{PMI(w_i, w_j)}{1 - Cosine(w_i, w_j)}$.

5   Graph Ranking Algorithm

A commonly used graph ranking algorithm by many studies [2,3] is PageRank [10], which was originally designed to compute the degree distribution of a directed graph. However, a document is usually represented as an undirected graph, since co-occurrence relations of words are bidirectional. Hence, a common way to use PageRank on an undirected graph is to simply cast each node into bidirectional. A problem with this approach is that the original non-personalised PageRank of an undirected graph equates to the degree distribution vector only if the graph is regular [18]. A text represented as an undirected graph is simply not a regular graph. Therefore, we propose to use a personalised vector that contains probability distribution of each node with a weighted PageRank [19].

Formally, in graph $G = (V, E)$, where V is the collection of all vertices, and E is the collection of all edges, let $C(w_i)$ be the set of vertices incident to w_i. The score of w_i is calculated by:

$$S(w_i) = (1 - d)pr(w_i) + d \times \sum_{w_j \in C(w_i)} \frac{sr(w_i, w_j)}{|out(w_j)|} S(w_j) \tag{5}$$

where $sr(w_i, w_j)$ is the *strength of relatedness* score between two words w_i and w_j calculated using Equation 2. $pr(w_i)$ is the probability distribution of word w_i, calculated as $\frac{freq(w_i)}{N}$ where $freq(w_i)$ is the occurrence frequency and N is the number of total valid words appearing in the text.

6   Keyword Generation

Keyword generation suggests potential keywords that do not appear in a given document. Intuitively, these words have very close semantic relatedness to the important words appearing in the document. As discussed in Section 4, a document D is constituted by a few important words W that bind the entire text. Each semantic related word $w \in W$ is represented as a point in word embedding vector space, thus forms a region. We hypothesise that words toward the centre of the region tend to be more abstractive, and provide high level description of W. One may think that a document could have multiple topics, and words representing each topic would form a cluster, thus our intuition would have a bias. In practice, however, keywords serve to present the highest level of abstraction, which is the core theme of an article rather than identify all the topics contained in the article.

Our generation approach uses a pipeline architecture built on top of our extraction approach. Given a document D, individual words are first ranked by our extraction algorithm, and the top ranked words W are extracted. The corresponding word embedding vector for each $w \in W$ is then extracted and saved in a list L and the mean vector v of L is computed. We apply the *approximate nearest neighbour* algorithm [6] to find the nearest vectors to v in the entire word embedding set, and saved in C. The corresponding words for each $c \in C$

are suggested keywords of D. Because the chosen word embedding vectors do not encode multi-word phrases, our generation approach is only able to suggest individual words.

7   Corpus-Aware Matching

A standard keyword exaction evaluation technique is *exact matching* using string comparison, which is unable to identify whether a pair of multi-word phrases have the same semantic meaning. In many cases, computer generated keywords can fail to match simply because they either contain or miss unnecessary words comparing with human assigned ones, such as *iPad* and *iPad tablet*. The problem we are interested is that how to identify whether a pair of phrases in which one phrase is a sub-phrase of another represent the same semantic meaning.

We propose a *corpus-aware matching* criteria, which excludes too-common and less meaningful words from the evaluation corpus when comparing two phrases. For example, our evaluation dataset is a collection of abstracts of computer science journal articles, words such as *algorithm, method, approach* are too common in this field, thus do not carry much meaning and not worth considering when comparing two multi-word phrases. For example, *backpropagation algorithm* and *backpropagation*, which refer to the same concept, are counted as correct match using proposed *corpus-aware matching* criteria, because *algorithm* is a common word in the evaluation dataset.

To identify the common and less meaningful words from a corpus, we follow the idea of Rose et al. [20] generating a per-corpus stop-list of words commonly used in the corpus. Note that these words are not standard stop-words. The intuition is that words frequently adjacent to, but not within, human assigned keywords are less likely to be meaningful, such as *approach, system, algorithm*, and *method* in computer science field. We separate a corpus into two parts, one is for evaluation, and another is for generating the stop-list in which only the top n most frequent words are saved.

8   Experiments

8.1   Experiment Setup

Dataset. We choose the publicly available dataset, Hulth2003[1], a collection of 2,000 abstracts of Computer Science journal articles, each of approximately 100-150 words, extracted from *Inspec*. Each document has two human assigned keyword lists - author assigned keywords (restricted to a certain number) and reader assigned ones (freely assigning), where 78% readers' assigned ones appear in the texts, in contrast to only 25% for authors' assigned keywords. The total number of keywords assigned by readers is 19,276, and 8,945 by authors, in which only 1,207 are in common. Moreover, we find that author assigned keywords tend to be more abstractive, often describing board fields, whereas the reader assigned ones tend to be more extractive.

---

[1] http://github.com/snkim/AutomaticKeyphraseExtraction

Evaluation Strategy. We first evaluate our extraction approach on three pre-trained word embedding vectors with four different weighting schemes. These compare different weighting scheme performance based on different word embedding vectors. Since word embeddings we used in this study are pre-trained, different learning objectives and training corpora can affect the performance and evaluation results. We therefore choose the best performing one for our next evaluation, which compares our proposed approach with three baseline and the state-of-the-art algorithms.

For the first and second evaluation, we selected 500 articles out of 2000, in which all reader assigned keywords appear in the corresponding text. We wish to determine the best performance of our extraction algorithm, since it is not able to identify any words that do not appear in the text. Secondly, we want a fair comparison with the state-of-the-art algorithm [21] on this dataset, which is also evaluated on 500 articles and only considering the keywords appearing in texts. We then use the remaining 1,500 documents to generate a stop-list, where the most frequent 30% words (exclusive stopwords) adjacent to assigned keywords are extracted and saved in the stop-list for *corpus-aware matching*. In the third evaluation, we evaluate the performance of our generation approach on the entire dataset - 2000 articles using the author assigned keywords.

Evaluation Matrix. For keyword extraction, we use two matching criteria: *exact matching* and *corpus-aware matching*. For keyword generation, we employ token matching using the automatically generated words against words in author assigned keywords. We use the *Precision, Recall,* and *F-measure* for all evaluations. The *Precision* is defined as: $precision = \frac{the\ number\ of\ correctly\ matched}{total\ number\ of\ extracted}$ *Recall* is defined as: $recall = \frac{the\ number\ of\ correctly\ matched}{total\ number\ of\ assigned}$ *F-measure* is defined as: $F = 2 \times \frac{precision \times recall}{precision + recall}$.

8.2   Baseline and the State-of-the-Art Algorithms

Since our approach is graph-based, we selected three graph-based ranking algorithms as the baseline: TextRank [2], SingleRank [3], and ExpandRank [3]. We use the re-implementations provided by Hasan and Ng [22]. TextRank first constructs an unweighted undirected graph to represent a text, and uses original PageRank for ranking. SingleRank is built on top of TextRank, but uses word co-occurrence frequencies as weights assigned to the edges of a graph. ExpendRank first finds k nearest documents using cosine similarity, then assigns weights to the edges using co-occurrence statistics of words collected from the k nearest neighbours and the document itself. The state-of-the-art algorithm on our chosen dataset is introduced by Liu et al. (2009) [21]. The algorithm firstly cluster candidate words based on their semantic relatedness using Wikipedia-based statistics. Then an exemplar term is extracted from each of the clusters, and finally multi-word phrases are formed by looking for longest n-grams from the document, and for the n-gram phrases contain an exemplar are extracted as keywords.

8.3   Experiment Results and Discussion

Our first experiment evaluates four different weighting schemes on three pre-trained word embedding vectors. Consistently using the ranking algorithm described in Section 5, and a window size of 10, we extracted the top 10 ranked phrases as the keywords compared with the gold standards using *exact match*. Although there is no significant difference, the *point-wise mutual information* coupled with *cosine similarity* outperforms other combinations on HLBL word embedding vectors. Therefore, we use this combination for all other evaluations.

Table 1. Comparison of three word embedding vectors with four weighting schemes

	SENNA			Turian			HLBL		
	Prec.	Rcl.	F-score	Prec.	Rcl.	F-score	Prec.	Rcl.	F-score
$\frac{dice}{Euclidean}$	37.6	51.3	43.4	37.4	51.2	43.2	37.8	51.7	43.7
$\frac{dice}{1-cosine}$	38.3	52.4	44.3	37.1	50.7	42.9	38.4	52.5	44.4
$\frac{pmi}{Euclidean}$	37.4	51.1	43.2	37.2	50.9	43.0	37.8	51.6	43.6
$\frac{pmi}{1-cosine}$	38.2	52.2	44.1	37.2	50.8	42.9	**38.7**	**52.8**	**44.7**

Table 2. Comparison with baseline algorithms and the state-of-the-art

		Exact Matching			Corpus-aware Matching		
		Prec.	Rcl.	F-score	Prec.	Rcl.	F-score
win = 2	TextRank	34.4	47.1	39.8	38.3	52.6	44.3
	SingleRank	34.0	47.7	40.3	38.7	52.9	44.7
	ExpandRank	34.9	47.7	40.3	38.7	52.9	44.7
	Our Approach	**37.8**	**51.6**	**43.6**	**41.8**	**57.1**	**48.3**
win = 5	TextRank	34.5	46.9	39.8	38.1	51.8	43.9
	SingleRank	34.6	47.1	39.9	38.3	52.2	44.2
	ExpandRank	34.6	47.1	39.9	38.3	52.2	44.2
	Our Approach	**38.0**	**52.0**	**43.9**	**42.1**	**57.6**	**48.7**
win = 10	TextRank	34.7	47.3	40.0	38.6	52.5	44.5
	SingleRank	34.4	46.8	39.6	38.4	52.2	44.2
	ExpandRank	34.4	46.9	39.7	38.5	52.4	44.3
	Our Approach	**38.7**	**52.8**	**44.7**	**42.7**	**58.3**	**49.3**
Topic Clustering Liu et al. [21]		35.0	**66.0**	45.7*	–	–	–
Our Approach Window = 10		**38.7**	52.8	44.7	42.7	58.3	49.3

*The authors used a customised word filter to remove some words that are too common to be keywords (yet not standard stop words). However the removal list is not publicly available, without that the overall F-score will drop 5% to 40.7.

Our second experiment compares our results with the baseline and the state-of-the-art algorithms using both *exact matching* and *corpus-aware matching*. We use the same tokeniser, POS tagger, and phrase formation technique. It is also worth noting that for TextRank, we modified the original phrase formation technique proposed by the authors to the one described in Equation 1. This is because by unifying phrase formation technique, effects are minimised, so we can clearly see the difference by comparing only the different algorithms. Finally, we extracted the top 10 ranked phrases as the keywords. Our ranking algorithm is personalised PageRank, however we also ran experiments using non-personalised PageRank, and the F-score uniformly drops about 0.6%. Our model would be the same as TextRank when we turn off our weighting feature and using non-personalised PageRank. Therefore, our proposed weighting scheme uniformly increases the performance about 4%. Overall, our approach produced superior performance over other baseline algorithms, and is very comparable to the state-of-the-art algorithm without using any customised word filter.

Using our *corpus-aware matching*, the overall performance of each algorithm increased about 5% on the F-score. The 5 most frequent words (stems) in the stop-list are: *use, new, includ, differ, develop*, other common words include *method, model, algorithm*. This matching strategy is a simple yet effective way to overcome the problem where two phrases, one a sub-phrase of the other, have the same semantic meaning in a domain specific corpus.

Table 3. Keyword Generation

	Appearing in text			Not appearing in text		
	Prec.	Rcl.	F-score	Prec.	Rcl.	F-score
keyword extraction	8.0	9.7	8.7	0.0	0.0	0.0
keyword generation	0.4	0.4	0.4	1.8	2.2	2.0

Our third experiment evaluates our keyword generation approach. Since the approach is not able to identify multi-word phrases, we tokenised all author assigned phrases into individual words. The extraction module only identified the top 10 ranked words, without the phrase formation process. Table 3 shows the evaluation results, which is lower than expected. Unlike the first two experiments, we only extracted top 10 ranked individual words.

The generation module only produced an F-score of 2.0. We found that the ground truth used in this experiment has many biases. All author assigned keywords are from a restricted vocabulary, whereas our approach uses an open vocabulary. The majority of matches typically fall into a few small groups of words in the restricted vocabulary. Another reason is that the word embedding vectors we used in this study are pre-trained over a general domain. However the evaluation dataset we used is a Computer Science specific corpus.

We randomly compared some of computer generated keywords with the original document, finding that many generated keywords do represent the core theme of the document. An example follows:

Text: *Is open source more or less secure? Networks dominate today's computing landscape and commercial technical protection is lagging behind attack technology. As a result, protection programme success depends more on prudent management decisions than on the selection of technical safeguards. The paper takes a management view of protection and seeks to reconcile the need for security with the limitations of technology.*

Author assigned keywords: *computer network, public domain software, security of data.* Generated keywords: *security, control, analysis, support, knowledge, intelligence, usage, complexity, problems, stability.*

9 Conclusion and Future Work

In this study, we have presented an approach for keyword identification. We have shown that extraction module outperforms baseline algorithms and is comparable to the state-of-the-art algorithm without any tuning. Moreover, our keyword generation approach employs word embedding vectors to suggest keywords not actually appearing in a text. We also show the *corpus-aware matching* evaluation is an effective way to overcome a common keyword evaluation problem.

We consider that the approach can be further improved by training the word embeddings over domain specific corpora. Our future work focuses on further investigating the background knowledge that word embeddings can encode, in order to further improve our approach.

Acknowledgments. This research was funded by an Australian Postgraduate Award, and Safety Top-Up Scholarships from the University of Western Australia, and Linkage Grant LP110100050 from the Australian Research Council.

References

1. Jones, K.S.: A statistical interpretation of term specificity and its application in retrieval. Journal of documentation **28**(1), 11–21 (1972)
2. Mihalcea, R., Tarau, P.: TextRank: Bringing Order into Texts. In: Conference on Empirical Methods in Natural Language Processing, Barcelona, Spain (2004)
3. Wan, X., Xiao, J.: Single document keyphrase extraction using neighborhood knowledge. AAAI **8**, 855–860 (2008)
4. Matsuo, Y., Ishizuka, M.: Keyword extraction from a single document using word co-occurrence statistical information. International Journal on Artificial Intelligence Tools **13**(01), 157–169 (2004)
5. Hasan, K.S., Ng, V.: Automatic keyphrase extraction: A survey of the state of the art. In: Proceedings of the 52nd Annual Meeting of the Association for Computational Linguistics (Volume 1: Long Papers), pp. 1262–1273 (2014)
6. Indyk, P., Motwani, R.: Approximate nearest neighbors: towards removing the curse of dimensionality. In: Proceedings of the Thirtieth Annual ACM Symposium on Theory of Computing, pp. 604–613. ACM (1998)
7. Collobert, R., Weston, J., Bottou, L., Karlen, M., Kavukcuoglu, K., Kuksa, P.: Natural language processing (almost) from scratch. The Journal of Machine Learning Research **12**, 2493–2537 (2011)
8. Mnih, A., Hinton, G.: Three new graphical models for statistical language modelling. In: Proceedings of the 24th International Conference on Machine Learning, pp. 641–648. ACM (2007)

9. Turian, J., Ratinov, L., Bengio, Y.: Word representations: a simple and general method for semi-supervised learning. In: Proceedings of the 48th Annual Meeting of the Association for Computational Linguistics, Association for Computational Linguistics, pp. 384–394 (2010)
10. Brin, S., Page, L.: The anatomy of a large-scale hypertextual web search engine. Computer networks and ISDN systems **30**(1), 107–117 (1998)
11. Liu, Z., Li, P., Zheng, Y., Sun, M.: Clustering to find exemplar terms for keyphrase extraction. In: Proceedings of the 2009 Conference on Empirical Methods in Natural Language Processing: Volume 1-Volume 1, Association for Computational Linguistics, pp. 257–266 (2009)
12. Grineva, M., Grinev, M., Lizorkin, D.: Extracting key terms from noisy and multi-theme documents. In: Proceedings of the 18th International Conference on World Wide Web, pp. 661–670. ACM (2009)
13. Wang, J., Liu, J., Wang, C.: Keyword extraction based on pagerank. In: Zhou, Z.-H., Li, H., Yang, Q. (eds.) PAKDD 2007. LNCS (LNAI), vol. 4426, pp. 857–864. Springer, Heidelberg (2007)
14. Mikolov, T., Sutskever, I., Chen, K., Corrado, G.S., Dean, J.: Distributed representations of words and phrases and their compositionality. In: Advances in Neural Information Processing Systems, pp. 3111–3119 (2013)
15. Liu, Z., Chen, X., Zheng, Y., Sun, M.: Automatic keyphrase extraction by bridging vocabulary gap. In: Proceedings of the Fifteenth Conference on Computational Natural Language Learning, Association for Computational Linguistics, pp. 135–144 (2011)
16. Brown, P.F., Pietra, V.J.D., Pietra, S.A.D., Mercer, R.L.: The mathematics of statistical machine translation: Parameter estimation. Computational linguistics **19**(2), 263–311 (1993)
17. Klein, D., Manning, C.D.: Accurate unlexicalized parsing. In: Proceedings of the 41st Annual Meeting on Association for Computational Linguistics-Volume 1, Association for Computational Linguistics, pp. 423–430 (2003)
18. Grolmusz, V.: A note on the pagerank of undirected graphs (2012). arXiv preprint arXiv:1205.1960
19. Xing, W., Ghorbani, A.: Weighted pagerank algorithm. In: 2004 Proceedings of the Second Annual Conference on Communication Networks and Services Research, pp. 305–314. IEEE (2004)
20. Rose, S., Engel, D., Cramer, N., Cowley, W.: Automatic keyword extraction from individual documents. Text Mining (2010)
21. Liu, Z., Huang, W., Zheng, Y., Sun, M.: Automatic keyphrase extraction via topic decomposition. In: Proceedings of the 2010 Conference on Empirical Methods in Natural Language Processing, Association for Computational Linguistics, pp. 366–376 (2010)
22. Hasan, K.S., Ng, V.: Conundrums in unsupervised keyphrase extraction: making sense of the state-of-the-art. In: Proceedings of the 23rd International Conference on Computational Linguistics: Posters, Association for Computational Linguistics, pp. 365–373 (2010)

Truth Discovery in Material Science Databases

Eve Bélisle[1]([✉]), Zi Huang[1], and Aimen Gheribi[2]

[1] University of Queensland, Brisbane, Australia
{uqzhuang,e.belisle}@uq.edu.au
[2] École Polytechnique de Montréal, Montréal, Canada
aimen.gheribi@polymtl.ca

Abstract. Instead of performing expensive experiments, it is common in industry to make predictions of important material properties based on some existing experimental results. Databases consisting of experimental observations are widely used in the field of Material Science Engineering. However, these databases are expected to be noisy since they rely on human measurements, and also because they are an amalgamation of various independent sources (research papers). Therefore, some conflicting information can be found between various sources. In this paper, we introduce a novel truth discovery approach to reduce the amount of noise and filter the incorrect conflicting information hidden in the scientific databases. Our method ranks the multiple data sources by considering the relationships between them, i.e., the amount of conflicting information and the amount of agreement, and as well eliminates the conflicting information. The scalable Gaussian process interpolation technique (SGP) is then applied to the clean dataset to make predictions of materials property. Comprehensive performance study has been done on a real life scientific database. With our new approach, we are able to highly improve the accuracy of SGP predictions and provide a more reliable database.

1 Introduction

In recent years, machine learning has been used extensively in the material engineering industry. Instead of performing expensive experiments, scientists rely on existing databases of experimental points fitted to various modelling techniques to make predictions on important material properties for the industry. Slag properties such as electrical conductivity, thermal conductivity, density, etc. play a key role in the metal industry [4]. If the databases do not already exist, the work simply involves a research in the literature, thus making it far less costly than performing actual experiments. Once a database of experimental points is assembled, one can use a machine learning model to fit the data and predict the desired properties in unknown areas.

One of the problems with databases consisting of experimental points is the human error involved in collecting the measurements. Furthermore, since these databases are assembled from different sources, some conflicting information

© Springer International Publishing Switzerland 2015
M.A. Sharaf et al. (Eds.): ADC 2015, LNCS 9093, pp. 269–280, 2015.
DOI: 10.1007/978-3-319-19548-3_22

between sources can alter the prediction accuracy of the chosen machine learning technique. In this work, we are looking at a way to improve the reliability of databases consisting of experimental points by analysing the conflicting information and attributing a quality measure to each source: the papers from which the points have been extracted, or authors. We calculate and compare the reliability of sources by using the amount of conflicting information for each source in combination with the amount of non-conflicting similarities with other sources. A level of reliability can then be attributed and the sources can be ranked, making it possible to choose between two conflicting data points. With this novel approach for analysing the data, a given database can be screened and improved by removing data points believed to be in error.

In order to test our approach, we used a database consisting of over 5000 data points on electrical conductivity (EC). EC predictions has been studied by various authors in the past using various modelling techniques [4,10,13]. We performed predictions using the Scalable Gaussian Process regression (SGP) [1], an interpolation technique developed in order to make on-line learning possible using Gaussian processes. This approach has proven to be superior for predicting material properties [2,12]. First, we evaluate the strength of SGP by testing how much conflicting information (noise) can be introduced and supported by this interpolation technique. Then we apply our new truth discovery approach to see how the predictions can be improved. We compare the prediction accuracy before pruning the database using our sources ranking truth discovery technique and after the database has been purged. On top of improving the predictions of machine learning techniques, the filtered database becomes more reliable when consulting existing information. Faced with conflicting data in an existing database, it can be confusing for a human being to decide which source is more reliable than an other. The process can involve time in research and reading and rely on a subjective evaluation. Our approach can therefore automatise this process and improve the quality of existing databases consisting of experimental points.

The rest of the paper is organized as follows. After discussing the related work in Section 2, we present our dataset and methods in Section 3. Experimental results are given and discussed in section 4 and we conclude in section 5.

2   Related Work

The topic of truth discovery is not new and has been extensively studied, especially in the domain of social networks and the world wide web, where many conflicting information can be found, and where the duplication of wrong information also becomes a problem. In their paper, Yin et al. [15] discuss the trustworthiness of websites by evaluating the amount of true information contained on the given website. The same authors propose a semi-supervised method for homogeneous network, again applied on web sources [16]. Kleinberg [9] also proposes a test algorithm to evaluate the quality of web pages according to their relationships with other pages. In our work we take inspiration from this approach by considering the amount of similar information linking our various sources together and how much they agree with each other, even though our sources are completely independent. Dong et al. [6] discuss truth discovery when accessing various sources of information, when the update history is known. They are evaluating the quality of sources over time and conducting a probability analysis. In another paper [7] they discuss the selection of sources when there is an overwhelming abundance of possible sources. A maximum likelihood approach is used by Wang et al. [14] to filter noisy social sensing data. Zhao et al. developed a probabilistic model for data steams, in order to evaluate sources quality in real time [17], applying their approach to weather forecast data.

None of the previous approaches have been applied on sets of experimental data points. In the field of data mining, Sheng et al. [11] address the problem of noisy labeling of data by carefully selecting a set of points where labelling will be repeated. Dekel et al. [5] are also proposing a way to prune low quality labels in a crowd.

In a previous publication, we evaluated SGP on prediction of electrical conductivity [2]. A different database was then used and in order to improve the predictions, based on the knowledge of the problem, the paper shows that using $ln(T*\sigma)$, where T is the temperature and σ is the value of electrical conductivity in Siemens per meter, provided a significant improvement on predictions. However, in the present work, we are using the non-logarithmic values of electrical conductivity in order to show the full range of errors.

3   Methods and Dataset

3.1   Dataset

The EC database used for this work has been collected from the literature. It consists of 5373 data points from 67 different sources. The inputs are chemical compositions from 8 oxide components in mole percent (SiO_2, Al_2O_3, MgO, CaO, MnO, PbO, FeO, Fe_2O_3) and a temperature in Kelvins. The value to be predicted is the corresponding electrical conductivity in Siemens per meter. See Table 1 for an example of data points taken from this database.

Here we consider each scientific paper from where the data points have been extracted to be independent sources of information. This database can be fed to a machine learning model in order to make predictions for new chemical compositions where the electrical conductivity is currently unknown. In Table 1, the last line is an example query that could be desired in the industry.

As an example taken from our dataset, Fig. 1 illustrates a series of conflicting pairs of data points between two sources. Here the values of EC, with input values extremely close in space and at the same temperature show large variations. Such differences are unacceptable when consulting in process design [8]. When consulting an existing database, faced to such variations in data, as it would be very confusing to decide which information is truthful and which should be discarded.

272   E. Bélisle et al.

Table 1. Sample data points for the electrical conductivity database. The input compositions are in mole percent and the electrical conductivity in Siemens per meter.

SiO$_2$	Al$_2$O$_3$	MgO	CaO	MnO	PbO	FeO	Fe$_2$O$_3$	T(K)	EC	Source
33.56	0	0	41.96	0	0	24.13	0.35	1573	48.00	1
33.3	0	0	0	0	66.7	0	0	1223	38.50	4
0	25.67	0	40.56	0	0	30.82	2.96	1673	94.00	6
50.00	0	0	25.00	25.00	0	0	0	1873	67.10	7
49.53	16.71	33.76	0	0	0	0	0	1923	18.79	31
25.31	0	18.33	9.01	0	0	46.37	0.99	1593	403.90	46
Example query point:										
54.24	5.32	40.43	0	0	0	0	0	1673	Predicted	N/A

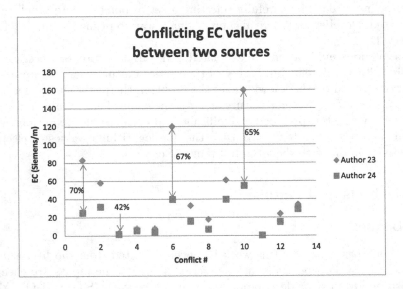

Fig. 1. Example of conflicting information found in our EC database. Each pair of conflict is shown on the X axis and the Y axis presents the values of EC.

3.2   Scalable Gaussian Process

A typical Gaussian process (GP) regression machine learning interpolation technique has a computational complexity of n^3 where n is the number of training data points. This is due to the necessary necessary matrices inversions used in this technique and it makes the calculations very tedious on the average desktop computer. The SGP interpolation technique has been introduced by the current authors in a previous published paper [1]. This approach improves the computational time of a traditional GP, by creating clusters of similar information as input queries, and then using a subset of the entire training database by choosing only the information close to a given query cluster. The calculations are therefore performed by small clusters, or batches, and the reduced training database is also compressed to remove similar information, improving drastically the overall computational time.

3.3 Truth Discovery by Sources Comparison

In an effort to reduce the noise in databases consisting of experimental point, we introduce a new method of truth discovery using the different sources (research papers in our case) and the amount of conflicting and similar but non-conflicting information between them to create a ranking of reliability.

In Fig. 2, we represent a sample of 13 of the 67 sources found in our database. One can see the amount of conflicting information over the amount of similarities. In this example, we want to consider source 50 as more reliable than source 23, because 50 has 3 similar data points agreeing with two other sources, whilst 23 has 18 conflicting data points, including 2 within its own data.

Of course, what is regarded as similar information and conflicting is entirely subjective and we had to define our own rules. In this work, we measure the distance between two data points in space using a custom distance equation, following this predicate to compare two points p and q:

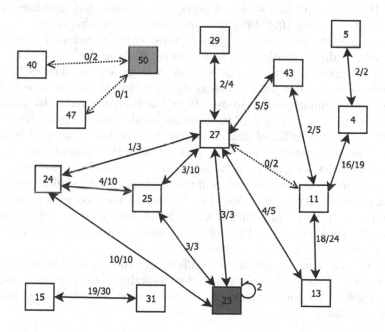

Fig. 2. Illustration of conflicting information between some sources of the EC database. The squares represent the different sources with their arbitrary numbering, the arrows represent similarities between two sources and the numbers on each arrow represent the amount of conflicts over the amount of similar data points. Dotted arrows indicate that there are no conflict, only agreements between two sources.

$$\forall i \in N : \left(\frac{|q_i - p_i|}{\sum_{j=1}^{N} q_j} < \epsilon \right) \wedge \left(\frac{|q_i - p_i|}{\sum_{j=1}^{N} p_j} < \epsilon \right) \wedge ((p_i = 0) \leftrightarrow (q_i = 0))$$

Where N is the total number of columns (dimensions), excluding the predicted column (EC) and ϵ is an arbitrary similarity constraint, we used a value of 5%. This equation has been introduced in our previous work [1], where it has proven to be a more accurate way of comparing material properties databases than using a simple euclidean distance. The reason being that we want points to be close in every dimension, as potential chemical interactions between the components can cause a very big difference in the predicted value. In other words, very close points in space can have a very big difference in their value of EC, caused by a small amount of a certain chemical component.

We define a conflict between two authors as two points that are similar, ie. relatively close in space ($\epsilon < 5\%$) but having a difference of more than twice the experimental error in the output used for prediction (EC). To find a reliable value of experimental error, we computed the average discrepancy within each source. That is, for each pair of similar points within a given author, we calculate the average difference in EC. In our case study, we find this average to be of approximately 5%, therefore we considered conflicting information values to be above a 10% difference in EC. By definition, the points that are close in space but where EC is under 10% difference are considered as agreements.

In our work we consider two types of similarities: direct and indirect. For a given source, direct similarities (agreements and conflicts) are the ones that can be found from its own data in relation to other sources. Let us define the list of sources with direct similarities as S_d. An example is illustrated in Fig. 3, showing direct similarities for source 23 by black arrows going to sources 24, 25 and 27. Here $S_d = 24, 25, 27$. On the other hand, indirect similarities are the similarities between our list of similar sources, S_d and other sources. The indirect similarities are illustrated by red arrows on Fig. 3. In this example, source 23 has indirect similarities with sources 24, 25, 27, 29, 43, 11 and 13. We consider that indirect similarities are an indication of the reliability of the similar sources. For example, source A could have a lot of conflicting information with source B, but if source B has also a lot of conflicts with a lot of other sources, this means that it may not be very reliable and therefore this information should be of less value than if B was considered very reliable.

Using the amount of similarities, conflicts, agreements and the total amount of data points for each author, we introduce a quality rate Q, giving an estimation of reliability for each source. For a given source, we calculate Q using the following equations:

$$Q = \frac{\alpha Q_d + \beta Q_i + \gamma Q_c}{\alpha + \beta + \gamma} \tag{1}$$

$$Q_d = \frac{1 - C_d - A_d}{P} \tag{2}$$

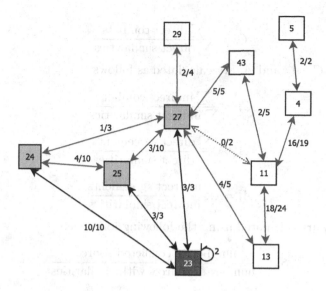

Fig. 3. Direct similarities of source 23 are shown by black arrows and indirect similarities by red arrows. Here sources 24, 25 and 27 contribute to the indirect similarities for source 23.

$$Q_i = \frac{1 - C_i - A_i}{P} \tag{3}$$

$$Q_c = \frac{1 - S_c - S_a}{P} \tag{4}$$

Where C_d is the ratio of direct conflicts, A_d is the ratio of direct agreements, C_i is the ratio of indirect conflicts, A_i is the ratio of indirect agreements, S_c is the ratio of authors with conflicting data, S_a is the ratio of sources with agreeing data and P is the number of data points for the given source. Three parameters are introduced in the formula: α, β and γ, allowing weights to be attributed to each type of conflicting information. In this work we have chosen the parameters $\alpha = 2$, $\beta = 1$ and $\gamma = 0.5$. We chose these values because we consider direct conflicts and similarities to be of the most influential on the reliability of a given source. A higher value of Q signifies a higher confidence level for a given author. Here in the case of only one direct conflicting source, for example sources 15 and 31 in Fig. 2, the source with the most amount of data will have a higher Q value. In Fig. 2, source 23 has two internal conflicts, meaning that two pairs of data points are conflicting within its own dataset. This is considered very unreliable and should have a big effect on Q. We chose to treat these as direct conflicts but it is not added to the total amount of similarities. This can mean that a given source could have a negative value of Q. The percentage of direct conflicts C_d is calculated as follows:

$$C_d = \frac{\sum \text{direct conflicts}}{\sum \text{direct similarities}} \tag{5}$$

Similarly, C_i, A_d and A_i are calculated as follows:

$$C_i = \frac{\sum \text{indirect conflicts}}{\sum \text{indirect similarities}} \tag{6}$$

$$A_d = \frac{\sum \text{direct agreements}}{\sum \text{direct similarities}} \tag{7}$$

$$A_i = \frac{\sum \text{indirect agreements}}{\sum \text{indirect similarities}} \tag{8}$$

S_c and S_a are calculated using the following formulae:

$$S_c = \frac{\text{number of conflicted sources}}{\text{number of sources with similarities}} \tag{9}$$

$$S_a = \frac{\text{number of agreeing sources}}{\text{number of sources with similarities}} \tag{10}$$

Note that the same sources can contribute to both S_c and S_a, as two sources can have conflicting and agreeing data simultaneously.

Once every source has been evaluated, we consider every pair of conflicting information and eliminate the data point where its source has a lower value of Q. This is applied recursively on the entire database until all the conflicted information has been eliminated. Table 2 shows an example of two conflicting data points and the values of Q for each source. After the process, the remaining database is pruned and reduced, eliminating noisy information in an attempt to get better predictions.

Table 2. Example of conflicting data points. Here source 48 would be chosen over 45 and the first data point would be elminated from the database.

SiO2	Al2O3	MgO	CaO	MnO	PbO	FeO	Fe2O3	T(K)	EC)	Source	Q
63.40	0	0	36.60	0	0	0	0	1873	16.10	45	0.004
61.38	0	0	38.62	0	0	0	0	1873	20.50	48	0.120

In order to test the prediction power of SGP, some artificial noise has been generated and introduced in the database. Section 4 presents the results of the predictions with various amount of introduced noisy data. In order to keep it realistic, the noisy data had to be close to the existing data points, but have possible conflicting values of EC. Therefore, these points have been produced by taking each existing source and creating a slightly modified version of each data point (randomly +/-5%) but with a possibly conflicting value of EC (randomly

+/- 50%). We then choose a random subset of all the generated noisy data and we introduce them in our database prior to testing. It is important to note that from this method, a random portion of the introduced points will not be conflicting information.

4   Results and Discussion

First, we test predictions with SGP, using a 10-fold cross-validation technique on the non-filtered database. In a 10-fold cross-validation, the entire database is split in 10 equal subsets. Each subset is then used as a testing set where the model is trained with the remaining 9 subsets. We repeat this procedure 10 times. The average error in percent and root mean square error are then computed over all the tests and this is what we are presenting in this section. Table 3 shows the influence of introduced noise on the predictions performed by SGP. Then, we test the same databases when applying our noise reduction technique introduced in Section 3.3. The results are presented as a graph of the RMSE in Fig. 4 and in Table 3.

From these results, one can see that SGP is extremely efficient at excluding a low to medium amount of noisy information. When applying our noise reduction technique on the original database, only 7% of data points were removed, and it explains the fact that there is no improvement on the original predictions. In order to test the robustness of SGP, up to 75% of noisy data points were introduced in the database. When introducing up to 20% of noisy data, the predictions remain acceptable with a RMSE of around 20%. This result is an example of the remarkable robustness of the SGP technique. The biggest effect can be seen around 40% of noise, where the RMSE jumps to 34%. Beyond

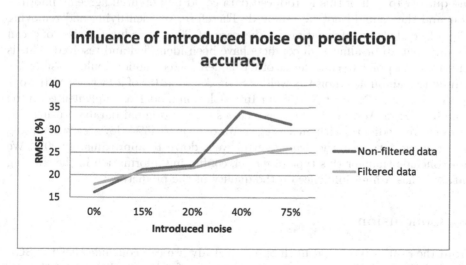

Fig. 4. Graph showing the influence of the amount of introduced noise on SGP prediction accuracy

Table 3. Influence of introduced noise on SGP predictions

Database	No filtering		Filtered database		
	Error (%)	RMSE	Filtered points (%)	Error (%)	RMSE
Original database	14.96	16.13	7	15.11	17.91
15% introduced noise	18.04	21.27	11	18.20	20.76
20% introduced noise	18.65	22.15	12	18.81	21.61
40% introduced noise	28.14	34.19	15	20.86	**24.38**
75% introduced noise	26.51	31.33	15	21.78	**25.96**

this amount, as it can be expected, the predictions are actually getting slightly better. This is because the noisy data is overtaking the actual real data and the SGP is actually over fitting. Nevertheless, our noise reduction technique is showing impressive results by keeping the error below the 22% mark. In Table 3, the pruned points column shows the amount of conflicting information that has been eliminated using our noise reduction technique. Here we can note that all the introduced noise is not completely removed during the filtering, and this is perfectly normal as some of the random noise can actually be non conflicting information. However, by removing the conflicts, we can improve the predictions by an impressive 10% on the RMSE for the case of 40% introduced noise, which is where we see the most effect on the SGP prediction accuracy.

Even if there is no a major improvement in the prediction accuracy under the bar of 20% introduced semi-noisy data, it is important to note that our filtering technique still managed to remove from 7% to 12% of conflicting information, meaning that the reliability of the database is improved when consulting existing data. Since the introduced data is random, it is not unreasonable to assume that one quarter to half of this introduced data could be potential agreeing information and thus should not be removed. Therefore, by identifying and removing 12% of conflicting information in the case of 20% introduced noise, one can assume that all or almost all conflicts have been identified and resolved. This is an important point to consider as other modelling techniques would possibly not handle this amount of noise as well as SGP. As a matter of a fact, we tested our approach using a Nearest Neighbour Interpolation model, as implemented in the XonGrid Excel Add-in [3]. The predictions on the original database using this model were mediocre, with an average error of around 45%. However, when using the training database, the average error went down to approximately 20%. We can conclude that for this type of model, conflicting information in the training database has a high influence on the quality of the predictions.

5   Conclusion

From the results presented in the present study, we can conclude that the SGP interpolation technique is robust when the ratio of noisy/conflicting data is relatively low. However, predictions start to deteriorate when more and more noisy data is involved. The proposed approach prunes the database using a novel noise reduction truth discovery technique, improving the predictions by 10%. This technique uses the amount of conflicting and agreeing information between different sources in order to resolve conflicts within the database. A filtered version of the database is then produced and the prediction accuracy and reliability is greatly improved. The new produced database can also be considered more reliable when consulting existing information, automatising the conflict resolution process.

Acknowledgments. We would like to thank Marcus Gallagher and Paul Voigt for stimulating discussions. The research is supported by ARC DP140103171.

References

1. Bélisle, E., Huang, Z., Gheribi, A.: Scalable gaussian process regression for prediction of material properties. In: Wang, H., Sharaf, M.A. (eds.) ADC 2014. LNCS, vol. 8506, pp. 38–49. Springer, Heidelberg (2014)
2. Bélisle, E., Huang, Z., Le Digabel, S., Gheribi, A.E.: Evaluation of machine learning interpolation techniques for prediction of physical properties. Computational Materials Science **98**, 170–177 (2015)
3. Besses, B.D.D.: Xongrid interpolation add-in (2015)
4. Birol, B., Polat, G., Saridede, M.: Estimation model for electrical conductivity of molten caf2-al2o3-cao slags based on optical basicity. JOM, pp. 1–9 (2014)
5. Dekel, O., Shamir, O.: Vox populi. Collecting high-quality labels from a crowd (2009)
6. Dong, X.L., Berti-Equille, L., Srivastava, D.: Truth discovery and copying detection in a dynamic world. Proceedings of the VLDB Endowment **2**(1), 562–573 (2009)
7. Dong, X.L., Saha, B., Srivastava, D.: Less is more: Selecting sources wisely for integration. In: Proceedings of the VLDB Endowment, vol. 6, pp. 37–48. VLDB Endowment (2012)
8. Gheribi, A., Audet, C., Digabel, S.L., Bélisle, E., Bale, C., Pelton, A.: Calculating optimal conditions for alloy and process design using thermodynamic and property databases, the factsage software and the mesh adaptive direct search algorithm. Calphad **36**, 135–143 (2012)
9. Kleinberg, J.M.: Authoritative sources in a hyperlinked environment. Journal of the ACM (JACM) **46**(5), 604–632 (1999)
10. Mualem, Y., Friedman, S.P.: Theoretical prediction of electrical conductivity in saturated and unsaturated soil. Water Resources Research **27**(10), 2771–2777 (1991)
11. Sheng, V.S., Provost, F., Ipeirotis, P.G.: Get another label? improving data quality and data mining using multiple, noisy labelers. In: Proceedings of the 14th ACM SIGKDD International Conference on Knowledge Discovery and Data Mining, pp. 614–622. ACM (2008)
12. Sourmail, T., Garcia-Mateo, C.: Critical assessment of models for predicting the ms temperature of steels. Computational Materials Science **34**(4), 323–334 (2005)
13. Tsuboi, H., Chutia, A., Lv, C., Zhu, Z., Onuma, H., Miura, R., Suzuki, A., Sahnoun, R., Koyama, M., Hatakeyama, N., Endou, A., Takaba, H., Carpio, C.A.D., Deka, R.C., Kubo, M., Miyamoto, A.: An electrical conductivity prediction simulator based on tb-qcmd and kmc. system development and applications. Journal of Molecular Structure: THEOCHEM, 903(1–3):11–22, Recent advances in the theoretical understanding of catalysis (2009)

14. Wang, D., Kaplan, L., Le, H., Abdelzaher, T.: On truth discovery in social sensing: A maximum likelihood estimation approach. In: Proceedings of the 11th International Conference on Information Processing in Sensor Networks, pp. 233–244. ACM (2012)

15. Yin, X., Han, J., Yu, P.: Truth discovery with multiple conflicting information providers on the web. IEEE Transactions on Knowledge and Data Engineering **20**(6), 796–808 (2008)

16. Yin, X., Tan, W.: Semi-supervised truth discovery. In: Proceedings of the 20th International Conference on World Wide Web, pp. 217–226. ACM (2011)

17. Zhao, Z., Cheng, J., Ng W.: Truth discovery in data streams: A single-pass probabilistic approach. In: Proceedings of the 23rd ACM International Conference on Conference on Information and Knowledge Management, pp. 1589–1598. ACM (2014)

Presto-RDF: SPARQL Querying over Big RDF Data

Mulugeta Mammo[⊠] and Srividya K. Bansal[⊠]

Arizona State University, Mesa, AZ, USA
{mmammo,srividya.bansal}@asu.edu

Abstract. There has been a rapid increase in the amount of Resource Description Framework (RDF) data on the web. The processing of large volumes of RDF data requires an efficient storage and query-processing engine that can scale well with the volume of data. In the past two and half years, however, heavy users of big data systems, like Facebook, noted limitations with the query performance of these big data systems and began to develop new distributed query engines for big data that do not rely on map-reduce. Facebook's Presto is one such example. This paper proposes an architecture based on Presto, called Presto-RDF, that can be used to process big RDF data. An evaluation of performance of Presto in processing big RDF data against Apache Hive is also presented. The results of the experiments show that Presto-RDF framework has a much higher performance than Apache Hive and native RDF store - 4Store and it can be used to process big RDF data.

Keywords: Database performance · Evaluation · Querying · Semantic web data

1   Introduction

Semantic Web is the web of data that provides a common framework and technologies for sharing data and reusing data in various applications and enterprises. Resource Description Framework (RDF) enables the representation of data as a set of linked statements, each of which consists of a subject, predicate, and object called a triple. RDF datasets, consisting of millions of triples, form a network of directed graph (DG) and are stored in systems called triple-stores. A query language standard, SPARQL, has also been developed to query RDF datasets. For the Semantic Web to work, both triple-stores and SPARQL query processing engines have to scale well with the size of data. This is especially true when the size of RDF data is too big such that it is difficult, if not impossible, for conventional triple-stores to work with. In the past few years, however, new advances have been made in the processing of large volumes of data sets, aka big data, which can be made to use for processing big RDF data. In this regard, various research studies that use big data technologies for RDF data processing have been published [1]–[3]. The initial attempts to address this issue focused on optimizing native RDF stores as well as conventional relational databases management systems. But as the volume of RDF data grew to exponential proportions, the limitations of these systems became apparent and researchers began to focus on using big data analysis tools, most notably Hadoop, to process RDF data. Various

© Springer International Publishing Switzerland 2015
M.A. Sharaf et al. (Eds.): ADC 2015, LNCS 9093, pp. 281–293, 2015.
DOI: 10.1007/978-3-319-19548-3_23

studies and benchmarks that evaluate these tools for RDF data processing have been published [1], [4]–[6].

In the past two and half-years, new trends in big data technology have emerged that use distributed in-memory query processing engines based on SQL syntax. Some of these tools include: Facebook Presto [7], Apache Shark, and Cloudera Impala. These tools promise to deliver high performance query execution than traditional Hadoop system like Hive. The motivation of this paper is to validate this claim for big RDF data – i.e. if these new in-memory query processing models work well to deliver faster response times for SPARQL queries, which must be translated to SQL. This paper investigates if there is a gain in query execution performance, compared to Hive and 4store, by storing big RDF data in HDFS and using in-memory processing engine instead of MapReduce. Specifically, it addresses the following questions:

- Is it feasible to store big RDF data in HDFS and get improved query execution time, compared to Hive and native RDF stores like 4store, by translating SPARQL queries into SQL and then using the Presto distributed SQL query processing engine to run the translated queries?
- How much improvement, in query response time, can be attained by using in-memory query processing engine, e.g. Presto, against native RDF stores, like 4store, and other query processing engines based on MapReduce, like Hive?
- How do different RDF storage schemes in HDFS affect the performance of SPARQL queries?
- Is it possible to construct an end-to-end distributed architecture to store and query RDF datasets?

The rest of the paper is organized as follows: Related work is presented in section 2. The architecture of Presto-RDF framework and RDF storage strategies are presented in section 3. Section 4 describes the experimental setup for performance evaluation of Presto-RDF and results. Section 5 presents conclusions and future work.

2   Related Work

This section presents a review of related works that propose and evaluate different distributed SPARQL query engines. It also presents a review of two systems, Apache Spark and Cloudera Impala, which are similar to Facebook Presto. Different RDF storage schemes are discussed in section 2.3.

2.1   Distributed SPARQL

A distributed SPARQL query engine based on Apache Jena ARQ has been proposed [11]. The query engine extends Jena ARQ and makes it distributed across a cluster of machines. The extension involves re-designing some parts of Jena ARQ. Document indexing and pre-computation joins were also used to optimize the design. The results of the experiments that were conducted showed that the distributed query engine scaled well with the size of RDF data but its overall performance was very poor. The query engine, unlike Facebook Presto, uses MapReduce. Scalable RDF stores have been proposed that efficiently perform distributed Merge and Sort-Merge [9].

Marcello Leida et al. [12] propose a query processing architecture that can be used to efficiently process RDF graphs that are distributed over a local data grid. The architecture has no single point of failure and no specialized nodes – which is a different than Hadoop. They propose a sophisticated non-memory query planning and execution algorithm based on streaming RDF triples. Presto uses a distributed in-memory query-processing algorithm.

Xin Wang et al. [13] discuss how the performance of a distributed SPARQL query processing can be optimized by applying methods from graph theory. The framework presented in this paper translates a SPARQL query into its equivalent SQL query, and hence the query optimization that is done by Presto is for the SQL query and not for the SPARQL query.

A distributed RDF query processing engine based on a message passing has been proposed [14]. The engine uses in-memory data structures to store indices for data blocks and dictionaries. Just like Presto, the query-processing engine avoids disk I/O operations. The authors experimented their design over several types of SPARQL queries and were able to get a significant performance gain (as compared to Hadoop). TriAD is a distributed RDF engine where communication is based on Message Passing Interface [10]. Researchers have also studied partitioning of SPARQL queries instead of RDF datasets for significant performance gain [8].

2.2 Apache Spark and Cloudera Impala

Apache Spark and Cloudera Impala are two open-sources systems that are very similar to Facebook Presto. Both Apache Spark and Cloudera Impala offer in-memory processing of queries over a cluster of machines. According to Apache, Apache Spark is a "fast and general engine for large-scale data processing". Spark uses advanced Directed Acyclic Graph (DAG) execution engine with cyclic data flow and in-memory processing to run programs up to 100 (for in-memory processing mode) or 10 times faster (for disk processing mode) than Hadoop MapReduce [8]. Cloudera Impala is an open-source massively parallel processing (MPP) engine for data stored in HDFS. Cloudera Impala is based on Cloudera's Distribution for Hadoop (CDH) and benefits from Hadoop's key features – scalability, flexibility, and fault tolerance. Cloudera Impala, just like Presto, uses Hive Metastore to store the metadata information of directories and files in HDFS.

2.3 RDF Triple Stores

RDF triples can be stored and accessed in Hadoop Distributed File System (HDFS) by creating a relational layer on top of HDFS that maps triples into relational schemas. Hive, for example, allows storing data in HDFS based on a relational schema that defined by the user. Though there are some discrepancies among researchers regarding the naming and classification of relational schemas for RDF data, most researchers classify these schemas in to three groups [1], [4], [5], [15]:

- Triple table – the entire RDF data is stored as a single table with three columns – subject, predicate and object. Each triple is stored as a row in this table.

- Property-table – triples are grouped together by predicate name. In this scheme, all triples with the same predicate are stored in a separate table. Some researchers call property tables vertical partitioning.
- Cluster-property tables – in this scheme triples are grouped into classes based on correlation and occurrence of predicates. A triple is stored in the table based on the predicate class it belongs to.

In this research study, we use Presto that is a distributed SQL query engine that runs on a cluster of machines controlled by a single coordinator with hundreds or thousands of worker nodes. Our literature review shows that there are no other studies done on Presto for semantic data processing. Presto is optimized for ad–hoc analysis and supports standard ANSI SQL, including complex queries, aggregation, joins, and window function [7]. The client sends SQL query using the Presto command line interface to the coordinator that would then parse, analyze and plan the query execution. The scheduler, component within the coordinator, connects together the execution pipeline and assigns and monitors work to worker nodes that are closer to the data. The client gets data from the output stage, one of the worker nodes, which in turn pulls data from the underlying stages. In this project we propose architecture for Presto to process big RDF data.

3   Presto-RDF

This section proposes architecture, called Presto–RDF, which can be used to store and query big RDF data using the Hadoop Distributed File System (HDFS) and Facebook Presto. It also presents RDF–Loader, one of the key components of the architecture, which is used to read, parse and store RDF triples.

3.1   Architecture

Presto–RDF consists of the following components: a command line interface (CLI), a SPARQL to SQL compiler (RQ2SQL), Facebook Presto, Hive Metastore, HDFS, and RDF–Loader. Figure 1 illustrates the different components of the architecture. RDF data that is extracted from the Semantic Web is parsed and loaded into HDFS using a custom–made RDF-loader, which will also store metadata information on Hive Thrift Server. When a user submits a SPARQL query over a command line interface, the query is processed by a custom–made SPARQL to SQL converter, RQ2SQL, that translates the SPARQL query into SQL which would then be submitted to Facebook Presto. Presto, using its Hive connector and Hive Thrift Server, runs the SQL against HDFS and returns the result back to the CLI.

3.2   RDF–Loader

The purpose of the RDF–Loader is to load, parse, and store RDF data in HDFS. RDF–Loader implements four different RDF storage schemes and creates external Hive tables whose metadata is stored in the Hive Thrift server. Before the

RDF–Loader is executed the raw RDF data to be first processed is loaded into HDFS using this command: *hadoop fs –put file hdfs–dir*

Once the raw RDF data is uploaded, RDF–Loader runs several MapReduce jobs and stores the output back into HDFS. The structure of the data is defined by the schema that can be specified by users of the system. In order for the RDF–Loader to run and process raw RDF, the following input parameters are required:

- *database* – is the name of the database that will be created.
- *target* – is the type of RDF storage structure, i.e. the type of schema. There are four options: *triples, vertical, wide,* and *horizontal.*
- *expand* – this option indicates if *qnames* are to be expanded.
- *server* – is the DNS name or IP address of the master node, NameNode, of the Hadoop cluster.
- *port* – is the port number Hadoop listens to connections.
- *input* – is the path of the HDFS directory that holds the raw RDF data.
- *output* – is the path of the HDFS directory the processed RDF data will be stored.
- *format* – defines the format of the output files as they are stored in HDFS. The current version of the Hive meta–store supports five different formats: SEQUENCEFILE, TEXTFILE, RCFILE, ORC, and AVRO. This study makes use of the TEXTFILE format.

The following sections discuss four different RDF storage strategies implemented by the RDF–Loader. The next section presents an analysis of performance of each of these storage strategies.

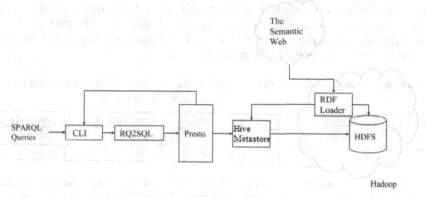

Fig. 1. Presto-RDF Architecture

3.3   Triple-Store Storage Strategy

In the triple-store storage scheme, an RDF triple is stored as is – resulting in a table with three columns: subject, predicate and object. If the raw RDF data has 30 million triples, the triple store strategy will have one table with 30 million rows.

The map–reduce algorithm that transforms the raw RDF data into the triples table is quite simple and shown in table 1.

```
map (String key, String value)
    // key: RDF file name
    // value: file contents
    for each triple in value
        emit_intermediate (subject + '\t ' + predicate, object)
reduce (String key, Iterator values)
    // key: subject and predicate delimited by tab
    // values: list of object values
    for each v in values
        emit (subject + '\t ' + predicate + object);
```

For an RDF dataset with n number of triples, the map algorithm has $O(n)$ running time while the reducer, which is called once for each unique subject, has $O(s*n)$ running time, where s is the number of unique subjects and o are the average number of object values per subject. Since, $s*o=n$, the total running time is $O(n)$.

3.4   Wide–Table Storage Strategy

In the wide table RDF storage scheme, the raw RDF data is parsed and stored as a single table having one column for subject values, and multiple predicate columns for object values. The resulting table has the following schema:

WideTable (String subject, String predicate_1, String predicate_2, ..., predicate_n)

Because it is unlikely that a subject has all the predicates found in the data set, this storage strategy will have a number of null values. For an RDF data set that has unique object values for a subject–predicate pair, this scheme would result in a table that has s number of rows, where s is the number of subjects in the data set. For example, given the following triple set in Table 1, the corresponding wide table representation for the triples would be as shown in Table 2.

Table 1. Example Triple set (a)

subject_1	predicate_1	object_1
subject_1	predicate_1	object_3
subject_2	predicate_2	object_3

Table 2. Wide table representation for example Triple set (a)

subject	predicate_1	predicate_2
subject_1	object_1	object_3
subject_2	null	object_3

If the dataset, however, contains multiple values for the same subject–predicate pair, the table will have multiple rows for the same subject. For example, given the following triple set in Table 3, the corresponding wide table representation for the triples would be as shown in Table 4.

Table 3. Example Triple set (b)

subject_1	predicate_1	object_1
subject_1	predicate_1	object_2
subject_2	predicate_2	object_3

Table 4. Wide table representation for example Triple set (b)

subject	predicate_1	predicate_2
subject_1	object_1	null
subject_1	object_2	null
subject_2	null	object_3

The algorithm for storing triples using the wide table storage scheme would get complicated if the data set contains subjects with multiple predicates (which is natural) and multiple object values for the same predicate. For example, given the following triple set in Table 5, the corresponding wide table representation for the triples would be as shown in Table 6.

The storage scheme, thus, forces new rows to be created for each unique subject–predicate pair. The map–reduce algorithm for the wide table storage scheme, as implemented in this study, is shown in the table below.

```
map (String key, String value)
    // key: RDF file name
    // value: file contents
    for each triple in value
        emit_intermediate (<subject, predicate>, <predicate, object>)
reduce (String key, Iterator values)
    // key: a <subject, predicate> pair
    // values: list of <predicate, object> pairs
    String subject = key.getSubject();
    String[] row = new String[1 + num_unique_predicates];
    int i = 0
    for each v in values
        row[i] = v.getObject();
        i++;
        emit (subject, row);
```

Table 5. Example Triple set (c)

subject_1	predicate_1	object_1
subject_1	predicate_1	object_2
subject_1	predicate_2	object_3
subject_2	predicate_2	object_4

Table 6. Wide table representation for example Triple set (c)

subject	predicate_1	predicate_2
subject_1	object_1	null
subject_1	object_2	null
subject_1	null	object_3
subject_2	null	object_4

3.5   Horizontal-Store Strategy

The horizontal storage scheme is similar to the wide table storage scheme in terms of the schema of the table. However, unlike the wide–table scheme, it optimizes the number of rows stored for subjects that have multiple object values for the same predicate. Given the example presented in the previous section in Table 5, the horizontal-store strategy stores the triples as shown in Table 7.

Table 7. Horizontal-store representation for example Triple set (c)

subject	predicate_1	predicate_2
subject_1	object_1	object_3
subject_1	object_2	null
subject_2	null	object_4

In this scheme, it is not necessary to create new rows for each unique subject–predicate pair. Instead, rows that are already created for the same subject, but for a different predicate will be used.

3.6   Vertical-Store Strategy

In the vertical storage scheme implemented in this research, the raw RDF data is partitioned into different tables based on the predicate values of the triples in the data with each table having two columns – the subject and object values of the triple. Thus, if the raw RDF data has 30 million triples that have 20 unique predicates, the vertical storage scheme will create 20 tables and stores the subject and object values of triples that share the same predicate in the same table. The map–reduce algorithm works with predicate as a key value and a pair of subject and object values as value:

```
map (String key, String value)
    // key: RDF file name
    // value: file contents
    for each triple in value
        emit_intermediate (predicate, <subject, object>);
reduce (String key, Iterator values)
    // key: predicate
    // values: list of <subject, predicate> pairs
    String table = key.replace_unwanted('_');
    MultipleOutputs<String, String> mos;
    for each v in values
        // create a directory table
        // write the subject, values inside the directory
        mos.write (v.getFirst(), v.getSecond(), table);
```

Because predicate values are URIs that contain non–alpha numeric characters, e.g. http://www.w3.org/1999/02/22–rdf–syntax–ns#, which cannot be used in naming directories, the reducer has to replace these characters with some other character, for example the underscore character, and creates the directory (which is considered as a table for the Hive Metastore). In the vertical storage scheme, for a raw RDF data that contains n number of triples, the mapper runs at $O(n)$ while the reducer runs at $O(p*x)$ where p and s are the number of unique predicates and subjects in the data set, respectively. In the worst case scenario, where there are as many unique predicates and subjects, the number of triples, the map-reduce algorithm for the vertical storage scheme runs at $O(n^2)$.

4   Benchmarking Presto-RDF

This section presents the experiments and the results conducted to benchmark the performance of Presto-RDF against Hive. A comparative measurement was also done on 4store – a native RDF store. Overall, two experimental setups were constructed for benchmarking the performance of Presto-RDF. The first setup was a 4-node cluster virtualized on a single 16GB memory machine. The second setup was 8-node cluster virtualized on the Windows Azure platform. The second setup was required because the experiments conducted used up the hard disk space and it was not possible to run queries on triples of more than 4 million. For the experiment, four benchmark queries from SP^{2}Bench [16], [17] were used and three different RDF storage schemes were evaluated – triple, vertical and horizontal stores. SP^{2}Bench is a SPARQL benchmark that is designed to test SPARQL queries over RDF triples stores as well as SPARQL–to–SQL re–write systems. SP^{2}Bench focuses on how well an RDF store supports the different SPARQL operators and their combination – known as operator constellations [16], [17]. SP^{2}Bench data model is based on the DBLP, http://www.informatik. uni–trier.de/~ley/db/, a computer science bibliography created in the 1980s and currently featuring more than 2.3 million articles. The SP^{2}Bench data generator can generate any number of triples based on what a user specifies. For the experiments conducted in this study, for example, triples of size 10, 20, and 30 million were generated.

4.1   Benchmarking Presto-RDF using 10, 20, and 30M Triples

The experiment was based on running four benchmark queries, from SP^{2}Bench with different degrees of complexity – query 1, 6, 8, and 11. The SP^{2}Bench use case is based on the DBLP. The experimental setup that was conducted involved setting up four and eight node clusters on Microsoft Windows Azure Platform. Each node in the cluster had a 2-core x86-64 processor, 14GB of memory, and 1TB of hard disk. Measurements were conducted for the four benchmark queries for 10, 20, and 30 million triples.

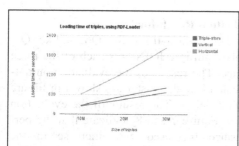

Fig. 2. Time Taken by RDF-Loader to Parse and Structure Raw RDF data

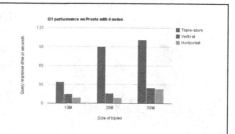

Fig. 3. Presto-RDF Query Processing Time of Ql over a 4 node Cluster

Loading Time

Once the RDF dataset is copied into HDFS, the RDF-Loader will parse and run a map-reduce job to convert the raw dataset to a structured dataset based on three storage schemas – triple-store, vertical and horizontal. The results of the measurement are shown in Figure 2. The performance of the RDF-loader has a linear relationship with the size of the triples. The horizontal store map-reduce algorithm always took much longer time than the triple-store and vertical store schemes. The result of running the above queries on Presto for a 4-node and 8-node cluster setup are shown in the figures below. For Q1, the vertical and horizontal stores have a much better performance than the triple-store schemas shown in Figure 3. This can be explained by looking into the SQL translations of the vertical and horizontal storage schemes – which have lesser rows involved in JOINs. This fact remains true when the number of nodes is increased from 4 to 8 – Figure 4. For Q1, increasing number of nodes resulted in performance improvement for the three storage schemes. Figure 5 below depicts the same.

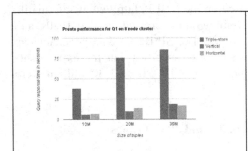

Fig. 4. Presto-RDF Query Processing Time of Q1 over 8 Node Cluster

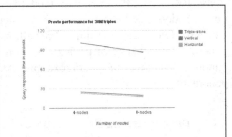

Fig. 5. Performance of Presto, with Increase in Number of Nodes, for Q1

Presto vs. Hive for Q1

Compared to Hive, Presto once again has a much higher performance. Figure 6 shows a comparison of Presto and Hive for 30M triples.

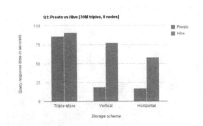

Fig. 6. Q1 Performance, for 30M Triples on 8 node Cluster

Evaluation Result for Q6

The SQL translations for Q6, unlike Q1, involve multiple JOINs for each of the three storage. The results of the evaluation on a 4-node and 8-node cluster are shown in Figure 7 and 8 below. The results of the evaluation above (Figure 7 and 8) indicate that the performance increased with increase in the number of nodes – see Figure 9 below. The vertical store, again, has a much better performance than the triple-store and horizontal store. Unlike Q1, however, where the horizontal store had a slightly better performance than the triple-store, the triple-store in Q6 had a slightly better performance than the horizontal store, especially as the size of the triples increases. This result can be explained by the fact that the horizontal store SQL for Q6, unlike the triple-store, involves multiple selections before making JOINs.

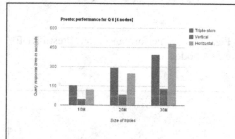

Fig. 7. Q6 Performance on Presto-RDF with 4 Nodes

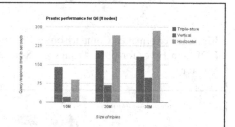

Fig. 8. Q6 Performance on Presto-RDF with 8 Nodes

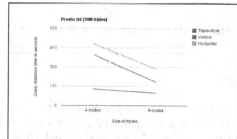

Fig. 9. Effect of Node Increase on Presto-RDF, for Q6 with 30M Triples

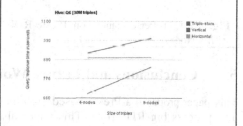

Fig. 10. Effect of Node Increase on Hive, for Q6 with 30M Triples

For Hive, unlike Presto-RDF, as the number of nodes was increased there was a drop in performance – which can be attributed to increase in replication across nodes and disk I/O operations – see figure 10.

Presto vs. Hive for Q6 and Q8
For Q6 as well, Presto-RDF has a much higher performance than Hive Evaluation. The SQL translations for Q8 involve multiple JOINs (just as the case were in Q6) and a UNION. The results have the same behavior as Q6 – the vertical store has a much better performance than the triple-store and horizontal stores, and Presto-RDF has a much higher performance than Hive. Figure 13 below shows the results of running the above queries over 10, 20 and 30M triples.

Because Q11 involves just one table that has less number of rows for the vertical and horizontal storage schemes than the triple-store (which is one table), the results shown above are expected. For 8 nodes, there is a performance improvement – see Figure 14.

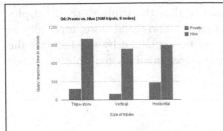

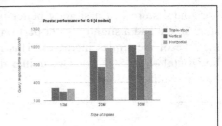

Fig. 11. Performance Comparison of Presto-RDF and Hive, for 30M Triples on 8-node Cluster

Fig. 12. Q8 performance, on Presto-RDF with 4 nodes

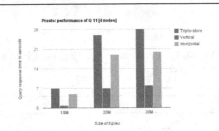

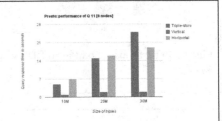

Fig. 13. Q11 Performance, on Presto-RDF with 4 Nodes

Fig. 14. Q11 Performance, on Presto-RDF with 8 Nodes

5   Conclusions and Future Work

This paper proposed a Presto-based architecture, Presto-RDF that can be used to store and process big RDF data. This paper also presented a comparative analysis of big RDF data using Presto, which uses in-memory query processing engine, and Hive, which uses Map Reduce to evaluate SQL queries. From the experiments conducted, following conclusions can be drawn:

- 4store has a much higher performance than Presto and Hive for small data sets. For bigger data sets (10M, 20M and 30M triples), however, 4store was simply unable to process the data and crashed. This is true when Presto, Hive and 4store are all tested with single-node setups.
- For all queries, Presto-RDF has a much higher performance than Hive.
- The vertical storage scheme has a consistent performance advantage than both the triple-store or horizontal storage schemes.
- As the size of data increases, the horizontal storage scheme performed relatively better than the triple-store scheme. This is unlike the articles reviewed during this research study, which ignore the horizontal scheme as being not efficient (because it has many null values).
- Increasing the number of nodes improved query performance in Presto but not in Hive. This can be explained by the fact that Hive replicates data across clusters and does IO operations – which increase as the size of nodes increase.

There are a number of areas to extend this study: this paper used a single benchmark, SP^{2}Bench. This work can be investigated on different benchmarks such as LUBM [18], BSBM [19], and DBPedia [6]. There are different optimization techniques that can be applied to the three storage schemas as well as to the RDF data directly. The RDF data is stored as a text file, which is not optimal. This work can be extended to test using RCFILE, ORC, and AVRO formats, which are better optimized than text file.

References

1. Luo, Y., Picalausa, F., Fletcher, G.H., Hidders, J., Vansummeren, S.: Storing and indexing massive RDF datasets. In: Semantic Search Over the Web, pp. 31–60. Springer (2012)
2. Cudré-Mauroux, P., Enchev, I., Fundatureanu, S., Groth, P., Haque, A., Harth, A., Keppmann, F.L., Miranker, D., Sequeda, J.F., Wylot, M.: NoSql databases for rdf: an empirical evaluation. In: Alani, H., Kagal, L., Fokoue, A., Groth, P., Biemann, C., Parreira, J.X., Aroyo, L., Noy, N., Welty, C., Janowicz, K. (eds.) ISWC 2013, Part II. LNCS, vol. 8219, pp. 310–325. Springer, Heidelberg (2013)
3. RDF, S.: Efficient RDF Storage and Retrieval in Jena2 (2003)
4. Sakr, S., Al-Naymat, G.: Relational processing of RDF queries: a survey. ACM SIGMOD Record 38(4), 23–28 (2010)
5. Abadi, D.J., Marcus, A., Madden, S.R., Hollenbach, K.: Scalable semantic web data management using vertical partitioning. In: Proc. of the Intl. Conf. on Very Large Data Bases, pp. 411–422 (2007)
6. Morsey, M., Lehmann, J., Auer, S., Ngonga Ngomo, A.-C.: DBpedia SPARQL benchmark – performance assessment with real queries on real data. In: Aroyo, L., Welty, C., Alani, H., Taylor, J., Bernstein, A., Kagal, L., Noy, N., Blomqvist, E. (eds.) ISWC 2011, Part I. LNCS, vol. 7031, pp. 454–469. Springer, Heidelberg (2011)
7. Presto: Interacting with petabytes of data at Facebook. https://www.facebook.com/notes/facebook-engineering/presto-interacting-with-petabytes-of-data-at-facebook/10151786197628920. (accessed: December 02, 2014)
8. Hammoud, M., etal.: DREAM: distributed RDF engine with adaptive query planner and minimal communication. In: Proc. of Intl. Conf. on Vary Large Databases (VLDB 2015)
9. Papailiou, N., Tsoumakos, D., Konstantinou, I., Karras, P., Koziris, N.: H2RDF+: an efficient data management system for big RDF graphs. In: Proceedings of SIGMOD Conference, pp. 909-912 (2014)
10. Gurajada, S., Seufert, S., Miliaraki, I., Theobald, M.: TriAD: a distributed shared-nothing RDF engine based on asynchronous message passing. In: Proceedings of SIGMOD Conference, pp. 289-300 (2014)
11. Kulkarni, P.: Distributed SPARQL query engine using MapReduce. In: Master of Science, Computer Science, School of Informatics, University of Edinburgh (2010)
12. Leida, M., Chu, A.: Distributed SPARQL query answering over RDF data streams. In: 2013 IEEE International Congress on Big Data (BigData Congress), pp. 369–378 (2013)
13. Wang, X., Tiropanis, T., Davis, H.C.: Evaluating graph traversal algorithms for distributed SPARQL query optimization. In: Pan, J.Z., Chen, H., Kim, H.-G., Li, J., Wu, Z., Horrocks, I., Mizoguchi, R., Wu, Z. (eds.) JIST 2011. LNCS, vol. 7185, pp. 210–225. Springer, Heidelberg (2012)
14. Dutta, A.K., Theobald, M., Schenkel, R.: A Distributed In-Memory SPARQL Query Processor based on Message Passing (2012)
15. Harth, A., Hose, K., Schenkel, R.: Linked Data Management. In: CRC Press (2014)
16. Schmidt, M., Hornung, T., Lausen, G., Pinkel, C.: SP^ 2Bench: a SPARQL performance benchmark. In: Data Engineering, ICDE 2009, pp. 222–233 (2009)
17. The SP2Bench SPARQL Performance Benchmark. http://dbis.informatik.uni-freiburg.de/forschung/projekte/SP2B/. (accessed: December 02, 2014)
18. Guo, Y., Pan, Z., Heflin, J.: LUBM: A benchmark for OWL knowledge base systems. Web Semantics: Science, Services & Agents on WWW 3(2), 158–182 (2005)
19. Berlin SPARQL Benchmark. http://wifo5-03.informatik.uni-mannheim.de/bizer/berlin sparqlbenchmark/. (accessed: December 02, 2014)

Detecting Spamming Groups in Social Media Based on Latent Graph

Qunyan Zhang, Chi Zhang, Peng Cai$^{(\boxtimes)}$, Weining Qian, and Aoying Zhou

Institute for Data Science and Engineering, ECNU-PINGAN Innovative Research Center for BigData, East China Normal University, Shanghai, China
{51121500043,chizhang}@ecnu.cn, pengcai2010@gmail.com,
{wnqian,ayzhou}@sei.ecnu.edu.cn

Abstract. Spammers in microblogging services aim to disseminate unuseful or misleading information, which leads to poor user experience and negative impact on the ecosystem of social media platform. Individual spammer detection, based on content and social network information, has been proposed to alleviate this predicament. However, most of the time spamming behavior is collaboratively conducted by a group of users, referred to as *spamming group*. In this paper, we propose to detect spamming groups in microblogging services. At the first step, we proposed RP-LDA to extract user features and find user groups within which users share similar retweeting behavior. Then, the degrees of individual users that are spammers are calculated by using a semi-supervised label propagation procedure. Finally, we determine the spamming groups using mixed membership distribution of users. Empirical studies over a real-life dataset demonstrate the effectiveness of our method and show that it can outperform the baseline.

Keywords: Spamming group · Latent graph · Social media

1 Introduction

Recently, we have seen the rise of new social marketing strategies and sophisticated automated promotion campaigns. Bot retweeting is one of the malicious activities occurring on Microblogging which is *collaboratively* and *purposefully* conducted by a group of users. The collaboration pattern is varied. One representative case is that a large number of criminal accounts registered and controlled by a single person or multiple persons for carrying on promotion activities such as information dissemination, advertising, speculation, etc. Another case is a large number of users organized by subscribing a "retweet timer" service which can help retweet automatically. As this kind of spamming that produces mass worthless information and misleads public opinion is harmful to user experience and still maintains its high visibility, detecting such fake retweeters become a pressing and challenging issue.

© Springer International Publishing Switzerland 2015
M.A. Sharaf et al. (Eds.): ADC 2015, LNCS 9093, pp. 294–305, 2015.
DOI: 10.1007/978-3-319-19548-3_24

While most of existing approaches focus on detecting criminal accounts in social media individually, limited research has been done to detect *spamming groups*, which is the focus of this paper. *Spamming group* refers the kind of activities that a group of users retweet together to promote some target tweets in a short period of time. It is hard to detect such misbehavior using indicators for detecting individual spammers, since then a single group member may no longer appear to behave abnormally by just looking at existing spamming indicators[15, 20]. Group based features are needed to characterize such accounts and groups. Previous work show that *distributed* and *bursty* are two important features for identifying group spamming[7]. Similar features to characterizing *transaction correlations* are used to detect spammers in online reviews[20].

The dynamics of retweeting velocity and collective #retweet of six hot tweets from Sina Weibo[1], the most popular microblogging service in China, is shown in Figure 1. Figure 1 (a) and (d) depict typical dynamics of tweets from celebrities or of newsworthy information, i.e. fast initial rise followed by a slow saturation in popularity. The short bursts of intense activity followed by a long period of inactivity in Figure 1 (b) and the smooth increasing of popularity in Figure 1 (e) show abnormal bursty charateristics, which can be easily detected based on retweet dynamics analysis[8]. However, Figure 1 (c) and (f) show two tweets whose retweets are generated by collaborated accounts from campaigns, which are difficult to be identified just based on retweet dynamics or user behaviors on a single tweet.

In this paper, not only the bursty and structure features of spamming groups, but also the semantic features of tweets from spamming groups are taken into consideration. Our work distinguishes itself from other group spamming methods, e.g. spamming review detection[15] and microblog spamming[8], in that: 1) We use co-retweeting relationships to capture the features of both bursty and structure characteristics of group spammers. 2) A semi-supervised label propagation process is used to solve the problem of lack of training data. 3) Latent graph is used to identify potential spamming groups based on *both* content of tweets and retweeting behavior, which provides more insightful results than the frequent-itemset-based method[15].

The main contributions of the paper are outlined as follows:

- We formalize the co-retweeting relationship and model the co-retweeter's profile using a unified model that integrates information from both retweet structure and tweet content.
- A graph-based semi-supervised learning algorithm with a label propagation procedure is adopted to identify spammers from non-spammers, which is effective for applications with few labeled spammers.
- Extensive empirical studies over a large real-life Sina Weibo dataset are conducted.

The remainder of this paper is organized as follows. In Section 2, we formally define the problem of social spammer detection in microblogging. In Section

---
[1] http://weibo.com/.

3, we show how to combine network structure and content information into a joint probabilistic word-link model. Section 4 is devoted to the detailed report of intensive study of the Sina Wiebo dataset. After the introduction of related work in Section 5, the last section is for concluding remarks and discussion.

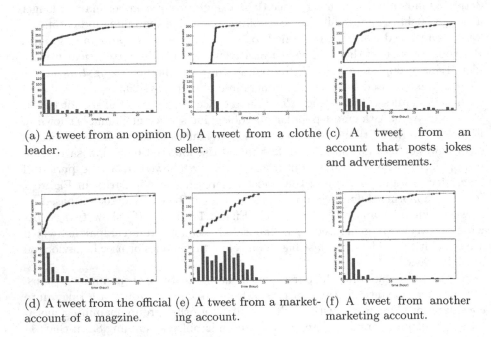

(a) A tweet from an opinion leader.

(b) A tweet from a clothe seller.

(c) A tweet from an account that posts jokes and advertisements.

(d) A tweet from the official account of a magzine.

(e) A tweet from a market- ing account.

(f) A tweet from another marketing account.

Fig. 1. Retweeting dynamics of six representative hotspot tweets in Sina Weibo

2   Problem Setting

In this section, we first introduce the notations used in the paper and then formally define the problem we study. The notations of this paper are listed in Table 1.

2.1   Co-Retweeting Relation

Let W be a set of tweets, U be a set of users, then a single tweet can be defined as $w_i = (c_i, l_i)$, where c_i is the textual content and l_i is the timeline of this tweet. The timeline is denoted as $l_i = \{(u_{i1}, t_{i1}), (u_{i2}, t_{i2}), ..., (u_{iL}, t_{iL})\}(t_{i1} < t_{i2} < ... < t_{iL})$. We say that every pair of users (u_{ip}, u_{iq}) are *co-retweeters* in tweet w_i, if $|t_{ip} - t_{iq}| < \Delta t$.

2.2 Spamming Group and Spamicity

In our problem, we assume that the users who retweet as co-retweeters frequently form a *co-retweeting group*. Considering the fact that in real world, a user can be active in many groups and the retweeting content may also be in various aspects, a reasonable assumption for user's membership should be mixed. Let $g_i \in G$ be a co-retweeting group, the distribution of groups conditioned on the user is $p(g_i|u)$. A user's feature is the vector of $p(g_i|u)$ on all the groups, where $\sum_{g_i \in G} p(g_i|u) = 1$. Assume S is the group of spammers, then the *spamicity* of a group can be defined as follows:

$$p(s = 1|g) = \sum_{u \in S} p(u|g) \tag{1}$$

Given the Weibo dataset, our task is to detect these co-retweeting groups automatically, and figure out the *spamicity* of each group. Groups ranked highly would be regard as the spamming groups. What's more, we are also interested in the topics of tweets each group usually retweeting. As the difficulty in judging web spam [5], and the huge number of users in Sina Weibo dataset, we define this problem as a semi-supervised learning problem.

Table 1. Notations of some frequently occuring variables

SYMBOL	DESCRIPTION	SYMBOL	DESCRIPTION
U	number of users	w	a word token
V	vocabulary size	c	co-retweeter
P	number of user profiles	α	the hyper-parameter for θ
N_p	number of words in the profile p	ϕ	membership parameter for each w
M_p	number of co-retweeters in profile p	ω	membership parameter for each c
K	number of groups	γ	the hyper-parameter for ω
u	user	θ_u	multinomial distribution of groups of u
l	timeline	β	the hyper-parameter for ϕ
g	hidden group variable	s	spamicity (1 spammer, 0 non-spammer)

3   Spammer Group Detecting Framework

In this part, we describe our spamming group detecting framework, *LPLDA*. Our framework mainly includes two parts: retweeter profile construction and graph based classification algorithm. In the first part, we proposed two models LDA-G and RP-LDA (Retweeter Profile LDA Model) for retweeter profile construction. The former just uses co-retweeting structure information while the latter integrates both the topic information of retweeted contents and the structure information simultaneously. The group of users is modeled as a latent variable in these two models. Then we use the label propagation algorithm to calculate the *spamicity* of remaining users based on the training data. Afterwards, the *spamicity* of every group is also gained.

3.1   Retweeter Profile LDA Model

Retweeter Profile Constructing. The profile p_i of a user u_i includes two parts: the co-retweeters p_{i_c} and the words p_{i_w} of retweets. Note that as the co-retweet relation is symmetric, for every pair of co-retweeters (u_a, u_b) in tweet w, u_b and u_a are added to each other's co-retweeter set p_{a_c} and p_{b_c} respectively. Words of w are both added to p_{a_w} and p_{b_w}. The construction process is shown in Figure 2 (a), tuples like {A,B} in third column represents A and B retweeted the same tweet within a specific time window which satisfy the definition of co-retweeting relation.

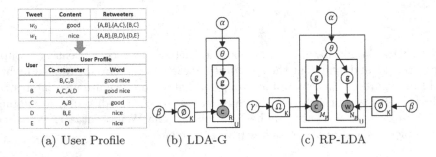

(a) User Profile   (b) LDA-G   (c) RP-LDA

Fig. 2. Latent topic model

LDA-G was presented as our baseline for retweeter profile construction by only taking co-retweeting structure into consideration. LDA-G in Figure 2 (b) resembles topic-based LDA (Latent Dirichlet Allocation)[4] model, with the retweeting network being analogous to the corpus, the users' co-retweeting profiles being analogous to documents; and the co-retweeters being analogous to words which is similar to [11]. While in our second model, RP-LDA not merely model the interaction between retweeters, but also capture the content information of tweets.

Table 2. Generative process for the RP-LDA model

For each user profile $p = 1, ..., U$
  Generate $\theta_p \sim Dir(\cdot|\alpha_\theta)$
  For each co-retweeter $m = 1, ..., M_p$
    Generate $g_m \in \{1, ..., K\} \sim Mult(\cdot|\theta_p)$
    Generate $c_m \in \{1, ..., U\} \sim Mult(\cdot|\Omega_{g_m})$
  For each word $n = 1, ..., N_p$
    Generate $g_n \in \{1, ..., K\} \sim Mult(\cdot|\theta_p)$
    Generate $w_n \in \{1, ..., V\} \sim Mult(\cdot|\beta_{g_n})$

Model Overview. A joint probabilistic model for spamming group detecting problem is designed in this paper, considering both content and connectivity. For our model in Figure 2 (c), we specify that a retweeting network contains a set of groups $G(g_1, g_2, \ldots, g_K)$ and each group in G is defined as a distribution on the retweeting group space, group assignments are modeled as a latent variable g for every co-retweeter and word in profile. The generative process of a user profile p is denoted as Table 2. In our work, we also use θ from LDA-G and RP-LDA as extracted features for users, which helps label propagation in the following sections.

3.2   Label Propagation

Having estimated the group membership parameter θ, we can easily construct a connected graph $G = (V, E)$ between users. Node $v \in V$ represents a user u_i, $w_{ij} = cosine(\theta_i, \theta_j)$ is the weight of edge e_{ij} between node v_i and v_j. In order to find the K nearest neighbours of every node quickly, LSH (Locality Sensitive Hashing) [17] is used in our experiments. The LPLDA algorithm is then used to help classifying unlabeled data using manually labeled data illustrated as follows.

Algorithm 1. Large-scale LPLDA Framework.

Input:
    Y: the $n * r$ labeling matrix, where y_{ij} represents the probability of user $u_i (i = 1...m)$ with label $r_i (j = 1...r)$;
    Y_L: the l labeled instance;
    Y_U: the u unlabeled instance;
    T: a $n * n$ matrix, with $\bar{t}_{ij}$ is the probability jumping from vertex x_i to vertex x_j; Z: topic distribution of each instance.
Output:
    label of each instance in Y_U;
1: initialize Y;
2: initialize the iteration index $t = 0$;
3: Let Y^0 be the initial soft labels attached to each vertex;
4: Let Y_L^0 be consistent with the labeling in the labeled data, where $y_{ij}^0 =$ the weight of the labeled instance if u_i has the label r_j;
5: using Locality Sensitive Hashing (LSH) to find the nearest neighbours of each instance.
6: **for** ($\forall u_i$ in $Y_L \cup Y_U$) **do**
7:     **for** ($\forall u_j$ in u_i's neighbours) **do**
8:         calculate $t_{ij} = \frac{w_{ij}}{\sum_j w_{ij}}$, where $w_{ij} = exp(-dist(u_i, u_j)^2/\sigma)$
9:     **end for**
10: **end for**
11: **repeat**
12:     propagate the labels of any vertex to nearby vertices by $Y^{t+1} = \bar{T}Y^t$;
13:     clamp the labeled data, that is, replace Y_L^{t+1} with Y_L^0;
14: **until** Y converges;
15: assign each unlabeled instance with a label with $\text{argmax}_j y_{ij}$

---

4   Experiments

4.1   Dataset

We now introduce the real-world Sina Weibo dataset used in our experiment. Our distributed crawler progressively crawls tweets, social networks and user

profiles via Sina API [2]. The total dataset contains 1.6 million users's time-
lines from August 2009 to the end of March 2014, including about 1.8 billion
tweets [19]. In this paper, we filter tweets from January 2010 to June 2011, in
which users are relatively active and many spammers are not moved at that
time. Only users who frequently retweet tweets with high retweet count should
be considered in this scenario, we got 86,603 users. To increase the ratio of
spammers, we further filter users by applying frequent itemset mining. 34,896
buckets with size ≥ 2 and support ≥ 5 were found. Finally 5,384 users in buckets
are labeled as follows: firstly, by observing frequent itemsets and their targeted
tweets, some buckets of users are easily to determine their labels. Secondly, some
heuristic rules like (1) Users with little personal information and log off now; (2)
User's name with explicit market purpose are applied for labeling spammers.
Thirdly, high duplicate tweets owners filtered out by previous work [22] tend to
be spammers; Finally, we manually labeled left users by checking their tweets
content, publishing time, platform and social network. Among them, 2,661 users
are spammers.

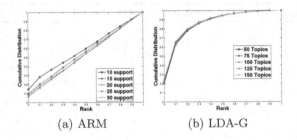

(a) ARM   (b) LDA-G

Fig. 3. Macro view of top-k performance

4.2   Effectiveness

In order to prove the effectiveness of topic model in finding co-occurrence relationship between retweeters. We compare LDA-G with traditional ARM (Association Rule Mining) methods.

The superiority of the group discovery algorithm is measured by suggesting
top-ranked co-retweeters to a user. Firstly, for each retweeter, we randomly with-
hold one co-retweeter c from his *original set* of co-retweeters to form the *training
set*. The training sets for all users form the training input for both algorithms.
Secondly, for each user u, we select $k - 1$ additional random users that were not
in user u's original set. The withheld co-retweeter c together with these $k - 1$
users form user u's *evaluation set*.

For user u, ARM assigns the score of each co-retweeter c based on the mined
rules, while LDA-G calculates the score for each of the k co-retweeters in the
evaluation set using:

$$\xi = \sum_g \theta_{ug}\phi_{gc}. \tag{2}$$

---

[2] http://open.weibo.com/

In Fig 3, the cumulative distribution of the probability of the withheld co-retweeter being found at top $k\%$ of the recommended list is plotted. The results show that the curve of LDA-G tends to be closer to 0 which means user can be better recommended by LDA-G model. Hence, in the paper, topic model is adopted for user profile construction.

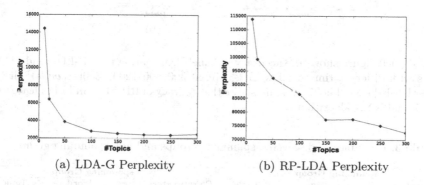

(a) LDA-G Perplexity   (b) RP-LDA Perplexity

Fig. 4. Perperlexity of LDA-G and RP-LDA

4.3   Group Discovery and User Classification

A result of group discovery using LDA-G and RP-LDA is shown in this section. Among 5,384 users, we randomly select 90% users as training dataset and 10% users as new users for inference. After the tf-idf filtering procedure, the vocabulary size $V=25950$. We set $\alpha = 50/K$, $\beta - 200/V$, $\gamma = 200/P$. Parameter estimation is implemented by Gibbs Sampling, and the number of iterations is set to 1000. Perplexcity of different topics of LDA-G and RP-LDA is illustrated in 4. 100 and 150 are selected for these two models separately.

Given the topic number, features of each user θ_u is extracted from the two models. Similarity between users is based on these features. In Figure 5(a), we compare classification accuracy at different time windows. For LDA-G, the value of Δt is important for feature extraction. when $\Delta t \geq 600$, growth rate of accuracy has flattened. However, RP-LDA's classification performance is not so sensitive to the value of Δt. In general, classification based on features extracted from RP-LDA performs better than that of LDA-G. In Figure 5(b), label propagation algorithm based on RP-LDA features achieved the highest accuracy rate at different labeled rates.

4.4   Spamming Group Discovery

By now, we have successfully classify users into two collections: spammers or non-spammers and all the users are clustered using proposed topic models. In Table 3, some typical groups of spammers and normal users found by our model are shown. Top ranked co-retweeters and words are listed in the table. We identify the aspects of each group according to co-retweeters and retweeting content.

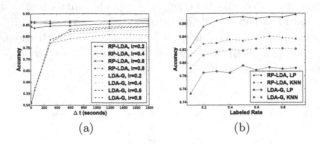

(a)   (b)

Fig. 5. The left figure shows the accuracy of Label Propagation for RP-LDA and LDA features with different time window $\Delta t \in \{10, 60, 300, 600, 1200, 1800(seconds)\}$, lr is short for labeled rate. The right figure shows the accuracy of RP-LDA and LDA features using LP and KNN algorithm.

Table 3. Examples of discovered spamming groups and non-spamming groups

Spamming Group			Non-spamming Group		
Co-retweeters	Word	Topic	Co-retweeters	Word	Topic
(家居微博, 微博热门, 口碑传播, 企业微博营销服务 口口相传, 微博营销资源) some market accounts	horoscope, love, egg, success skin	ads quotations	(秋中瑶, 叶子飞呀飞 学会梦游, live女王周 BC_nearfuture, 万万万代) fans of a singer	Zhou Bichang concert, album music, song	music
(专家微博, 微博主联盟, 微博红人联盟, 微博欢迎您 微博原创, 微博公关) some weibo promoters	revelation, time, suddenly, world, whether, rank	hot news	(Crystal-晶晶, 岚楚天国的梦, Virgo, HOHO-Chen) normal users	family, house safety, health, society	family society
(小粟旬星闻, 户田慧梨香星闻, 张东健星闻, 生田斗真星闻 最爱Maybelline) some star news reporters	advertisement, ha-ha, hee-hee, idea	fun	(小V, 夏小沫Sandy, 阁子不想做球球_IMJ, 有理说不清就不说算了) normal users	place, lament safe, intimate, airport, police, countrymen	news disaster
(baby嘉珍Niki, baby心语Irene, baby婉慧love, best欣荣May) robots	breast, slim, woman, wife, underwear	female	(五星级的厕所, 魏索楠, 蒙娜丽莎的男人, 没钱找媳妇) normal users	culture, CCTV, public, benefit	culture society

4.5   Group Spammer Indicator

Here, we discuss some group features, GTW, GETF and GCS. The definition of these features can de found in [15,16]. Members in a spamming group are likely to have worked together in retweeting for the target tweet during a short time interval. The definition of GTW says that a group g of retweeters retweet on a tweet within a short burst of time is more prone to be spamming (attaining a value close to 1) as shown in Figure 6 (a). GETF captures the time frame as how early a spamming group retweet a tweet. It attains a value of 0 as retweeters retweeted then are not considered to be early any more. From Figure 6 (b), the dashed curve tend to be closer to zero. GCS captures context of group spamming, the spammers sometimes pretend to act as normal users, attaching some prepared reviews when retweeting tweeters, while this kind of reviews usually have high similarity in content as shown in Figure 6 (c). This section tries to analyze whether spamming groups behave strangely based on a set of unusual behaviors.

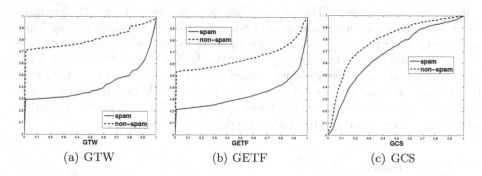

| (a) GTW | (b) GETF | (c) GCS |

Fig. 6. Behavioral distribution. Cumulative % of spam and non-spam groups vs. feature value.

5   Related Work

Spammer Detection in Microblogging. In recent years, much research work has focused on the study of spammer detection of microblogging communities. Most of the exsiting approaches rely on URL blacklists [7,10,18] and machine learning method to classify spam users from non-spam users [1,12,13,21]. Supervised learning methods can generally be divided into two categories. One is content based analysis. In [22], An improved locality-sensitive hashing based method is used for detecting near duplicated tweets. Duplicate and near duplicate tweets which are almost certainly fake tweets created by spammers who auto-tweet at regular intervals. Users with high duplicate tweets rate are cheated as candidate spammers and their behaviors are compared with normal users. The other category of methods is to detect spammers via social network analysis. Link farming in the Twitter network was investigated in [9], which shows that spammers can successfully acquire a number of normal followers, especially those referred to as social capitalists who tend to increase their social capital by following back anyone who follows them. Topic Models [2,3] are also widely used in spammer detection in recent years.

Spammer Group Detection in Social Network. Prior works on social spammer focused on detecting fake reviews and individual spammers. Few work has been done on detecting group spammers. A.Mukherjee et. al presented a novel relation-based model, called GSRank [15], to detect group spammers in product reviews. [6] built an automatic classification framework on detect spam campaigns that manipulate multiple accounts to spread spam on Twitter by using the features combining both content and behavior to distinguish spam campaigns from legitimate ones. However, well-labeled data is essential to these supervised classification. Also, URL based clustering was used in [7] to find candidate groups in [7]. In [15], frequent itemset mining method was applied to find a set of candidate groups which is time consuming and may lose some latent relationship among users. The problem has been pointed out in [20] that indicator based classification sometimes is not so stable since spammers may learn and change their tactics and equip with those indicators. while, group structure and transaction correlations are harder to fake. In [6,14], only tweets with URL were considered and clustered into campaigns based on shared final URLs. In this paper, topic model is used to find out latent collaborate retweeting campaigns and semi-supervised method was applied to label each user as spammer on non-spammer according to his/her neighbours' labels.

6   Conclusions

We propose to detect spamming group in Microblogging. The retweet network and content information are integrated by RP-LDA to find the candidate spamming group. To reduce the effort of data annotation, semi-supervised label propagation learning method is adopted to estimate the probability whether the tweeter is a spammer. We evaluated our approach on the real dataset from Sina Weibo, i.e. the largest Microblogging service provider in China. Experimental results demonstrate the effectiveness of our method and show that it can outperform the baseline.

Acknowledgments. This work is partially supported by National Hightech R&D Program (863 Program) under grant number 2012AA011003, and National Science Foundation of China under grant number 61432006 and 61170086. The authors would also like to thank Ping An Technology (Shenzhen) Co., Ltd. for the support of this research.

References

1. Benevenuto, F., Rodrigues, T., Magno, G., Almeida, V.A.F.: Detecting spammers on twitter. In: CEAS (2010)
2. Bíró, I., Siklósi, D., Szabó, J., Benczúr, A.A.: Linked latent dirichlet allocation in web spam filtering. In: AIRWeb, pp. 37–40 (2009)
3. Bíró, I., Szabó, J., Benczúr, A.A.: Latent dirichlet allocation in web spam filtering. In: AIRWeb, pp. 29–32 (2008)
4. Blei, D.M., Ng, A.Y., Jordan, M.I.: Latent dirichlet allocation. Journal of Machine Learning Research **3**, 993–1022 (2003)
5. Castillo, C., Donato, D., Becchetti, L., Boldi, P., Leonardi, S., Santini, M., Vigna, S.: A reference collection for web spam. SIGIR **40**(2), 11–24 (2006)
6. Chu, Z., Widjaja, I., Wang, H.: Detecting social spam campaigns on twitter. In: Bao, F., Samarati, P., Zhou, J. (eds.) ACNS 2012. LNCS, vol. 7341, pp. 455–472. Springer, Heidelberg (2012)
7. Gao, H., Hu, J., Wilson, C., Li, Z., Chen, Y., Zhao, B. Y.: Detecting and characterizing social spam campaigns. In: CCS, pp. 681–683 (2010)
8. Ghosh, R., Surachawala, T., Lerman, K.: Entropy-based classification of 'retweeting' activity on twitter (2011). CoRR, abs/1106.0346

9. Ghosh, S., Viswanath, B., Kooti, F., Sharma, N.K., Korlam, G., Benevenuto, F., Ganguly, N., Gummadi, P. K.: Understanding and combating link farming in the twitter social network. In: WWW, pp. 61–70 (2012)

10. Grier, C., Thomas, K., Paxson, V., Zhang, C. M.: @spam: the underground on 140 characters or less. In: CCS, pp. 27–37 (2010)

11. Henderson, K., Eliassi-Rad, T.: Applying latent dirichlet allocation to group discovery in large graphs. In: SAC, pp. 1456–1461 (2009)

12. Hu, X., Tang, J., Zhang, Y., Liu, H.: Social spammer detection in microblogging. In: IJCAI (2013)

13. Lee, K., Caverlee, J., Webb, S.: Uncovering social spammers: social honeypots + machine learning. In: SIGIR, pp. 435–442 (2010)

14. Li, F., Hsieh, M., An empirical study of clustering behavior of spammers and group-based anti-spam strategies. In: CEAS (2006)

15. Mukherjee, A., Liu, B., Glance, N.S.: Spotting fake reviewer groups in consumer reviews. In: WWW, pp. 191–200 (2012)

16. Mukherjee, A., Liu, B., Wang, J., Glance, N.S., Jindal, N.: Detecting group review spam. In: WWW, pp. 93–94 (2011)

17. Slaney, M., Casey, M.: Locality-sensitive hashing for finding nearest neighbors. IEEE, Signal Processing Magazine 25(2), 128–131 (2008). (lecture notes)

18. Thomas, K., Grier, C., Ma, J., Paxson, V., Song, D.: Design and evaluation of a real-time URL spam filtering service. In: S&P, pp. 447–462 (2011)

19. Xia, F., Zhang, Q., Wang, C., Qian, W., Zhou, A.: On the rise and fall of sina weibo: Analysis based on a fixed user group. In: SSEPM (2015)

20. Xu, C., Zhang, J., Chang, K., Long, C.: Uncovering collusive spammers in chinese review websites. In: CIKM, pp. 979–988 (2013)

21. Yang, C., Harkreader, R.C., Zhang, J., Shin, S., Gu, G.: Analyzing spammers' social networks for fun and profit: a case study of cyber criminal ecosystem on twitter. In: WWW, pp. 71–80 (2012)

22. Zhang, Q., Ma, H., Qian, W., Zhou, A.: Duplicate detection for identifying social spam in microblogs. In: BigData Congress, pp. 141–148 (2013)

Demo Papers

A Framework of Enriching Business Processes Life-Cycle with Tagging Information

Zakaria Maamar[1], Sherif Sakr[2,4(✉)],
Ahmed Barnawi[3], and Seyed-Mehdi-Reza Beheshti[2]

[1] Zayed University, Dubai, UAE
zakaria.maamar@zu.ac.ae
[2] University of New South Wales, Sydney, Australia
{ssakr,sbeheshti}@cse.unsw.edu.au
[3] King Abdulaziz University, Jeddah, Saudi Arabia
ambarnawi@kau.edu.sa
[4] King Saud Bin Abdulaziz University for Health Sciences, Riyadh, Saudi Arabia

Abstract. In this demonstration, we present a framework for enriching business processes with tags specialized into *social, resource, location*, and *temporal*. Using the framework, business-process engineers and end-users (i.e., executors) provide the tags with the necessary details which are then automatically propagated from one tag to another, when appropriate. At design time phase of a business process, the propagation of relations between tags reflects *unidirectional-transfer-of-final-details, unidirectional-transfer-of-partial-details*, and *bidirectional-transfer-of-partial-details* while at run-time the propagation of relations reflects *strong-trigger, weak-trigger*, and *meet-in-the-middle trigger*. Our provides an elegant mechanism for monitoring business processes which is more user-driven than traditional approaches which heavily rely on log analysis mechanisms.

1 Overview

With the increasing reliance on Internet and Web technologies in every aspect of our life, we are witnessing an increasing interest in blending social computing with various disciplines. The social enterprise is a typical example of this blend since maintaining "social" contacts with customers, suppliers, competitors, and partners is highly rewarding [2]. In practice, tags are considered as one of the important facets of Web 2.0 acceptance. In principle, tagging is defined *"as the practice of creating and managing labels (or tags) that categorize content using simple keywords"* [1]. We design a framework that aims at enriching the business process with different types of user-driven tagging information [3]. For example, a *social* tag can be used to indicate that a person sought the recommendations of a peer when executing a task, which should help in the future maintain a proper support network of contacts for this person. A *technological* tag can be

---

[1] www.practicalecommerce.com/articles/589-What-Are-Tags-And-What-Is-Tagging

© Springer International Publishing Switzerland 2015
M.A. Sharaf et al. (Eds.): ADC 2015, LNCS 9093, pp. 309–313, 2015.
DOI: 10.1007/978-3-319-19548-3_25

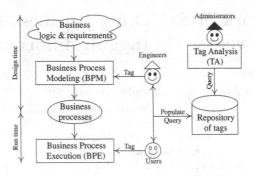

Fig. 1. Tag-based approach's general architecture

used to indicate the satisfaction level of a person with a certain technology when executing a task. In addition to categorizing tags, our framework goes one step further by establishing relations between tags in a way that automatically and logically propagate them between related tasks and ease their management in the case of changes that affect dependent tags. The use of tags is reported in different publications and is already offered through some popular Web sites like delicious.com and flickr.com. However few initiatives, in fact only two (namely [1] and [5]) to the best of our knowledge, consider tag connection.

2   Tag-Based Framework for Business Process Life Cycle

2.1   Architecture

Fig. 1 illustrates our tag-based framework which consists of three main components: BP Modeling (BPM), BP Execution (BPE), and Tag Analysis (TA). The three components are all connected to a repository of tags: tagRepository.

The BPM component is designed for business process engineers so they create new BPs and/or edit and delete existing ones. In particular, the engineers can access the tagRepository in two modes: (i) population mode that consists of creating (editing as well) tags when modeling BPs and (ii) query mode that consists of searching for specific details over the saved tags that could help model BPs. The BPE component is designed for users so that they execute BPs upon receipt from the BPM. Similar to engineers, users can access the tagRepository in two modes: (i) population mode that consists of creating (editing as well) tags when executing the BPs and (ii) query mode that consists of searching for specific details over the saved tags that could help execute BPs. The TA component is designed for administrators who run queries over the tagRepository so that they can mine engineers and users' tags. The mining could help identify patterns between BPs, prevent problems prior to BPs execution, develop design and/or execution recommendations, etc. In principle, we differentiate between BP mining and tag mining. BP mining relies on logs that are well-structured and capture execution (not design) details automatically without human assistance. Contrarily tag mining relies on human assistance to populate tags with unstructured details related to BPs at both design- and run-time. Moreover, execution logs are typically analyzed after BP execution has been completed, whereas tags can be analyzed during and after BP execution; this could even provide guidance towards modifying BPs at run-time in response to actual or predicted failures.

2.2   Types of Tags

For the sake for enriching business processes with tagging information, our framework supports different types of tags that cover various details as follows [3]:

1. $Social$ tag (S-tag): it reports on the interactions that an engineer/user initiates with persons affiliated or not with the organization during the design/execution of BPs.
2. $Resource$ tag ($\mathcal{E}$-tag): it reports on the means (software, hardware) used during the design/execution of BPs. This could be related to the satisfaction level with the resources in terms of performance, reliability, and availability.
3. $Location$ tag ($\mathcal{L}$-tag): it reports on where the design/execution of a BP takes place. Examples of locations could be at work, outside the office, shopping mall, etc. This tag suggests options on where a BP can be designed/executed.
4. $Temporal$ tag ($\mathcal{T}$-tag): it reports on when the design/execution of a BP takes place. Examples include during business hours, after business hours, etc.

It should be noted that other types of could be flexibly considered in our framework without any impact on the proposed architecture.

2.3   Relationships Between Tags

Our framework connects same tags anchored to different tasks t_i and t_j using data dependencies that are classified as *prerequisite* (preReq), *parallel prerequisite* (parPreReq), and *parallel* (par); these dependencies are data driven in the sense of treating one task's outputs as inputs for other tasks. In a preReq dependency relevant outputs of t_i are sent to t_j, whereas in a parPreReq dependency intermediary outputs of t_i are sent to t_j. In a par dependency, t_i and t_j execute in parallel without an explicit dependency between them, but they may need to synchronize their executions at a certain stage. For more details about our defined rules for automated propagation of the tags between tasks, we refer the reader to [3].

3   Implementation and Demonstration

The implementation of our framework extends Yaoqiang BPMN editor[2] and has been integrated with SUPER Framework (Social-based Business Process Management Framework) [4]. In particular, our extensions permit to anchor tags to tasks, connect tags together, store tags persistently, etc. (Fig. 2 for screenshots).

In our demonstration, we will present the tagging process for both of the design phase[3] and the runtime phase[4]. Our demonstration will show different

---

[2] www.sourceforge.net/projects/bpmn
[3] Design-time video: www.youtube.com/watch?v=02uNQam9Z7Q
[4] Run-time video: www.youtube.com/watch?v=1chT-iD07iM&feature=youtu.be

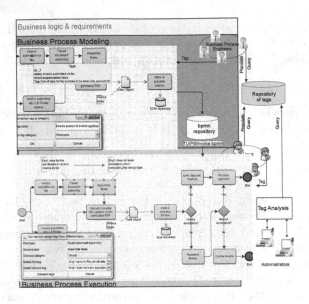

Fig. 2. System's architecture and screenshot

use cases. For example, in use case 1 (top part of Fig. 2), a Business Process Engineer (BPE) leaves the first tag to other BPEs in order to warn them that checking invoice acceptability is the next step. Tags anchored to tasks are stored in appropriate repositories, i.e., BPMN and tag. Use case 2 (bottom part of Fig. 2) shows how a user may connect two tags anchored to different tasks using specific trigger relations. Concretely speaking a partial list of persons that are included in the first tag (i.e., Accountant$_1$ as per could be contacted again in order to complete the tag of faxed document assembly task. Depending on the type of trigger relation (whether strong, week, or meet-in-the middle) and content of the first tag the execution of assemble faxes task may be completed successfully or may need to be reviewed as per the cut-off date of receiving fax invoices. Through proper synchronization of tags anchored to faxed document assembly task and assemble faxes task it becomes possible to prevent failure by ensuring that appropriate details are passed along. Having in mind that the execution of the same business process takes usually several instances at the same time such as "on the fly tag synchronization" may significantly enrich business process execution reducing the time needed to solve difficulties that could occur.

References

1. Garcia-Castro, L.J., Hepp, M., Garcia, A.: Tags4Tags: using tagging to consolidate tags. In: Bhowmick, S.S., Küng, J., Wagner, R. (eds.) DEXA 2009. LNCS, vol. 5690, pp. 619–628. Springer, Heidelberg (2009)
2. Maamar, Z.: Enterprise 2.0 - research challenges and opportunities. In: WEBIST (2014)

3. Maamar, Z., Narendra, N.C., Kajan, E., Pljaskovic, A., Boukhebouze, M.: Using tags for business process enrichment. In: ICIST (2014)
4. Maamar, Z., Sakr, S., Noura Faci, A.B., Boukhebouze, M.: SUPER: social-based business process management framework. In: ICSOC (2014)
5. Tanasescu, V., Streibel, O.: Extreme tagging: emergent semantics through the tagging of tags. In: ESOE (2007)

SocialTrail:Recommending Social Trajectories from Location-Based Social Networks

Qinzhe Zhang[1,2]([∅]), Litao Yu[1,2], and Guodong Long[1,2]

[1] The University of Queensland, Brisbane, Australia
Qinzhe.Zhang@student.uts.edu.au
[2] The University of Technology Sydney, Ultimo, Australia
l.yu4@uq.edu.au, Guodong.Long@uts.edu.au

Abstract. Trajectory recommendation plays an important role for travel planning. Most existing systems are mainly designed for spot recommendation without the understanding of the overall trip and tend to utilize homogeneous data only (e.g., geo-tagged images). Furthermore, they focus on the popularity of locations and fail to consider other important factors like traveling time and sequence, etc. In this paper, we propose a novel system that can not only integrate geo-tagged images and check-in data to discover meaningful social trajectories to enrich the travel information, but also take both temporal and spatial factors into consideration to make trajectory recommendation more accurately.

1 Introduction

Nowadays, thanks to the popularity of mobile devices, location-based social networks (LBSNs) which enable mobile users to share their activities with geographic locations attached have become more and more prevalent. For instance, Panoramio,allows users to share geo-tagged images. Another famous one is Foursquare, in which users can "check-in" venues and provide ratings or comments on the venues.

The widespread use of LBSNs has engendered a great interest of discovering *social trajectories* from LBSNs for travel recommendation. We define a *social trajectory* as a sequence of locations/areas mined from online social data. Compared with traditional "editor-selected" trajectories, social trajectories would be more effective in building travel plans. With exploiting the wisdom of a vast number of users, they not only suggest famous landmarks in a city, but also recommend interesting venues which are popular among local residents, such as restaurants, bars, etc.EveryTrail[1], a travel trail sharing website, allows users to browse trajectories shared by others. However, it cannot recommend personalized trajectories to meet various travel requirements. In other words, the travelling trajectory could be stable as it was planned by the expert but not social media users. There are also some studies [1,2] mine trips from geo-tagged images for recommendation. However, they have the following limitations.

---

[1] http://www.everytrail.com

© Springer International Publishing Switzerland 2015
M.A. Sharaf et al. (Eds.): ADC 2015, LNCS 9093, pp. 314–317, 2015.
DOI: 10.1007/978-3-319-19548-3_26

First, only considering geo-tagged images is insufficient, as popular venues for daily-life purposes like dining and shopping are often missing in geo-tagged images [3]. As such, the mined trajectories may not contain locally-popular venues. Second, their recommendation algorithms do not consider the temporal consistence among locations when constructing trajectories, which affect the quality of the recommended trajectories.

We propose a novel social trajectory recommendation system called **Social-Trail**. It integrates the social data extracted from two LBSNs, Panoramio and Foursquare. As the former focuses on geo-tagged images of tourist attractions while the latter captures local preferences, SocialTrail can discover more comprehensive social trajectories containing both famous landmarks and locally-popular spots.On top of the raw social data, SocialTrail firstly mines Areas of Interest (AOI) from tourism attractions and local preferences, called *Tourism Areas of Interest* (TAOI) and *Local Areas of Interest* (LAOI) respectively. Given a query specified by a user with the departure location/time and arriving location/time, SocialTrail then recommends a social trajectory formed by a sequence of AOIs. For supporting effective trajectory recommendation, SocialTrail not only takes into account the spatial consistence of AOIs, but also considers whether the "best-visiting hours" of recommended AOIs match the travel duration. Compared with existing systems, SocialTrail has the following new features:

- Social trajectory is a new concept as the trajectory is constructed based on the location information extracted from social data.
- SocialTrail recommends social trajectories that consist of sequences of AOIs, providing more visiting options in each way point on trajectories.
- Social trajectories in SocialTrail are derived from integrating heterogeneous social data including geo-tagged images and check-ins.
- The recommendation algorithm in SocialTrail is spatial-, temporal- and popularity-aware.

2 The SocialTrail System

Figure 1 shows an overview of our SocialTrail system. The system first collects social data from two LBSNs, i.e., Panoramio and Foursquare. Then, it discovers areas of interests by using clustering techniques on top of the collected social data. In the recommendation component, given a query that specifies the travel requirement, i.e., start location/time and end location/time, the system recommends a social trajectory consisting of TAOIs and LAOIs which match the query in terms of spatial/temporal/popularity relevance.

Collecting Geo-tagged Images and Check-Ins. The integration of geo-tagged images and check-ins can provide more comprehensive social trajectories in the recommendation. We crawled about 10,000 geo-tagged images from Panoramio and downloaded the Foursquare dataset[2] for New York City.

---

[2] http://www.public.asu.edu/~hgao16/dataset.html

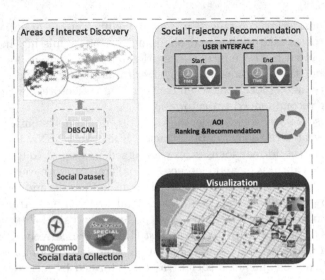

Fig. 1. System Framework

We obtained about 20,000 venues from 20 millions check-ins. The integration of geo-tagged images and check-in data do not only provide the tourist attractions but also the preferences of local residents.

Discovering Areas of Interest. To effectively discover AOIs from geo-tagged images and checkins respectively, we adopted the DBSCAN algorithm on them separately, since this clustering approach can find the clusters starting from the estimated density distribution of corresponding data points. Each cluster represents an AOI.

Recommending Social Trajectories. The recommendation is incrementally generated by simultaneously evaluating the importance of spatial/temporal information and overall AOI popularity for an AOI, and formulating the sequence of optimal AOIs as trajectories. Usually, as one of the most important factors, the overall AOI popularity is often considered by tourists. However, solely considering it may neglect some other attributes such as temporal/spatial information which also plays significant roles in trajectory recommendation. The synthesis of three factors on spatial, temporal and popularity information in our system can bring better matched trajectories than existing systems. For instance, the recommended routes in our system can effectively reduce the deviation from the main direction specified by the start/end locations. At the same time, the ares with high popularities but in an inappropriate traveling time can be filtered. Specifically for temporal information, we use time vectors to represent the temporal popularity for each AOI, in which each entry in time vector indicates the popularity level in the corresponding time stamp. Thus the recommended trajectories may vary with different traveling time periods. Such recommendations can undoubtedly cater the travelers' needs in different time intervals, as shown in Figure 2.

Fig. 2. Different recommendations for different time intervals

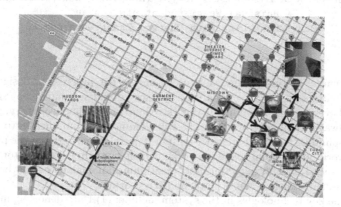

Fig. 3. Recommended Social Trajectory

3   USER SCENARIOS

Scenario 1: For the new comers to the city, when they click any AOI on the map, the available places and their links in the area can be displayed. If there are geo-tagged images associated to places, they can also used to visualize the places in the area.

Scenario 2: A user can specify the start location/time and the arriving location/time to get the recommended trajectory. Locations can be inputted by clicking any two AOIs on the map. The system then provides a sequence of AOIs, where each AOI include a variety of visiting places in the area and can also be visualized by associated images, as shown in Figure 3.

References

1. Arase, Y., Xie, X., Hara, T., Nishio, S.: Mining people's trips from large scale geo-tagged photos. In: ACM MM (2010)
2. Cheng, A.-J., Chen, Y.-Y., Huang, Y.-T., Hsu, W.H., Mark Liao, H.-Y.: Personalized travel recommendation by mining people attributes from community-contributed photos. In: ACM MM (2011)
3. Liu, J., Huang, Z., Chen, L., Shen, H.T., Yan, Z.: Discovering areas of interest with geo-tagged images and check-ins. In: ACM MM (2012)

SocialAnalysis: A Real-Time Query and Mining System from Social Media Data Streams

Haishuai Wang[✉], Peng Zhang, Ling Chen, and Chengqi Zhang

University of Technology Sydney, Ultimo, NSW 2007, Australia
Haishuai.Wang@student.uts.edu.au,
{Peng.Zhang,Ling.Chen,Chengqi.Zhang}@uts.edu.au

Abstract. In this paper, we present our recent progress of designing a real-time system, *SocialAnalysis*, to discover and summarize emergent social events from social media data streams. In social networks era, people always frequently post messages or comments about their activities and opinions. Hence, there exist temporal correlations between the physical world and virtual social networks, which can help us to monitor and track social events, detecting and positioning anomalous events before their outbreakings, so as to provide early warning.

The key technologies in the system include: (1) Data denoising methods based on multi-features, which screens out the query-related event data from massive background data. (2) Abnormal events detection methods based on statistical learning, which can detect anomalies by analyzing and mining a series of observations and statistics on the time axis. (3) Geographical position recognition, which is used to recognize regions where abnormal events may happen.

1 Introduction

Online social networks have become a major platform for instant information spreading and sharing [1,2]. For example, the event that a 7.3-magnitude earthquake hit northwest China on February 12, 2014 was first reported by the Sina Weibo (http://weibo.com/) in less than 1 minute. Besides, malicious information is also widely spread on social networks, which raises the question of how to discover negative information propagation.

Motivated by the above question, we aim to develop data stream detection methods on Sina Weibo. To date, most real-time event detection models, such as "TwitInfo" from MIT, "SensePlace2" from Pennsylvania State University and "Skynet Search" from Peking University have been designed to monitor and analyze public sentiments on social networks [3–6]. However, these works are merely focused on keyword search and cannot provide real-time event organization such as event location mining and event early warning.

In this paper, we build up a social media data stream query and mining system on the Sina Weibo. We analyze the Weibo data to recommend relevant events, find abnormal events and locate the event on the BAIDU map. The components of the system are: 1) a real-time event detection model for given

© Springer International Publishing Switzerland 2015
M.A. Sharaf et al. (Eds.): ADC 2015, LNCS 9093, pp. 318–322, 2015.
DOI: 10.1007/978-3-319-19548-3_27

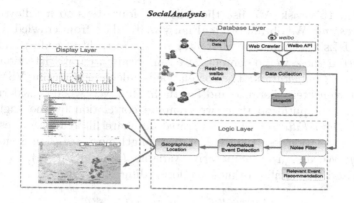

Fig. 1. An illustration of the system workflow

queries, 2) an abnormal event outbreak detection model, 3) a topic summarization for outbreak event detection, and 4) a location mining model from the outbreak events.

When designing the *SocailAnalysis* system, we face three challenges: *Challenge 1*, how to obtain weibo data with respect to given stream queries; *Challenge 2*, how to find abnormal time points where the number of twitters reach its peak; *Challenge 3*, how to find locations of events by combining three kinds of location data, i.e., content-based location data, posting location data and user registration location data.

To address *Challenge 1*, we first use the Sina API and Web crawlers to collect data, and then use event correlation analysis to store relevant Weibo data. To solve *Challenge 2*, we design an event peak identification algorithm to highlight abnormal time points. To tackle *Challenge 3*, we identify locations from content-based location data and use different location weights to discover the genuine locations. The post location and registration location data can be obtained by using the Sina API functions.

2   The System Architecture and Technical Details

Fig. 1 shows the workflow of the system *SocialAnalysis*. The system contains four components: 1) Data collection. We use both a Web page crawler and Sina API functions to collect data. 2) Noisy text filtering. Event correlation analysis is used to filter texts which are irrelevant to an event. 3) Anomalous event detection. The text data are statistically analyzed to detect anomalous time points. 4) Geographical location detection. Based on correlated Weibo contents, exacting and analyzing the locations of the event.

2.1   Event Correlation Analysis

Event correlation analysis aims to remove irrelevant data with respect to given stream queries and rank the query results. Therefore, the filtering process plays

a key role in this task. We use the following four steps to handle this part:
1) *Pre-processing:* We obtain original noisy weibo data from crawled pages. By removing URLs, insignificance character strings (e.g., html codes) and special symbols within these weibo contents, we can obtain accurate detection results.
2) *Correlation Computation:* The correlation is calculated by using VSM, String matching, Levenshtein distance and SimHash. However, when considering both accuracy and computation time, we compute correlation by the angle cosine method. 3) *Data Filtering:* The generated data are filtered by corresponding rules, e.g., correlation less than 0.2 or text length no more than 8 characters.
4) *Ranking:* We design a comparative variable L, as in Eq. (1), to integrate correlation and the number of message forwarding.

$$L = \beta * \frac{count}{Maxcount} + (1 - \beta) * \frac{sim}{Maxsim} \tag{1}$$

where *count* is the number of message forwarding, *Maxcount* is the maximal number of forward messages, *sim* is the correlation between a weibo text and the query words, and *Maxsim* is the maximal correlation result in the computation result. $\beta > 0$ controls the weight of the two parts, and we set β equals 0.5 in this demo.

Based on the event correlation analysis, we design a relevant event recommendation model. This model is to rank and return N events from the hot event database according to search keywords in two steps: *a)* to construct a hot event database. We use the same technique with data collection model to get hot events, and *b)* to recommend hot events which are relevant to query keywords. We use a scoring function in Eq. (2) to measure the similarity between searched keywords and hot events. Following the scoring function, we choose top-K events as the recommend events. Users can retrieve events they are interested in by the following model.

$$\mathcal{F}(keywords, hot\text{-}events) = \delta * \rho + (1 - \delta) * \varepsilon \tag{2}$$

where δ is weight parameter, ρ is the similarity between keywords and hot events under the *Levenshtein* distance, ε is 1 if keywords are contained in hot events, otherwise 0.

2.2 Real-time Anomalous Events Detection

The aims of this part are 1) to visualize the number of weibo message related to an event, and 2) to discover abnormal time points by a Peak-finding algorithm. For time-critical monitoring applications, peak-finding algorithm adopts moving average, moving variance and sliding window methods.

Peak-finding algorithm first gets the number of candidate weibo messages and stores the data into an array ζ. The size of ζ is equal to the size of observation window. Then, the algorithm initializes the mean (μ) and variance (σ) of data in the array ζ. If ζ_i in ζ satisfies Eq. (3), where τ is a threshold and set to 1.5 based on the crawled data and the number of peaks, there will be a peak range around ζ_i $(start, end)$. The start index is $i - 1$, and we can get a pseudo index by using greedy algorithms.

$$\frac{|\zeta_i - \mu|}{\sigma} > \tau \quad and \quad \zeta_i > \zeta_{i-1} \quad (i > 1) \tag{3}$$

2.3 Location Mining

The location of weibo data can be categorized into three classes: 1) the user registration location, i.e., a user attribute; 2) the weibo position location, i.e., location data attached to weibo texts sent from mobile devices; and 3) the content-based location, i.e., the location name entity in weibo messages.

In this system, we extract content-based location data after preprocessing. We discard weibo messages which there is no location information can be found, then choose the most relevant one for the event detection. For each location l_i, we add a weight η_i, as shown in Eq. (4), to sort the locations and select the most relevant location.

$$\eta_i = \frac{\sum\limits_{j=1}^{m} (\mathcal{F}_{index}[i] - \mathcal{G}_{index}[j])^2}{\sum\limits_{i=1}^{n} \sum\limits_{j=1}^{m} (\mathcal{F}_{index}[i] - \mathcal{G}_{index}[j])^2} \tag{4}$$

where $\mathcal{F}_{index}$ and $\mathcal{G}_{index}$ are the index position of location and keywords in the weibo messages respectively, and m is the number of keywords.

Based on the above method, we can get the location of each weibo, and use hierarchical clustering method to estimate the accurate event location from all the locations.

3 Demonstration

We demonstrate the *SocialAnalysis* system from the viewpoints of hot event recommendation, anomalous event detection and geographical location detection. When users input a keyword, the system will query database to return relevant events for recommendation, the trend of the event, abnormal time points, and the location of the event.

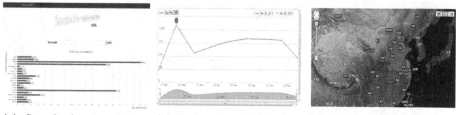

(a) Search box and hot event recommendation.   (b) Weibo event detection.   (c) Location demonstration.

Fig. 2. Demonstration of the system

As the limitation of space, we only show the main modules of our system. Fig. 2(a) shows the search page and event recommendation. We use histogram to visualized the degree of hot events. After users submit a keyword, the detected relevant curve is demonstrated in Fig. 2(b), where the abnormal point is highlighted by a red circle. Fig. 2(c) shows the location of the event.

4   Conclusion

In this paper, we present a real-time query and mining system, which aims to analyze and mine social network data for finding bursts of events. We solve the challenges, i.e. data collection, abnormal event detection and location identification. Our system can detect events rapidly, estimate locations of the events and provide early warnings.

References

1. Sakaki, T., Okazaki, M., Matsuo, Y.: Tweet analysis for real-time event detection and earthquake reporting system development. TKDE **25**(4), 919–931 (2013)
2. Zhang, P., Zhou, C., Wang, P., Gao, B., Zhu, X., Guo, L.: E-tree: An efficient indexing structure for ensemble models on data streams. TKDE **27**, 461–474 (2015)
3. Lee, R., Wakamiya, S., Sumiya, K.: Discovery of unusual regional social activities using geo-tagged microblogs. World Wide Web **14**(4), 321–349 (2011)
4. Marcus, A., Bernstein, M., Badar, O.: Twitinfo: aggregating and visualizing microblogs for event exploration. In: Conference on Human Factors in Computing Systems, pp. 227–236 (2011)
5. MacEachren, A., Robinson, A., Jaiswal, A.: Geo-twitter analytics: applications in crisis management. In: 25th International Cartographic Conference, pp. 3–8 (2011)
6. Shan, D., Zhao, W., Chen, R.: Eventsearch: a system for event discovery and retrieval on multi-type historical data. In: KDD, pp. 1564–1567 (2012)

SCIT: A Schema Change Interpretation Tool for Dynamic-Schema Data Warehouses

Rihan Hai[1](✉), Vasileios Theodorou[2], Maik Thiele[1], and Wolfgang Lehner[1]

[1] Technische Universität Dresden, Dresden, Germany
{rihan.hai,maik.thiele,wolfgang.lehner}@tu-dresden.de
[2] Universitat Politècnica de Catalunya, Barcelona, Spain
vasileios@essi.upc.edu

Abstract. Data Warehouses (DW) have to continuously adapt to evolving business requirements, which implies structure modification (schema changes) and data migration requirements in the system design. However, it is challenging for designers to control the performance and cost overhead of different schema change implementations. In this paper, we demonstrate SCIT, a tool for DW designers to test and implement different logical design alternatives in a two-fold manner. As a main functionality, SCIT translates common DW schema modifications into directly executable SQL scripts for relational database systems, facilitating design and testing automation. At the same time, SCIT assesses changes and recommends alternative design decisions to help designers improve logical designs and avoid common dimensional modeling pitfalls and mistakes. This paper serves as a walk-through of the system features, showcasing the interaction with the tool's user interface in order to easily and effectively modify DW schemata and enable schema change analysis.

1 Introduction

Data Warehouses (DW) have traditionally been designed to work on carefully modeled, predefined schemata, developed in the first phase of the overall system design. However, in a modern and highly volatile business environment, with changing market needs, company policies and technological advances, neither the data structure nor the analytical needs remain stable. It has recently been stressed that the incessant pressure of schema change is impacting every database [1]. Moreover, schema changes pose a threat to the accuracy of current applications and queries regarding the consistency of query answering, deciding schema equivalence, schema mapping composition and inversion. In addition, DW designers often lack the intuition of the performance and cost impact of different schema change implementations, while at the same time some dimensional modeling pitfalls may even compromise the original DW system [2]. Nonetheless, to our best knowledge, there is no existing implemented tool facilitating effectiveness and automation in DW schema change operations and analysis.

We thereby present SCIT, a flexible schema change interpretation tool for DW designers, supporting the improvement of logical system modeling. In its

© Springer International Publishing Switzerland 2015
M.A. Sharaf et al. (Eds.): ADC 2015, LNCS 9093, pp. 323–327, 2015.
DOI: 10.1007/978-3-319-19548-3_28

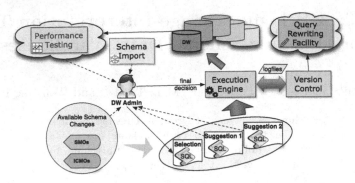

Fig. 1. SCIT framework

design, SCIT balances the need to interpret a wide range of schema changes with the imposed complexity of providing design alternatives for special dimensional structures. The main contributions of SCIT are summarized as follows:

- SCIT facilitates the interpretation of DW schema changes in a practical implementation where users are guided through a sequence of button clicks on an intuitive user interface. A representative set of thirty structural changes and eight basic integrity constraint changes, as well as their combinations, are supported by our tool, which automatically interprets them and generates executable SQL scripts containing equivalent Data Definition Language (DDL) statements. DW designers benefit from this time-saving function, especially during massive schema modification implementation tasks.
- SCIT employs an iterative mechanism to facilitate the design of alternative schemata. At every step, changes are assessed. And according to the satisfaction of specific dimensional structure conditions, SCIT recommends possible alternative designs based on best-practices. By comparing and simulating such alternatives, DW designers can obtain enhanced solutions regarding performance and cost.
- SCIT offers schema evolution traceability by maintaining files, which record the sequence of schema changes. Thus, it enables schema evolution analysis where designers can further explore potential issues, e.g., identifying and rewriting queries invalidated by schema changes and maintaining structural understandability.

2   System Overview

In this section, we introduce the functions and underlying mechanisms that make SCIT an efficient and flexible schema change interpretation tool. A high level view of our system architecture can be seen in Fig. 1 showing the process of using SCIT.

We build our tool to operate on a relational star schema, since DW systems are usually implemented following the star schema pattern, and according to Kimbal's recommendation [2], detailed and atomic information should be loaded into a star schema database. Below we describe the main system functionalities, showing how SCIT takes a user selection from a wide range of schema changes as input and generates equivalent SQL scripts as output, as well as intelligently providing design alternatives for improving logical system design:

Supported Schema Changes. SCIT supports a comprehensive set of Schema Modification Operators (SMOs) and Integrity Constraint Modification Operators (ICMOs) to represent atomic structural schema changes and keys or other dependencies, respectively. Interested readers can refer to [1,3] for formal representation and characteristics of those operators. The enhancement of SCIT is that we extend the original set from 18 [1] to 38 operators (30 atomic structural changes and 8 integrity constraint changes) and their combinations. This set includes operators for special design alternatives, e.g., pivot a row-based table into an Entity-Attribute-Value (EAV) model (column-based model).

Special Design Alternatives. Based on [2,4], SCIT supports a representative, practical set of twelve common design alternatives corresponding to specific dimensional structures, e.g., mini-dimensions for fast-changing attributes, EAV model for highly sparse dimensions and bridge tables for ragged variable depth hierarchies and multi-valued attributes. When schema changes satisfy certain predefined conditions, SCIT recommends the implementation of such alternatives to users, while automatically generating the corresponding executable SQL script for each suggestion, which supports users to choose desirable design regarding further performance test results or specific system requirements. As an example, let us consider a schema change scenario where a user inputs *add new attributes* as desired schema change and the following conditions also stand: (1) new attributes are text attributes (e.g., comment fields) and (2) the target table is a fact table. Then, SCIT will suggest the usage of *Garbage dimension* by putting the text attributes in a new dimension table, adding a new column in the fact table and connecting the dimension table to the target fact table by a foreign key constraint. In case these attributes are highly sparse, SCIT additionally suggests the application of an EAV model to the new dimension table. Designers are then able to conduct performance testing to compare the original design and alternative recommendations.

Schema Change Interpretation. Considering the required processes after schema modifications, such as rewriting invalid queries, which requires information preserving, necessary actions are automatically taken by the system. All thirty atomic structural implementation changes are embedded with foreign keys, which hides dimensions or facts supposed to be dropped, by cutting all existing foreign key constraints and renaming the target table. Taking *split a dimension table vertically* as an example, each new sub-table will get a newly generated surrogate key as primary key, as well as equivalent constraints mapped from the original dimension table. In the original table, all the foreign key constraints are dropped and the table is renamed to be hidden from the schema.

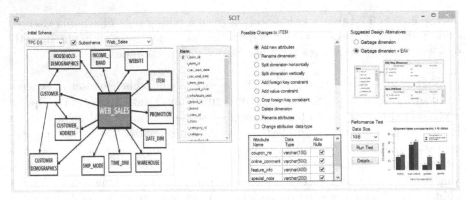

Fig. 2. SCIT User Interface

3   Demonstration Scenario

The demonstration will illustrate the basic use of SCIT as an effective schema change interpretation tool, highlighting its role as an alternative design adviser. The following scenarios will be running on Microsoft SQL server 2014:

Initial Schema Set Up. We use the TPC-DS benchmark[1] for our demonstration, which has a relatively complex schema of a retail model, as our initial schema. Visualization of the whole or partial schema is presented on the main screen (shown in the left section of Fig.2). By clicking on one dimension/fact table, users can obtain basic structural information of the chosen element(s).

Schema Change Selection. In this scenario we demonstrate the implementation of certain schema changes. Possible structural or integrity constraint changes regarding the chosen dimension/fact will be automatically listed in the middle of the main screen. After choosing a desired schema modification from the list, users need to further input corresponding parameters, as shown in the middle section of Fig.2 (attribute name, data type, etc.).

Special Scenarios for Design Alternatives. In this scenario we select changes that trigger the suggestions of alternative designs. Users will observe the visualized sub-schema of each alternative and performance test results with configurable size of data loaded into the RDBMS. We measure the query elapsed time for the original design and a chosen alternative. Thus, we illustrate how users can obtain an estimation of performance comparison between alternative designs. In the right section of Fig.2, we show the performance comparison between a basic schema change implementation and SCIT suggested design alternative (Garbage dimension + EAV) based on 1GB data.

Acknowledgments. This research has been funded by the European Commission through the Erasmus Mundus Joint Doctorate "Information Technologies for Business Intelligence - Doctoral College" (IT4BI-DC).

---

[1] http://www.tpc.org/tpcds/

References

1. Curino, Carlo, et al.: Automating the database schema evolution process. The VLDB Journal The International Journal on Very Large Data Bases **22**(1), 73–98 (2013)
2. Kimball, R., Ross, M.: The data warehouse toolkit: The definitive guide to dimensional modeling. John Wiley & Sons (2013)
3. Rahm, E., Bernstein, P.A.: A survey of approaches to automatic schema matching. The VLDB Journal **10**(4), 334–350 (2001)
4. Ballard, C., et al.: Dimensional Modeling: In a Business Intelligence Environment. IBM Redbooks (2012)

SEMI: A Scalable Entity Matching System Based on MapReduce

Pingfu Chao[1,2], Yuming Li[1,2], Zhu Gao[2], Junhua Fang[1,2],
Xiaofeng He[1,2], and Rong Zhang[1,2(✉)]

[1] Institute for Data Science and Engineering, East China Normal University,
Shanghai, China
[2] Shanghai Key Laboratory of Trustworthy Computing,
East China Normal University, Shanghai, China
{51121500001,51141500019,10132510331,52131500020}@ecnu.cn,
{xfhe,rzhang}@sei.ecnu.edu.cn

Abstract. MapReduce framework provides a new platform for data integration on distributed environment. We demonstrate a MapReduce-based entity resolution framework which efficiently solves the matching problem for structured, semi-structured and unstructured entities. We propose a random-based data representation method for reducing network transmission; we implement our design on MapReduce and design two solutions for reducing redundant comparisons. Our demo provides an easy-to-use platform for entity matching and performance analysis. We also compare the performance of our algorithm with the state-of-the-art blocking-based methods.

1 Introduction

Entity resolution, which aims to identify entities referring to the same items, is introduced to clean the dirty data, such as typos, duplicated records and so on. However, in the scenario of C2C (Customer to Customer) online markets, as the rarity of descriptions, missing of uniform schema or intended errors, tradition entity resolution methods can not get good match performance. But these online platforms have provided massive unstructured information generated by users, such as reviews. These reviews provide us the chance for retrieving useful information to identify the entities.

Entity resolution is a compute-intensive job since it needs to compare all possible pairs in the data set. It often spends days or even months to make a matching process on a huge data set. The state-of-the-art distributed entity resolution methods, such as Dedoop[2] and Document Similarity Self Join [1] (DSSJ), introduce blocking-based matching methods to improve the processing speed and implement them on MapReduce framework, a parallel computing framework with high usability and flexibility. By means of this distributed framework, those methods achieve good performance on handling big size of data, but they still have performance bottlenecks dealing with noisy/enormous data sets. In this paper, we propose an efficient random-based matching algorithm

© Springer International Publishing Switzerland 2015
M.A. Sharaf et al. (Eds.): ADC 2015, LNCS 9093, pp. 328–332, 2015.
DOI: 10.1007/978-3-319-19548-3_29

for entity matching on MapReduce framework. We will compare our method with the blocking-based methods and demonstrate our competitive power on computation efficiency and transmission cost.

2 System Overview

Framework of random-based entity resolution system **SEMI** is shown in Fig. 1. It is comprised of three parts: Data Preprocessing, Entity Matching and Redundancy Elimination. We complete all those tasks distributively on MapReduce. SEMI can process variety of data formats including structured, semi-structured and unstructured data. The final output of SEMI is a set of entity pairs whose similarity is above a certain threshold θ, $0 \leq \theta \leq 1$. Alternatively, we can output the top N pairs of highest similarity.

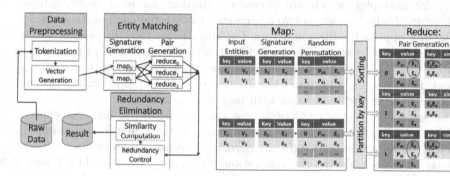

Fig. 1. System Framework   **Fig. 2.** Matching Example on MapReduce

Data Preprocessing. It takes three steps to preprocess the source data. Initially, we tag and tokenize the input entities by the *Part-Of-Speech Tagger*[4]. Then we start stop word removing. The rest tokens are sorted by their frequency in ascending order and form a token dictionary. In the end, we generate a k-dimension vector V_u for each entity u as system input. The nth dimension in V_u represents the frequency of the nth token in the dictionary occurred in entity u. So the output of data preprocessing step is a set of (*key, value*) pairs with the entity ID E_u as its key and its k-dimension vector V_u as its value. In order to differentiate data coming from different parts, we can weight the vector elements differently. For instance, every item in online shopping, the tokens from the structured part, such as *name, colour*, are more important than those from the unstructured parts such as *review content*.

Entity Matching. Dimensional reduction is important for user to reduce the transmission cost. We realize this by using Local Sensitive Hashing (LSH) to transform the high-dimensional feature vector into a low-dimensional binary signature. Then we calculate the *hamming distance* between every two signatures to estimate the cosine similarity between them, which is proved to be reasonable in [3]. Since it is expensive to compare every possible pair in the entity set. We introduce a random-based pair generation method to reduce the high comparison cost.

Fig. 2 illustrates the workflow of our matching job. We use one MapReduce job to finish our matching task. In the map phase, we collect all input (E_u, V_u) and transform the k-dimension vector V_u into a set of bit streams S_u of length d, with $d \ll k$. Then we randomly permute the d bits signature S_u t times and get t permuted signatures (all entities must use the same permutating rule when starting one round of permutation). We output each permutation as (t, P_{ut}, E_u), which contains the series number (t) of this permutation and its entity ID (E_u) in the map phase. In the reduce phase, each reducer receives signatures of the same series number, it sorts all signatures and generates pairs between each signature and its m nearest neighbors. The output of our matching step is the IDs and signatures of each entity pair $(E_u E_v, S_u S_v)$.

Redundancy Elimination. As we can see in Fig. 2, there are duplicated pairs $(E_0 E_4$, for example), which are pervasive problems for most blocking-based matching methods. We design two approaches to save the computation cost.

The first approach adds a new round of MapReduce jobs. In the map phase, we group the pairs by keys as shown in Fig. 2, since all duplicated pairs may have the same *key*. Then we compute similarity once for each group, thus reducing the cost. However, the additional MapReduce job can introduce extra cost that cannot be neglected when dealing with huge data set.

The second approach involves a new random algorithm into the existing matching job. For the map phase in matching job, the signature is permuted t times and therefore we get t different permutations for each entity.

In this approach, during the generation of permutations, we add a f bits of similarity hashing check bit for each permutation to record the information of previous generated permutations of the same signature. Before the reduce phase, if the check bit of two entities which are supposed to be paired are the same, it means that these two entities are possibly paired in another reducer. Therefore, we cancel this pair generation in a possibility of α, $0 \leq \alpha \leq 1$. This method has an advantage in processing speed, but it can only reduce the redundancy but not eliminate it. It is more suitable for processing huge data set.

3   Demonstration

Our demo is based on Hadoop-1.2.1 and Java-1.7.0. The interface of our demo will help to connect the Hadoop cluster, choose the matching methods, and configure parameters. We also provide a file explorer to preview the input data and output results.

After the matching process, a log analyzer is provided to visualize and compare the performance of different jobs implemented by different systems or different parameter settings or different optimization algorithms. We can show the result in different formats, such as line chart, histogram, pie chart, etc.

4   Experiments

We run experiments on the CiteSeerX data set. It contains nearly 1.32 Million citations of total size 2.89 GB in XML format including structured features, such as *record ID, author, title, date, page, volume, publisher, etc*, and also unstructured features, like *abstract*. We use *accuracy* and *run-time* metrics for performance evaluation. The baseline methods are Document Similarity Self-Join (DSSJ) and Dedoop, which are the state-of-the-art methods. We run experiments on a 22-node HP blade cluster. Each node has two Intel Xeon processors E5335 2.00GHz with four cores and one thread per core, 16GB of RAM, and two 1TB hard disks. In order to measure the accuracy, we use a validation set which contains 200 records.

Table 1. Accuracy Comparison

Name	Top 10	Top 20	Top 50
DSSJ	90%	95%	94%
Ours	90%	100%	94%
Dedoop	100%	100%	100%

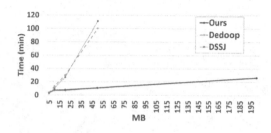

Fig. 3. Run-time Comparison

We run the three approaches and find out the top 10, 20 and 50 similar pairs. Table 1 shows the accuracy of the test. Since Dedoop compares all possible pairs of entities and calculates cosine similarity directly, the result of Dedoop are always correct but with high computation cost as seen in Fig.3. Meanwhile, we can see that our system achieves better accuracy than DSSJ method with much less computation cost seen in Fig.3. At the same time, we evaluate their processing speed using the same parameters as shown in Fig. 3. Since Dedoop and DSSJ will generate enormous size of pairs, it can be a big burden on network transmission and is hard to be processed in memory. In our experiment, when the data set exceeds 200MB, the transmission data generated by Dedoop or DSSJ is up to several terabyte, therefore it obstructs the matching process and causes job fails. However, our algorithm is significantly faster than Dedoop, and far more stable even dealing with gigabytes of input data.

Acknowledgment. This work is partially supported by National Basic Research Program of China (Grant No. 2012CB316200), National Science Foundation of China (Grant No.61232002, 61402177 and 61103039), and the Key Program of Natural Science Foundation of Yunnan Province under grant No. 2014FA023.

References

1. Baraglia, R., De Francisci Morales, G., Lucchese, C.: Document similarity self-join with mapreduce. In: Proc. of ICDM, pp. 731–736 (2010)
2. Kolb, L., Thor, A., Rahm, E.: Dedoop: efficient deduplication with hadoop. Proc. of VLDB **5**(12), 1878–1881 (2012)

3. Ravichandran, D., Pantel, P., Hovy, E.: Randomized algorithms and nlp: using locality sensitive hash function for high speed noun clustering. In: Proc. of ACL, pp. 622–629 (2005)
4. Toutanova, K., Manning, C.D.: Enriching the knowledge sources used in a maximum entropy part-of-speech tagger. In: Proc. of EMNLP, pp. 63–70 (2000)

Author Index

Printed in the United States
By Bookmasters